From Hydrology to Hydroarchaeology in the Ancient Mediterranean

From Hydrology to Hydroarchaeology in the Ancient Mediterranean

Edited by

Giovanni Polizzi, Vincent Ollivier, Sophie Bouffier

Archaeopress Archaeology

ARCHAEOPRESS PUBLISHING LTD
Summertown Pavilion
18-24 Middle Way
Summertown
Oxford OX2 7LG
www.archaeopress.com

ISBN 978-1-80327-374-7
ISBN 978-1-80327-375-4 (e-Pdf)

© the individual authors and Archaeopress 2022

Cover: A public cistern of the Solunto agora. Photo by Giovanni Polizzi

All rights reserved. No part of this book may be reproduced, or transmitted, in any form or by any means, electronic, mechanical, photocopying or otherwise, without the prior written permission of the copyright owners.

This book is available direct from Archaeopress or from our website www.archaeopress.com

Contents

Introduction 1
Sophie Bouffier and Vincent Ollivier

The Impact of Climate, Resource Availability, Natural Disturbances and Human Subsistence Strategies on Sicilian Landscape Dynamics During the Holocene 8
Salvatore Pasta, Giuseppe D'Amore, Cipriano Di Maggio, Gaetano Di Pasquale, Vincenza Forgia, Alessandro Incarbona, Giuliana Madonia, César Morales-Molino, Silvio Giuseppe Rotolo, Luca Sineo, Claudia Speciale, Attilio Sulli, Willy Tinner and Matteo Vacchi

Analyse historique des variations du débit provoqué par les séismes pendant les siècles XVe–XXe: le cas de Termini Imerese (Sicile centro-septentrionale) 61
Patrizia Bova, Antonio Contino and Giuseppe Esposito

Le risorse idriche nel territorio di Alesa 76
Aurelio Burgio

***Aquae caldae*. Note sparse sul termalismo e lo sfruttamento delle acque sulfuree nel territorio a est di Agrigento tra archeologia e paletnologia** 91
Luca Zambito

Archaeology and Hydrogeology in Sicily: Solunt and Tindari 102
Giovanni Polizzi, Vincent Ollivier, Olivier Bellier, Edwige Pons-Branchu and Michel Fontugne

Baia (Bacoli-NA): l'acqua e il suo utilizzo nel complesso delle Terme romane 127
Daniele De Simone

Natural Risks and Water Management in Delphi 143
Amélie Perrier, Isabelle Moretti and Luigi Piccardi

Introduction

Sophie Bouffier[1] and Vincent Ollivier[2]

[1] Aix Marseille Univ, CNRS, CCJ, Aix-en-Provence, France
[2] CNRS, Aix Marseille Univ, Minist Culture, LAMPEA, Aix-en-Provence, France

This work appears as part of an interdisciplinary programme, supported by the Foundation A*Midex between 2018–2021, which encourages cross-interdisciplinary initiatives on complex issues to advance knowledge in fields previously exploited by parallel disciplines, and to build knowledge shared by people from different backgrounds.[1] It follows on from a previous programme that led to the creation of an international network on water issues in pre-Roman societies: HYDRΩMED.[2] Within this framework, a scientific meeting was organised at Aix-Marseille University in May 2019 to strengthen links between archaeology and environmental sciences around water resources in ancient times, in terms of selected topics.

Context

Research on water has been generally conceived of within an environmental framework and focused on the issue of the contemporary risks and deficits of a resource threatened by climate change and anthropogenic over-exploitation, giving rise to latent crisis situations or open war in some parts of the world,[3] including the Middle East, as highlighted in certain international reports.[4] However, except for work on the Palaeolithic, or periods prior to the appearance of the human species, we too rarely have taken into account the natural habitat of ancient societies, faced with hydric situations comparable to those we know today. Thus, some teams have conducted pioneering work on a conceptual level, essentially on the palaeoclimate and on the very long duration, as, for example, the CEREGE project at Aix-Marseille University, with the work of Joël Guiot and his colleagues from the European Pollen DataBase:[5] the stated goals are clearly helping the understanding of contemporary climate change using the Quaternary palynological data.

In the Mediterranean area, the development of societies has always been intimately tied to the management of water resources. Very early on, from the 6th millennium BC, in Mesopotamia the first hydraulic facilities appeared, with the intention, among other things, of irrigating cereal crops.[6] These modifications to the hydrographic network, such as the digging of canals and the deviation of rivers by humans, produced major changes in the environment, itself subject to external forces (climate, eustasy, tectonics). This changing environment has had a significant impact on the establishment of occupation from the Neolithic, as shown, for example, in the articles by Ollivier[7] on the region of the Lesser Caucasus (Georgia, Armenia and Azerbaijan), as well as on water and land use management strategies.[8]

In addition, there is the vulnerability of populations, from the Mediterranean to the Caspian Sea basin, to seismic risk. This is often tragically illustrated,[9] to cite only recent earthquakes. The convergence of the Eurasian, African and Arabian plates creates an important regional North–South shortening, shaping landscapes and exposing prehistoric and historical societies to seismic hazards. These events affect also the hydraulic structures and methods of water supply and management.

In the field of archaeology *stricto sensu*, for a long time research has focused on two different and sometimes complementary aspects: ancient architecture and technological knowledge, political management of water systems.[10] The interest has been primarily focused on hydraulic facilities from an architectural point of view, and the technological knowledge that ancient civilizations, particularly the Mesopotamians and Romans, were able to develop in their own way, or from knowledge gained from societies that had preceded them in the Mediterranean. Most of the projects and international conferences, including

---

[1] The project leading to this publication has received funding from Excellence Initiative of Aix-Marseille University – A*Midex, a French 'Investissements d'Avenir' programme.
[2] https://hydromed.hypotheses.org/.
[3] Lasserre and Descroix 2011.
[4] Cf., e.g., the study of the World Resource Institute, American National Center for Environmental Research: Luo *et al.* 2015.
[5] http://www.europeanpollendatabase.net/index.php.
[6] Sherratt 1980.
[7] Ollivier 2022, Ollivier *et al.* 2018.
[8] Ricci *et al.* 2012.
[9] Notably the 1988 Armenian earthquake, in Spitak, Ms 6.7, Westaway, 1990; the 1908 earthquake in Messina, Sicily, Ms 7.1; the 1977 earthquake in Vrancea, Romania, Ms 7.2.
[10] Bouffier 2019: 5–9.

those organised through the *Cura Aquarum* network by the Foundation Frontinus-Gesellschaft or the Deutsche Wasserhistorische Gesellschaft, have focused on technical, especially architectural, systems developed by the Romans.[11] The network HYDRΩMED highlighted pre-Roman water cultures, seeking to fill a documentary gap about cultures that have been neglected by traditional historiography.[12] Apart from the history of ancient technology, political history has been favoured, especially in the wake of Karl August Wittfogel,[13] as evidenced by publications on politics of the Assyrian Kings,[14] or the special issue of the *Annales. Histoire, Sciences Sociales* dedicated to *Politique et contrôle de l'eau dans le Moyen-Orient ancien*.[15]

The first joint investigations occurred between archaeologists, historians and environmentalists concerning the great rivers exploited for their drinking or irrigation water, notably those of the Tigris, Euphrates, and Nile.[16] It was between prehistorians that the closest and earliest dialogue was developed: on the site of Aix-Marseille, where work in geomorphology on the rhythms of sedimentation in water courses under climate and/or anthropogenic control in Mediterranean mountain areas initiated the first long-term collaborations.[17] Other partnerships in the fields of coastal and fluvial geomorphology, with the work of Philippe Leveau, Mireille Provansal, and Christophe Morhange,[18] and naval and maritime archaeology, by Giulia Boetto and Valérie Andrieu-Ponel, have also been developed recently.[19]

Collaboration is today more regular and seems both natural and necessary, but it is not sufficiently structured and leaves still a whole section of research undeveloped. In hydraulic topics, there have been some first initiatives from the end of the 1990s.[20] The interdisciplinary perspective gave archaeologists the opportunity to access closer readings on the problems faced by ancient users and the solutions adopted in response to variable resources due to climatic and geomorphological reasons. The water work developed in the arid and semi-arid areas, ranging from the valleys of the Tigris and Euphrates, passing through the Valley of the Nile as far as the shores of the Aegean Sea, underlined that in these regions, hydraulics developed under different natural constraints. In Alexandria, for example, these experiences gave birth to inventions, technology, and then to a science – a prelude to the Alexandrine innovations in the 3rd century BC. The confrontation between rational Greek thought and the hydraulic know-how of the East gave birth to a new science that began with Archimedes' hydrostatics and the first devices functioning with water pressure. Archimedes (287–212 BC) introduced into science various hydraulic machines, such as the shaduf, in use for millennia, and the screw. In the same epoch, the *sakieh* for irrigation, or the bucket chain, spread in Egypt. The Hellenistic world would rely on this major innovation to develop its agriculture by extending irrigated land. At the same time, archaeological research focused on the more unusual aspects of the history of techniques, such as running water networks[21] or evacuation practices,[22] in connection with the current issues of environmental pollution, and land irrigation.[23]

Additionally, the HYDRΩMED programme has encouraged an international network to emerge, establish partnerships, and evaluate hydraulic knowledge, even if over only a limited period, the 1st millennium BC, and in a limited geographical area: the Mediterranean. The network thus highlighted what was necessary to encourage and develop. Five international meetings have focused on four main themes:[24] the climatology and hydrology of the Mediterranean in the 1st millennium BC; ancient hydraulic facilities and exploitation of natural resources; history of science and technology in relation to economic and political history; and cults and cultures of water and waters. These conferences evaluated our knowledge of the different pre-Roman societies – including the Greek, Phoenician-Carthaginian, and Iberian worlds – in political, technological, and cultural areas. They also underlined insufficient exchanges between geosciences and archaeology, especially in the restitution of the natural conditions of past societies and the resources available, as highlighted by the 3rd international symposium (March 2016, *Geoarchaeological and palaeoenvironmental approaches to water resources management in Antiquity*),[25] which documented the findings of several researchers from France, Germany, England, Scotland, Spain and Austria, specialists in palaeoenvironmental (paleoclimatology, paleohydrology) and geoarchaeological issues

---

[11] Bouffier 1919.
[12] Bouffier and Fumadó Ortega 2020; Robinson, Bouffier and Fumadó Ortega 2019; Fumadó Ortega and Bouffier 2019; Bouffier, Belvedere and Vassallo 2019.
[13] Wittfogel 1957.
[14] Masetti-Rouaux and Defendenti 2020; Bagg 2000.
[15] 2002/3; 57e année.
[16] Hairy 2010.
[17] See the summary in the festschrift to Maurice Jorda led by Miramont 2004; and also, Provansal 1995; Bruneton 1999; Ollivier 2006.
[18] Leveau and Provansal 1993; Leveau 2011; Morhange 1994.
[19] Boetto *et al.* 2012.
[20] The French School of Athens' programme on ancient water at Delos, 1999–2004: Brunet 2008; Desruelles and Fouache 2015; at Delphi, Perrier *infra*, the programme on water at Syracuse: Bouffier *et al.* 2018; in the Roman field, Hélène Dessales' work (2013).

[21] Abadie-Reynal *et al.* 2011.
[22] Bouffier and Brunet 2020; Nenna 2010.
[23] Bouffier 2002.
[24] Bouffier, Belvedere and Vassallo 2019; Bouffier and Fumadó Ortega 2020; Fumadó Ortega and Bouffier 2020; Robinson, Bouffier and Fumadó Ortega 2019.
[25] https://calenda.org/357764?formatage=print&lang=en. The meeting remains unpublished.

associated with the management of hydric resources. In terms of the Mediterranean climate, for example, Joël Guiot's work suggested mathematical modelling from pollen data, but for the ancient Mediterranean climate, and ongoing discussions show a complex situation, as data is too rare, with archaeologists from these regions still mostly unaware of palynological sampling.

Thus the *Watertraces* project has been conceived as part of closer links between environmental sciences and archaeology, which began at Aix-Marseille University with one-off collaborations, especially on Mediterranean fluvial-deltaic environments and the Mediterranean ports and environmental mountain changes. This has brought together historians and archaeologists, as well as hydrology, geology, and tectonic specialists, who worked on hydrologic and hydraulic problems – with investigations common to the humanities, social sciences and geosciences, sometimes using common methods and tools, but also specific ones. The aim was to understand the functioning of water resources in past societies, so as to draw lessons for our current behaviour, and the protection of a resource which has been declared a future scarcity. The interdisciplinary dialogue relied on specific methods: data archives, prospecting, excavations and samples, as well as physicochemical analyses which gave rise to confrontations between archaeologists, historians and environmentalists.

Project objectives

The project's aim was to highlight the resources around which human societies choose to settle, and their coping strategies when faced with a resource that is either too rare or too abundant. Investigations focused along three axes:

1. To procure water: the state of the resource

Near to which resources do human societies settle? What quality of water do they look for? What quantity of water do they use? Has the resource varied over the course of history? In what proportions? Why? For example, the formation of carbonated deposits in old aquatic systems is a source of data on climate and the gradual change in source water properties, but also a source for understanding the functioning of hydraulic systems: why were certain structures built the way they were? What knowledge do they reveal and how have they evolved? Can we identify knowledge transfer? How have water systems evolved? How long did they work? We asked all these questions when comparing water systems from different parts of the Mediterranean and the areas surrounding it. The interdisciplinary teams worked on several Mediterranean sites that offered different geomorphological and hydrological situations, such as sedimentary and semi-lagoon environments (Egyptian Alexandria, Sicilian Syracuse, Arpi in southern Italy), karstic ones (Solunto, Syracuse, Taormina in Sicily, Nimes in France, Locres Epizephyrii in southern Italy), clay and marshy soils (Arpi in southern Italy), karst, dolomite, marls, sandstones and terra rossa (Loron in Croatia). Some of their results are presented in this volume (Solunto).

2. Water storage: waterproofing processes of hydraulic facilities

The aim here was to study, from a technical point of view, the process of waterproofing hydraulic facilities by means of hydraulic mortars, a field still insufficiently developed in archaeology. The presence and type of coatings need investigating to find cases of permanency and ruptures of technologies over time. Archaeologists are often faced with hydraulic installations that have been used over a very long period, without it being possible to date the construction and clarify the chronology of human occupation. The good functioning of the construction, through regular maintenance and repair, obliterates previous traces. The nature and composition of waterproofing mortars could provide clues for dating, but also give us information about the materials and the techniques used, that might reveal the artisans' know-how, circulation of materials, and technologies. Archaeological research is very underdeveloped on these aspects. For the Roman epoch, the architect Vitruvius left a work of incomparable value, providing us with a technical presentation on hydraulic mortars, concentrating only on *pozzolana*, whereas for earlier periods, and in the majority of Roman provinces, a crushed terracotta mortar was mainly used, so that archaeologists associate almost systematically crushed terracotta mortar with hydraulic structures. Actually these crushed terracotta mortars can be used in several contexts, being mainly characterised by their superior mechanical resistance, and thus considerable durability. The project under discussion, in addition to an in-depth literature review about hydraulic mortars, will allow advances to be made in terms of the physical characteristics of mortar samples from sites (porosity, capillary action, porometry, etc.), their mineralogical characteristics, certain possible organic additions (fats, oils, etc.) capable of providing waterproofing properties to these materials, chalk-crushed terracotta mortar reactions (*pozzolanic* reaction), and the study of their impact on the properties of these materials.

The *Watertraces* project programmed sample collection and systematic analysis at the sites studied within its framework, proposing further identification protocols available to all. The aim was to develop petrographical and chemical studies to determine the composition of mortars, the provenance of granulates, and the hydraulicity factors of these mortars. In partnership with the *Centre interdisciplinaire pour la conservation et la restauration du Patrimoine* (CICRP, Marseilles) studies have been conducted to identify the materials and production techniques by observing them through stereoscopic or optical microscopes, complemented by chemical and particle size analyses, by mineralogical characterisations by X-ray diffraction on powder, and observations and analyses basic to the electronic scanning microscope, coupled with an energy dispersive spectrometer; whereas dating can be given by the street furniture found in stratigraphy at the site. A 'Mortars' database has been initiated on the French Archaeological Platform ArkeoGis[26] that provides the scientific community with the wide range of mortars and techniques developed by ancient societies.

3. Water Loss? The concretions, seismic risks, the natural vicissitudes of hydraulic facilities

One of the realities that archaeologists often have to face is the deterioration of facilities due to both anthropogenic and natural causes, particularly tectonic ones. The observations between geologists and archaeologists, which began with an example from Sicily, with the traces left by the earthquakes of the 17th century, particularly the 1693 Noto earthquake, on the hydraulic facilities and attested to by archives of the era, should encourage the highlighting of recurring phenomena in the history of local societies. Among the structural drivers that can cause damage to hydraulic infrastructures (erosion, floods, defects, wear and tear, defective maintenance, etc.),[27] the seismic component is one of the most radical and widespread within the geographical area included in our programme. Occasionally, main and/or secondary faults can promote hydrothermal lifts supplying sites (i.e. Solunto[28] or Germisara[29]), potentially steering the organisation of hydraulic infrastructures and urban morphology. At other times they can partly control the route of aqueducts or accentuate the loss of water from such structures, as we can find at Syracuse.

This present volume, therefore, proposes some current reflections on these questions, as well as some results to date from this programme, as well as showing work in progress and offering new ways of thinking. Centre stage is given to seven interdisciplinary teams, who have worked in various fields.

As a concise introduction to the relevant work from Sicily, Salvatore Pasta and his coauthors present the most recent discoveries and hypotheses concerning the factors driving human subsistence economy and landscape shaping during the Holocene, while explaining the role played by different disciplines (geology, paleogeography, climatology, palynology, archaeobotany, archaeozoology, anthropology, archaeology) in reconstructing the palaeoenvironment of the largest island in the Mediterranean. They do not overlook that the historical approach also makes it possible to highlight past episodes of unsustainable land-use and mankind's action on the environment, while the paleoclimatic studies emphasise the favourable/unfavourable conditions for the development of forests and vegetation long before the historical period. The seismic events appear to have played a major role in the socio-economic evolution of populations, particularly in the Aeolian islands. A wide assessment of the presence of man since his late arrival in Sicily, within a time lapse of 20.0 to 15.0 cal ka BP, assesses the development of the island's settlement process.

Several contributions here, mostly focused on Sicily, highlight sites that have been the object of interdisciplinary projects between archaeologists and environmentalists.

Antonio Contino, Patrizia Bova,[30] and Giuseppe Esposito look at northern-central Sicily, especially the Trabia-Termini Imerese mountains. The geological study, combined with the historical approach, sheds light on water resources in the Augustan colony of Termini Imerese. The ancient town was supplied via an aqueduct that brought potable water from the springs of Brocato-Fridda and Favara-Scamaccio. The team presents the history of this aqueduct, in association with climatic vicissitudes that forced the authorities to overhaul and restore it several times during its history until the output from the Favara-Scamaccio springs ceased. This case-study is paralleled with another aqueduct at Galermi (Syracuse), restored on several occasions and studied by an interdisciplinary team from Aix-Marseille University.[31] Both monuments show how past hydraulic installations evolved and were transformed by local populations to meet their needs.

Aurelio Burgio's contribution looks at the countryside around the ancient city of Alesa, which is known, fortunately, thanks to an archaeological survey and the extant epigraphy. The fact that prospection teams are now interdisciplinary makes it possible to cross-reference data and raise questions that archaeologists have not been used to asking, or could not answer.

---

[26] The 'Mortars' database has been initiated on the Publication Platform ArkeoGis: https://app.arkeogis.org/#/database/562
[27] Passchier *et al.* 2013.
[28] Polizzi *et al.* 2017.
[29] Ollivier *et al.* 2008.
[30] Very sadly, Patrizia Bova left us in 2021 and will not see the publication of her chapter; we would like here, of course, to salute her memory.
[31] Bouffier and Wateau 2019.

Luca Zambito examines the Agrigento region, richly endowed with sulphurous springs and mineral deposits which ancient populations exploited for thermal and therapeutic purposes, and where ancient groups established cults, and passed down their legends.

Giovanni Polizzi, Vincent Ollivier, and their colleagues, present the ancient cities of Solunto, on an intensely fractured dolomitic limestone massif, and Tindari, situated on metamorphic rocks, limestones and flysch, affected by deep karst/pseudo-karstification processes in the fractured areas. Previously considered to be exclusively from the collection of rainwater through systems of tanks, the water supply to the site appears, in fact, to be much more abundant and more complex. This seems to have resulted from the exploitation of systems of tectonic fractures guiding hydrothermal lifts, and to have provided additional and sustainable thermal mineral water via a significant number of the identified hydraulic infrastructures.

Two further papers extend beyond this geographical area – one relating to Campanian Baia, characterised by its volcanic geological situation, and the other to Apollo's famous sanctuary at Delphi.

Daniele De Simone's ongoing PhD is exploring Baia's thermal Baths, a huge architectural and engineering complex that has exploited the numerous hydrothermal springs of the Phlegrean fields from at least the 3rd/2nd c. BC. Notably, the area lacks drinking water and the Romans were obliged to construct an entire water collection, storage, and distribution system – initially relying on rainwater, then constructing a huge aqueduct, the 'Acqua Augusta' pipeline, in the 1st c. BC. The chapter starts with the geomorphology of the area, highlighting the expedients used by Roman designers (from the Late Republican era to the entire 4th c. AD., to exploit these exceptional resources.

Crossing to the Greek Cyclades, Amélie Perrier, Isabelle Moretti, and Luigi Piccardi focus on the natural calamities experienced by the celebrated Sanctuary at Delphi, which suffered from many disasters, such as earthquakes and landslides. The team from the French School of Archaeology at Athens, with its long association with the site, start by looking at the architectural remains, and using a geological approach, show how the buildings were damaged, how the ancient Greeks approached the natural hazards, and how they responded to water-related risks by exploiting them and channelling the water to their own ends.

Thus this present volume, modestly aims to shed light on the new paths that have emerged and linked our teams of classical archaeologists. If perhaps not exactly new in terms of several of the fields we explore, then these paths have been insufficiently followed in the past, and the creative dialogue between ourselves and geomorphologists have allowed us to understand many situations we have found obscure, especially in the field of hydraulics.

Bibliography

Abadie-Reynal, C., Provost, S., and Vipard, P. 2011. *Les réseaux d'eau courante dans l'Antiquité. Réparations, modifications, réutilisations, abandon, récupération.* Rennes: Presses Universitaires de Rennes.

Bagg, A.M. 2000. *Assyrische Wasserbauten: Landwirtschaftliche Wasserbauten im Kernland Assyriens zwischen der 2. Hälfte des 2. und der 1. Hälfte des 1. Jahrtausends v. Chr.* C Mainz am Rhein: P. von Zabern.

Boetto, G., Radić Rossi, I., Marlier, S., Brusio, Z., Huguet, C., Capelli, C., Guibal, F., Greck, S., Cenzon-Salvayre, C., Andrieu-Ponel, V. and Dumas, V. 2012. Résultats d'un projet de recherche franco-croate. *Archaeonautica* 17: 105–151.

Bouffier, S. 2019. Introduction, in S. Bouffier, O. Belvedere, S. Vassallo (eds) *Gérer l'eau en Méditerranée au premier millénaire av. J.-C.*: 5–9. Aix-en-Provence: Presses Universitaires de Provence : 5-9.

Bouffier, S.C. 2002. Eau et campagne en Sicile grecque : observations préliminaires à l'étude, in *IN BINOS ACTUS LUMINA. Rivista di studi e ricerche sull'idraulica storica e la storia della tecnica, Atti del Convegno Internazionale di Studi su Metodologie per lo studio della scienza idraulica antica* (Ravenne, 13–15 mai 1999): 27–35. Sarzana: Agora Edizioni.

Bouffier, S., Belvedere, O. and Vassallo, S. (eds) 2019. *Gérer l'eau en Méditerranée au premier millénaire av. J.-C.* Aix-en-Provence: Presses Universitaires de Provence.

Bouffier, S. and Brunet, M. 2020. L'eau dans les cités grecques antiques : approvisionnement et salubrité, in S. Bouffier and I. Fumadó Ortega (eds), *L'eau dans tous ses états*: 175–193. Aix-en-Provence: Presses Universitaires de Provence.

Bouffier, S., Dumas, V., Lenhardt, P. and Paillet, J.-L. 2018. HYDROSYRA Project. Some reflections on the typology of the Ancient Aqueduct of Galermi (Syracuse, Italy), in J. Berking (ed.) *Water Management in Ancient civilizations*: 289–309. Berlin, Excellence Cluster Topoi, Workshop 2, 11–12 February 2016, 2018, ISSN (online) 2366-665X; doi.org/10.17171/3-53.Bouffier, S. and Fumadó Ortega, I. (eds) 2020. *L'eau dans tous ses états*. Aix-en-Provence: Presses Universitairse de Provence.

Bouffier, S. and Wateau, F. 2019. Galermi au fil de son eau : un aqueduc syracusain dans la reconfiguration de son territoire, *Développement durable et territoires*

[En ligne], Vol. 10, n°3 | Décembre 2019, mis en ligne le 20 décembre 2019, consulté le 22 janvier 2020; http://journals.openedition.org/developpementdurable/16046; doi.org/10.4000/developpementdurable.16046.

Brunet, M., 2008. La gestion de l'eau en milieu urbain et rural à Délos dans l'Antiquité, in E. Hermon (ed.) *Vers une gestion intégrée de l'eau dans l'empire romain*. Actes du Colloque International Université Laval. Laval: Michael Shamansky Bookseller.

Bruneton, H. 1999. *Evolution holocène d'un hydrosystème nord-méditerranéen et de son environnement géomorphologique. Les plaines d'Arles à l'interface entre le massif des Alpilles et le Rhône*. Thèse de Doctorat de Géographie Physique, Université de Provence.

Desruelles, S. and Fouache, E. 2015. Chapitre XV. L'eau dans la ville antique de Délos (Cyclades, Grèce). Des ressources aux aménagements, in N. Carcaud and G. Arnaud-Fassetta (eds) *La géoarchéologie française au XXIe siècle*: 205–214. Paris, Ed. du CNRS.

Dessales, H. 2013. *Le partage de l'eau. Fontaines et distribution hydraulique dans l'habitat urbain de l'Italie romaine*. Rome: École française de Rome.

Fumadó Ortega, I. and Bouffier, S. (eds) 2020. *Mortiers et hydraulique en Méditerranée antique*. Aix-en-Provence: Presses Universitaires de Provence.

Hairy, I. (ed.) 2010. *Du Nil à Alexandrie, Histoires d'Eaux, Hauterive*. Alexandrie: Ed. Harpocrates.

Karakhanian, A., Djrbashian, R., Trifonov, V., Philip, H., Arakelian, S. and Avagian, A. 2002. Holocene-historical volcanism and active faults as natural risk factors for Armenia and adjacent countries. *Journal of Volcanology and Geothermal Research* 113: 319–344.

Lasserre, F. and Descroix, L. 2011. *Eaux et territoires. Tensions, coopérations et géopolitique de l'eau*. Québec: Presses de l'université du Québec.

Leveau, P. 2011. Les études de cas sur des milieux palustres et fluviaux en Basse-Provence. De l'anthropisation à la prévision environnementale. *Méditerranée* 117: 17–23.Leveau, P. and Provansal, M. (eds) 1998. Archéologie et paléopaysages. *Méditerranée* 1998/4.

Luo, T., Young, R. and P. Reig 2015. Aqueduct Projected Water Stress Country Rankings. Technical Note. Washington, D.C.: World Resources Institute. Available online at: www.wri.org/publication/aqueduct-projected-water-stress-country-rankings.

Masetti-Rouault, M.G. and Defendenti, F. 2020. Eau pour l'Assyrie : les rois assyriens et leur politique hydrique, in S. Bouffier and I. Fumadó Ortega (eds) *L'eau dans tous ses états*: 105–116. Aix-en-Provence: Presses Universitaire de Provence.

Miramont, C. (ed) 2004. Géosystèmes montagnards et méditerranéens. Un mélange offert à Maurice Jorda. *Méditerranée* 102(1–2): 71–84.

Morhange, C. 1994. *La mobilité récente des littoraux provençaux, éléments d'analyse géomorphologique*, Thèse de doctorat en géographie physique, Université de Provence, http://tel.archives-ouvertes.fr/tel-00685442/fr

Nenna, M.-D. 2010. L'eau et l'hygiène dans l'Alexandrie antique, in I. Hairy (ed.) *Du Nil à Alexandrie, Histoires d'Eaux, Hauterive*: 490-501. Alexandrie: Ed. Harpocrates.

Ollivier, V. 2006. *Continuités, instabilités et ruptures morphogéniques en Provence depuis la dernière glaciation. Travertinisation, détritisme et incisions sur le piémont sud du Grand Luberon (Vaucluse, France). Relations avec les changements climatiques et l'anthropisation*. Thèse de doctorat de Géographie Physique, Université de Provence U1.

Ollivier, V. 2022. Multiple geomorphic factors and responses in the landscape reconstruction of Aknashen archaeological site (Arax Valley, Armenia). Oxford: Archaeopress Archaeology, 61-66.

Ollivier V., Fontugne M., Hamon C., Decaix A., Hatté C. and Jalabadze M. 2018. Neolithic water management and flooding in the Lesser Caucasus (Georgia). *Quaternary Science Reviews* 197: 267–287.

Ollivier, V., Roiron, P., Balasescu, A., Nahapetyan, S., Gabrielyan, Y. and J.-L. Guendon 2008. Milieux, processus, faciès et dynamiques morphosédimentaires des formations travertineuses quaternaires en relation avec les changements climatiques et les occupations humaines entre Méditerranée et Caucase. *Studii de Preistorie (Etudes de Préhistoire), revue du Musée National d'Histoire de la Roumanie* 5: 15–35.

Passchier, C.W., Wiplinger, G., Güngör, T., Kessener, P. and Sürmelihindi, G. 2013. Normal fault displacement dislocating a Roman aqueduct of Ephesos, western Turkey. *Terra Nova* 25: 292–297.

Politique et contrôle de l'eau dans le Moyen-Orient ancien. Annales. Histoire, Sciences Sociales 2002/3; 57e année.

Polizzi, G., Ollivier, V., Fumadó Ortega, I. and S. Bouffier 2017. Archéologie et hydrogéologie : de nouvelles données concernant la gestion de l'eau dans la cité de Solonte (Nord-Ouest Sicile, Italie). *Chronique des activités archéologiques de l'École française de Rome* [En ligne], Sicile, mis en ligne le 12 avril 2017; http://cefr.revues.org/1705; doi.org/ 10.4000/cefr.1705.

Provansal, M. 1995, The role of climate in landscape morphogenesis since the Bronze Age in Provence, southeastern France. *The Holocene* 5(3): 348–353.

Ricci, A., Helwing, B. and T. Aliyev 2012. The Neolithic on the Move: High Resolution Settlement Dynamics Investigations and Their Impact on Archaeological Landscape Studies in Southwest Azerbaijan. *eTopoi, journal for Ancient Studies*, Special Volume 3: 369–375.

Robinson, B., Bouffier, S. and Fumadó Ortega, I. (eds) 2019. *Ancient Waterlands*. Aix-en-Provence: Presses Universitaires de Provence.

Sherratt, A. 1980. Water, Soil and Seasonality in Early Cereal Cultivation. *World Archaeology* 11.3 (Water Management): 313–330.

Westaway, R. 1990. Seismicity and tectonic deformation rate in Soviet Armenia: implications for local earthquake hazard and evolution of adjacent regions. *Tectonics* 9: 477–503.

Wittfogel, K. 1957. *Oriental Despotism: A Comparative Study of Total Power*. New Haven, CT: Yale University Press.

The Impact of Climate, Resource Availability, Natural Disturbances and Human Subsistence Strategies on Sicilian Landscape Dynamics During the Holocene

Salvatore Pasta,[1] Giuseppe D'Amore,[8] Cipriano Di Maggio,[2] Gaetano Di Pasquale,[3] Vincenza Forgia,[9] Alessandro Incarbona,[2] Giuliana Madonia,[2] César Morales-Molino,[4] Silvio Giuseppe Rotolo,[2,5] Luca Sineo,[6] Claudia Speciale,[7] Attilio Sulli,[2] Willy Tinner[4] and Matteo Vacchi[10]

[1] Institute of Bioscience and BioResources (IBBR), Italian National Council of Research (CNR), Unit of Palermo, Corso Calatafimi 414, I-90129 Palermo, Italy
[2] Department of Earth and Marine di Sciences (DiSTeM), University of Palermo, via Archirafi 22, I-90123 Palermo, Italy
[3] Department of Agricultural Sciences, University of Naples 'Federico II', via Università 100, I-80055 Portici, Italy
[4] Institute of Plant Sciences and Oeschger Centre for Climate Change Research, University of Bern, Altenbergrain 21, CH-3013 Bern, Switzerland
[5] National Institute of Geophysics and Volcanology (INGV), Unit of Palermo, Ugo La Malfa 153, I-90146 Palermo, Italy
[6] Department of Biological, Chemical and Pharmaceutical Sciences and Technologies (STEBICEF), University of Palermo, via Archirafi 18, I-90123 Palermo, Italy
[7] Department of Historical Studies, University of Gothenburg, Renströmsgatan 6; 41255 Göteborg, Sweden
[8] Institute of Archaeo-Anthropological Studies (ISA), via delle Cascine 46, I-50018 Scandicci, Italy
[9] via San Giuseppe 24, I-90018, Termini Imerese, Italy
[10] Department of Earth Sciences (DST), University of Pisa, via Santa Maria 53, I-56126 Pisa, Italy

Abstract: This paper presents a multidisciplinary summary of the most recent discoveries and hypotheses concerning factors driving the human subsistence economy and landscape shaping in Sicily during the Holocene. A number of scientific papers have recently pointed out the key role played by paleogeography, resource (water, food) availability and natural disturbances (volcanic eruptions, tsunamis) in local human activities. Modern anthropology and archaeology increasingly use biological remains (e.g. soils, bones, wood, plant macroremains, pollen) to better understand how human communities managed to survive and spread. Likewise, refined reconstructions of past human demographic fluxes and socio-economic structures may enable a better understanding of landscape dynamics. Specifically, this historical perspective on the management of natural resources allows the finding of past episodes of unsustainable land use (e.g. forest destruction, overgrazing), thus providing a useful basis for future nature protection and maintaining sustainable ecosystem services.

Keywords: palaeoclimatic patterns, marine geology, palaeogeography, volcanic and seismic activity, environmental constraints, vegetation history, human impact, anthropology, palynology, charcoal and wood analysis, archaeological sciences

Introduction

Due to its ancient settlement history and long-lasting and almost continuously high demographic density, Sicily, the largest Mediterranean island, offers many opportunities to investigate the complex and dynamic relationships between landscape evolution, anthropogenic disturbance and resource availability. In this contribution, considering its geographic position in the midst of the Mediterranean Sea, its high geo-topographical and pedo-climatic variability, resulting in marked environmental gradients, its habitat patchiness and the complexity of its human history,[1] earth and life scientists were invited to share their knowledge and opinions with anthropologists and archaeologists to provide an up-to-date, multi-disciplinary and dynamic overview of the most recent hypotheses and data concerning the driving factors of Sicilian human peopling. Authors were explicitly requested to focus on the complex balance between resources and disturbances. Hence this paper will provide some clues on how and to what extent factors like food and water availability, geomorphology, sea level, climate, local vegetation patterns and dynamics or volcanic hazards may have affected the routes and speed of colonisation, the location of settlements, or the subsistence strategies of human communities living in Sicily from the Palaeolithic to the Common Era (CE).

Geologists and anthropologists were explicitly asked to provide a synthetic overview of current knowledge on coastal and regional landscape dynamics over the last 20,000 years, in order to check if the most updated information supports the hypothesis of the two main paths of colonisation of Sicily, namely Europe and north Africa.

---
[1] Guarino and Pasta 2017.

To help readers we have created four additional maps (Figures 8–11) to show the exact location of the place names cited in the text. We are confident that any possibly missing information can be easily retraced by consulting the papers quoted throughout the text. Ages are given in millions or thousands of years (calibrated) Before Present (Ma and ka BP, respectively), with 'present' corresponding to 1950 CE. For comparison with archaeology, we frequently shift to BCE and CE (Before Common Era, Common Era), respectively. Also in this case the dates must be considered as calibrated, unless specified otherwise by adding 'uncal.'.

We are aware that the present synthesis sounds rather ambitious due to the very wide time lapse considered and that many questions remained unanswered, namely the management of water resources – which was actually the topic of the congress but is as yet not well studied enough to provide a full overview at the regional scale. Although the congress focused on the 1st millennium BCE, we decided to revisit the long period between the Neolithic – with the dawn of productive economy – and the Iron Age, so as to give a reasoned overview of the most recent literature on Sicilian land use history. In fact, during this phase and, in particular, with the Mesolithic-Neolithic transition, the paradigms of humans/nature relationship changed forever. For the same reasons, we decided to provide some useful hints to stimulate further debate and address future research projects on the major events affecting the evolution of Sicilian landscapes over last three millennia.

Evolution of the physical landscape of Sicily over the last 20,000 years (C. Di Maggio and A. Sulli)

Sicily is a very young island that did not yet exist during the Early Pliocene (5.3–3.6 Ma), when its already deformed successions of Mesozoic-Paleogene rocks constituted a subsiding seabed (intramountain, syntectonic and foreland basin) affected by accumulations of pelagic deposits (locally known as Trubi formation). Subsequently, crustal shortening, thickening and consequent isostatic compensation due to the convergence between the European (Corsican-Sardinian block) and the African plates triggered uplift. More specifically, according to Di Maggio et al. 2017 (and references therein), between the Late Pliocene and Early Pleistocene (3.6–2.6 Ma) and the Early Calabrian (1.5 Ma), uplift was responsible for the gradual emersion of the current northern areas of Sicily, while its current central-southern areas were still below sea level, forming a subsiding foredeep basin. Clastic deposits (locally known as Marnoso-Arenacea, Monte Narbone and Agrigento formations) filled this basin. From the Early Calabrian (1.5 Ma) to the present, with the southward progressive migration of the Sicilian chain-foredeep-foreland system, the central-southern areas were also involved in the uplift and began to emerge, causing a gradual southward shift of the shoreline. At the same time, many blocks of the previously emerged northernmost areas began to sink below sea level, due to extensional faults related to the opening of the Tyrrhenian Sea. Starting from the Middle Pleistocene (around 0.8 Ma), some of the drowned blocks began to rise again and gradually re-emerged, producing the coastal plains of Bonagia, Castelluzzo, Partinico-Castellammare del Golfo, Carini and Palermo-Bagheria. The average speed of those uplift movements is generally between 0.2–0.4 mm yr^{-1}, but high-angle extensional to transtensive or reverse to transpressive faults may modify these values at the local scale.[2]

Uplift is the main cause of Sicily's long-term geomorphological evolution.[3] Its interaction with sea-level changes and coastal processes is responsible for the flat setting of the southern coastal areas and the more recently emerged small northern coastal plains, both characterised by a staircase of raised marine terraces. In the older relief of central Sicily, where the marine terraces have been dismantled and the underlying (mainly Neogene) clayey rocks (foredeep successions) have been exposed, competition between tectonic uplift and river down-cutting is responsible for the genesis of an alternation of deep V-shaped valleys and rounded hills prone to rapid and constant changes due to landslides and water erosion. In the mountain areas of northern Sicily affected by erosion for much longer, where down-cutting processes have dismantled even the foredeep deposits and exposed the deeper Mesozoic-Paleogene (mostly carbonate) rocks, the development of karst has slowed down the river deepening, reducing the frequency of V-shaped valleys and canyons and allowing the relief to reach higher altitudes.

From the Pliocene to Early Pleistocene, the current Hyblaean plateau was still below sea level. During the Early Pleistocene, punctual uplift, driven mainly by magmatic processes and secondly by flexural bulging[4] typical to this foreland area, involved stepwise emergence, triggering, indirectly, both river deepening, with the development of numerous canyons, and downward migration of the coastal processes, with the genesis of a Middle/Upper Pleistocene staircase of raised marine terraces along the coastal areas. Just emerged, the Hyblaean plateau represented an island separated from Sicily by a sea-arm submerging the current plains of Catania and Gela, where the foredeep was developing. Due to strong marine to continental deposition[5] compensating the flexural subsidence,

---

[2] Catalano et al. 2013 and references therein; Gasparo Morticelli et al. 2017.
[3] Di Maggio et al. 2017.
[4] Henriquet et al. 2019.
[5] Longhitano and Colella 2003 and references therein.

the bottom of this sea (submerged foredeep) finally emerged in the Late Pleistocene, connecting the Hyblaean Plateau to Sicily and giving rise to the current emerged foreland basin.

Climate is the main driver of geomorphological changes affecting Sicily over the short term (from the Late Pleistocene to present). During the last glacial maximum (hereafter LGM, i.e. ~20 ka BP), the physical landscape of Sicily was broadly similar to that of today, with a harsh and abrupt mountain relief in the north areas, an undulating hilly relief in the central areas, and a larger flat relief than the current one (see next paragraph, Figure 1) in the south areas.[6]

Present-day regional equilibrium line altitudes (ELAs), i.e. the line between the ice ablation and accumulation zones, are between c. 3,400–3,600 m a.s.l.,[7] whilst during the Pleistocene cold stages, they were estimated to be c. 2,000–2,500 m a.s.l.[8] As suggested by Neri et al. 1994 and Carveni et al. 2016 (and references therein), these values allowed the existence of glaciers on Mt Etna (3,329 m a.s.l.), by far the highest mountain on the island. Particularly, through geomorphological, stratigraphic and volcanological analyses, Carveni et al. 2016 recognised some glacial landforms, which lead back to the LGM. Moreover, the highest uplands of the Madonie (Carbonara Massif) and Nebrodi Mountains (Monte Soro area), both above 1,700–1,800 m a.s.l., but below the regional ELAs of that time, were covered with perennial snow.[9]

Di Maggio et al. 2009 (and references therein) and Agate et al. 2017 report the occurrence of large talus slopes and talus cones at the foot of the main rock scarps, which occur from mountain to coastal areas. These landforms are produced by accumulation of thick, stratified slope deposits, variously cemented during the last glacial cycle (LGC, i.e. ~110.8–11.7 ka BP). Their genesis implies strong physical weathering, producing large quantities of coarse scree and frequent debris/rock falls affecting the steep slopes, both favoured by a cold climate. Their stratification is indicative of runoff processes, mainly due to prevailing conditions of aridity. These conditions are also testified by the colluvial deposits with fossil mammals dated to the Late Upper Pleistocene, widely occurring in Sicily. Thin intercalations of paleosoils in stratified slope deposits and reworked soils in colluvial deposits indicate short climate oscillations in humid/temperate sense, which allowed soil formation.

Still during the LGC, in mountain and inland hilly areas, both the strong deposition rates due to higher production of scree and the lower capacity of the streams to transport and remove material due to poor discharges, linked to conditions of prevailing aridity typical of each glacial cycle,[10] were responsible for the filling of the river valleys, as indicated by both the alluvial river terraces of the Corleone area[11] and the filled valleys of the Madonie area[12]. Moreover, the large amounts of scree accumulated over clayey slopes have triggered numerous and large landslides.[13]

Conversely, the river segments crossing the coastal areas underwent vertical erosion due to sea-level lowering, producing river canyons and V-shaped valleys and creating steep river slopes subjected to landslides and soil erosion. Additionally, sea-level lowering induced the emersion of sandy deposits cropping out on previously submerged continental platforms (Figure 1), favouring the genesis and the development of wind-fed coastal and climbing dunes.[14]

With the transition from LGM to the Holocene, higher values of temperature and moisture reduced scree production and induced an increase in stream discharge, triggering processes of river down-cutting along the mountain and inland hilly areas. The growing local relief (i.e. the elevation range between the lowest and highest altitudes) triggered several landslides, particularly frequent on the clayey slopes (e.g. Servizio Geologico d'Italia 2010; 2011). Along the coastal areas, the sea-level rise involved the drowning and subsequent filling of river valleys, with the creation of large alluvial plains periodically affected by flooding.

Little data are available on the reconstruction of the ages in which the barrier lagoon/marsh systems of the Sicilian coastal areas began to form. These systems are controlled by multiple processes working along the coasts, the main ones being the relative sea-level changes and net sedimentation rates along the coasts. Available geological data obtained through different approaches, combined with multi-dating techniques, agree in suggesting that the primary cause in the genesis of the present-day systems of barrier lagoon/marsh is the slowdown in the Holocenic rise in sea-levels. Most of the systems of barrier, lagoon, and marsh along oceanic and gulf coasts of the USA, Brazil, Northern Australia, South Africa, and North Europe, actually started to form between c. 8.0–5.0 ka BP.[15] Specifically, the barrier lagoon/marsh systems of the Mediterranean[16] and, particularly, the Italian coastal

---

[6] Di Maggio et al. 2017.
[7] Messerli 1980
[8] Hughes and Woodward 2017 and references therein.
[9] Hughes and Woodward 2017.
[10] E.g. Lambert et al. 2008.
[11] Servizio Geologico d'Italia 2010
[12] Servizio Geologico d'Italia 2011
[13] E.g. Portella Colla and Piano Zucchi landslides; see Servizio Geologico d'Italia 2011 and references therein.
[14] Di Maggio et al. 2009 and references therein; Agate et al. 2017.
[15] McBride et al. 2013, and references therein; Benallack et al. 2016.
[16] Marco-Barba et al. 2013.

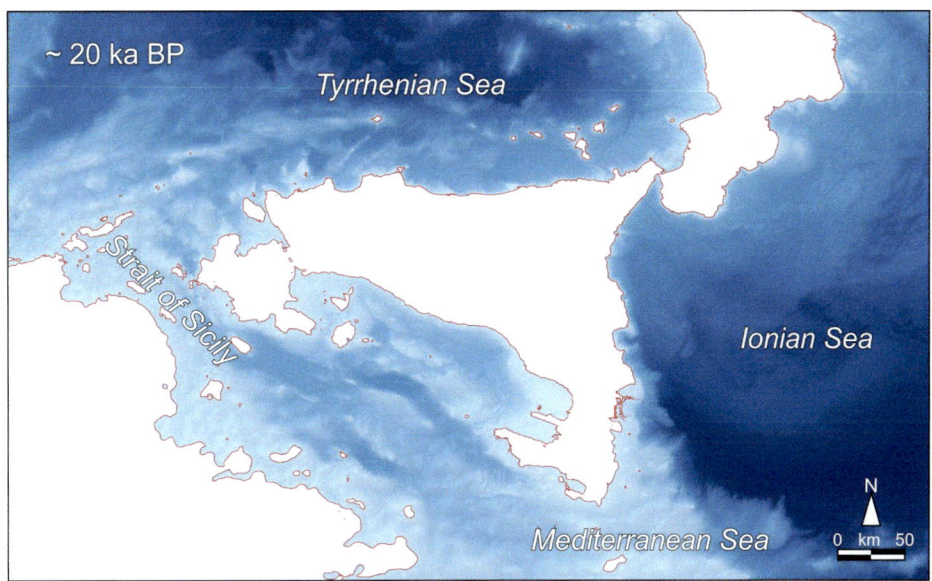
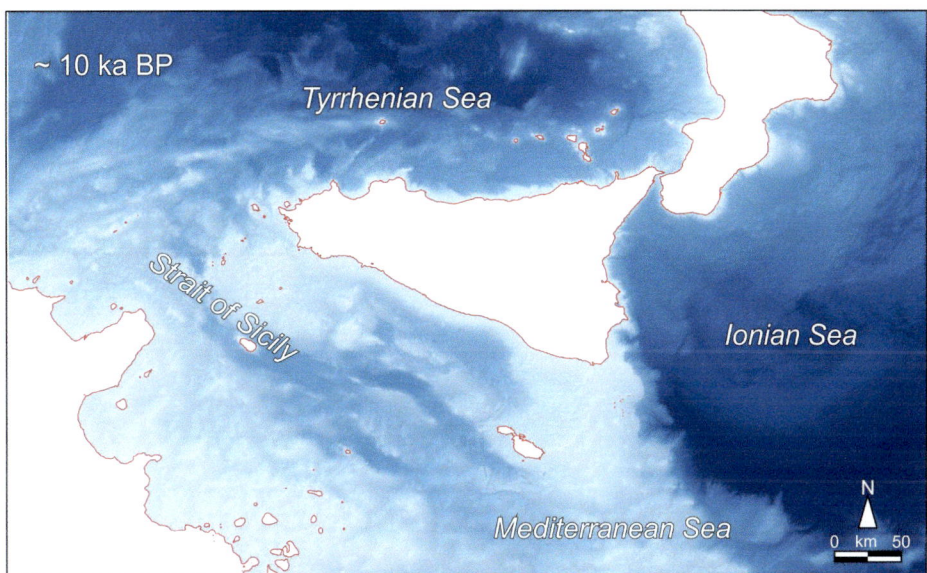
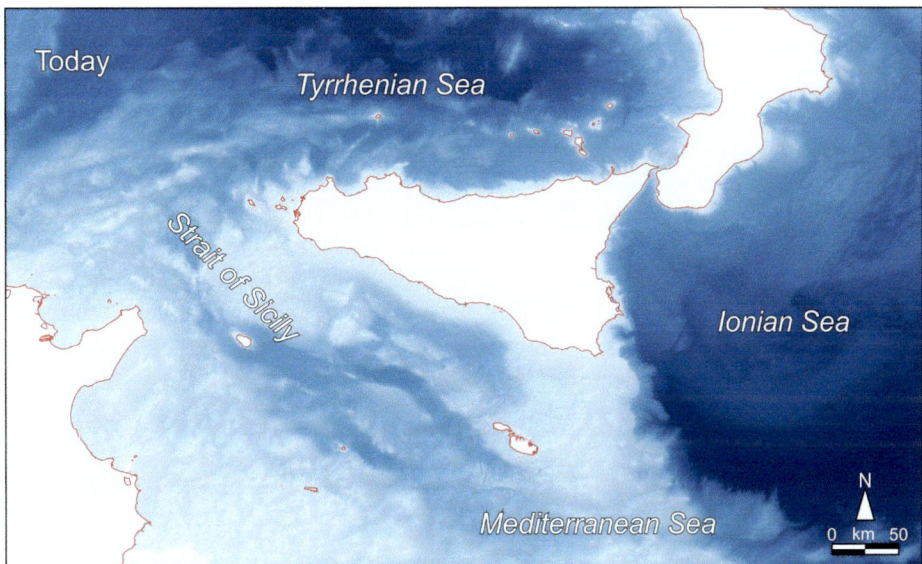

Figure 1: Paleo-geographical reconstruction of Sicily and its surrounding at ~20 ka and ~10 ka BP. Red lines indicate the approximate location of the paleo-shorelines on the basis of the present bathymetry (source: EMODnet Bathymetry Consortium, Thierry et al. 2019).

plains,[17] began to develop between c. 8.7–5.0 ka BP. It is reasonable to assume that the Sicilian coastal barriers, lagoons, and marshes began to form during the Holocenic deceleration in sea-level rise, c. 8.0–5.0 ka BP, similar to many parts of the world. On the other hand, AMS ^{14}C age determinations,[18] both on archaeological remains (~4.6 ka BP) found within coastal dune deposits of the barrier lagoon system of Ganzirri (Messina) and on underlying beach conglomerates (~5.9 ka BP), confirm this hypothesis. Historians provide additional information about when people abandoned the coasts near these lagoons to escape malaria. Sicilian wetlands, i. e. river mouths, coastal marshes and lagoons, but also some wetlands located in central Sicily – like those once occurring near Vallelunga Pratameno and Villarosa, now disappeared – represented a danger to human life for millennia, due to the occurrence of this disease. For this reason, most of the territories near the coastal wetlands remained uninhabited until 150–100 years ago, when most were reclaimed for agricultural purposes.

During the Holocene, milder climate conditions generally favoured soil formation, inhibiting erosion processes. However, whenever climate deterioration has occurred (see the paragraph on palaeoclimate reconstruction), physical weathering, soil erosion and landslides have increased their power. As a consequence of deforestation, particularly intense under Roman rule, runoff produced strong soil erosion.[19]

Sea-level changes and the Sicilian coastland during the last 20,000 years (M. Vacchi)

The subject of the evolution of the largest Mediterranean island after the LGM has been central to several multiproxy investigations carried out along Sicilian coastlines. In particular, sea-level changes and the palaeogeographic reconstruction of the past shorelines have been the focus of several publications.[20] These studies are of great archaeological interest – coastlines and coastal plains have always played an important role as natural landfall sites and cultural exchange.[21]

Global eustatic sea-level rose more than 100 m since 17.0 ka BP, and c. 56 m between 11.7–7.0 ka BP (Figure 1, 2D), with an average rate of 1.2 cm/year.[22] This estimate is one order of magnitude higher than the highest tectonic uplift rates in Sicily (<2.0 mm/year).[23] Thus, eustatic sea-level rise was a relevant factor for the dynamics of Sicilian coastal landscape over the Early/Middle Holocene.

At the end of the LGM, with the sea-level 120–140 m below that of the present day (Figure 1),[24] Sicily remained connected with the Italian peninsula for at least 500 years in the period between 21.5–20.0 ka BP.[25] This land bridge likely facilitated the first colonisation of the island by anatomically modern humans.[26] The progressive melting of the Northern Hemisphere ice sheets produced a very rapid sea-level rise until ~7.0 ka BP,[27] which resulted in the progressive development of the Messina Strait and the consequent separation of Sicily from the Italian mainland[28] and islands such as Malta. Since that period, the evolution of the Sicilian coasts has been driven by the complex interplay between eustatic, isostatic and tectonic movements. The latter were of particular importance in the north-eastern portion of the island, where the activity of some major faults significantly affected coastal evolution during the last millennia. For instance, at Milazzo[29] and around the Taormina area[30] the coupled activity of long-term and co-seismic uplifts, with rates up to ~2 mm a^{-1}, played a major role in the Middle to Late Holocene (last 6.0 ka BP) evolution of local coasts.

A more complex tectonic pattern was recorded along the coastal stretch between Catania and Siracusa.[31] Here, the results of archaeological surveys and coring activities in coastal lagoons provided a long record of sea-level evolution.[32] The oldest data indicate that sea-level rose from -36.9±1.1 m at ~9.6 ka BP to -6.2±1.0 m at ~6.6 ka BP and finally to -2.3±1.0 m at ~4.2 ka BP. Late Holocene data show variability, most likely related to the differential uplift trend affecting the area.[33] Data from Augusta and Priolo indicate that sea-level reached the present value around ~3.7 ka BP. Conversely, archaeological data indicate a different pattern, with the sea-level placed between -3 and -1.2 m at ~3.7 ka BP and at -1.5 m±0.3 m at ~2.6 ka BP. These data probably need to be corroborated by further field investigation: as far as we know, their partial disagreement could also depend on the differential speed of local tectonic uplift. This area was also affected by a series of extreme wave inundations (storms and tsunami), at least over the last 4000 years. Geomorphological and sedimentary evidence of these extreme events was found south

---

[17] Amorosi et al. 2013; D'Orefice et al. 2020, and references therein.
[18] Antonioli et al. 2006b, and references therein.
[19] Brandolini et al. 2019, and references therein.
[20] Scicchitano et al. 2008; Antonioli et al. 2014; Lo Presti et al. 2019.
[21] Vacchi et al. 2019; Revelles et al. 2019.
[22] Lambeck et al. 2014; 2016.
[23] Di Stefano and Branca 2002; Antonioli et al. 2006b.
[24] Lambeck et al. 2016; Thierry et al. 2019.
[25] Antonioli et al. 2014.
[26] Antonioli et al. 2016; Lo Presti et al. 2019.
[27] Lambeck et al. 2014; Vacchi et al. 2018.
[28] Antonioli et al. 2014.
[29] Scicchitano et al. 2011.
[30] Antonioli et al. 2003; De Guidi et al. 2003; Spampinato et al. 2012.
[31] Scicchitano et al. 2008; Spampinato et al. 2011.
[32] Vacchi et al. 2016.
[33] Spampinato et al. 2011.

of Siracusa[34] and in the Augusta Bay.[35] However, the current elevation of the last interglacial shoreline (~125 ka BP) suggests a much higher long-term tectonic stability in this portion of the Sicilian coast.[36]

The tectonic component is, conversely, considered negligible along the western coast of Sicily. In fact this area is considered tectonically stable at least since 125 ka BP.[37] Most of the available sea-level data concerning this sector of the island were extracted from cores retrieved from the Stagnone di Marsala lagoon[38] and from fossil vermetid reefs in Capo San Vito and near Palermo.[39] Additional sea-level insights were obtained from fossil marine shells collected on Marettimo Island and near Palermo[40] and from archaeological investigations in the Punic town of Mozia.[41] All these data allowed the reconstruction of sea-level evolution since 9.5 ka BP. At this time, the sea-level likely varied between ~–27 and ~–25 m at ~9.5 ka BP. A hiatus in the data did not allow an assessment of sea-level evolution in the Middle Holocene. Conversely, the sedimentary data from the Stagnone di Marsala lagoon, coupled with the fossil vermetid reef, did provide an opportunity to reconstruct with some confidence evolution in sea-level over the last three millennia. At ~2.8 ka BP, the sea-level was at -1.1±0.6m, while it rose to -0.4±0.6 m at ~2.5 ka BP. The entire sea-level variation in the last 1 ka BP was within ~0.3 m.

Climate reconstruction for the last 12,000 years (A. Incarbona and G. Madonia)

The Holocene is the current interglacial, starting *c.* 11.7 ka BP,[42] and is either divided into an early phase of climatic optimum, between 11.7–5.0 ka BP, and a later 'neo-glacial' stage, characterised by the advance of alpine glaciers,[43] or in three phases, with the Early Holocene lasting 11.7–8.2 ka BP, the Middle Holocene 8.2–5.0 ka BP, and the Late Holocene, which corresponds to the 'neo-glacial' in the two-phase system.[44]

Holocene marine sediments around Sicily testify to a climatic evolution that is in agreement with other Mediterranean Sea sites, both in terms of trends and RCCs. Alkenone-derived sea-surface temperature (SST) data, $\delta^{18}O$ profiles from planktonic foraminifera, and micropaleontological evidence from cores recovered in the southernmost Tyrrhenian Sea and the Sicily Channel, show the occurrence of a distinct thermal maximum for the Early to Middle Holocene and a subsequent cooling trend,[45] like in other central-western Mediterranean locations.[46] SST variations in the southernmost sector of the Tyrrhenian Sea (north Sicily offshore) (Figure 2A) and net primary productivity changes in the Sicily Channel (Figure 2B) show the full suborbital climatic variability of Holocene RCCs,[47] as observed by IRD deposition in the North Atlantic (Figure 2C).[48] IRD peaks (Bc 0–8) are associated, within dating uncertainties (Figure 2A–C), with decreased SST and increased primary productivity. This correlation suggests a long-distance connection between the Atlantic Ocean and the central Mediterranean Sea, probably similar to the North Atlantic Oscillation (NAO), whose positive phases are related to the arrival of more prolonged and stronger storm tracks to the southern central Mediterranean.[49]

Focusing on historical times, it is well established that the Little Ice Age (LIA, 1450–1850 CE) behaved as a RCCs event in the marine environment (Bc 0 in Figure 2C), once again characterised by cooling and enhanced primary productivity.[50] While a period of terrestrial drought is suggested by documentary evidence,[51] palaeoclimatic evidence from the Pantelleria island points to increased humidity,[52] exemplifying that moisture changes were variable in space. More recently, the full sequence of climatic variability since 1 CE (Roman Period, Medieval Warm Period and Little Ice Age) was documented by Mg/Ca-derived SST fluctuations in the Sicily Channel.[53] These historical climatic fluctuations perfectly overlap with the timing of those recorded in the Alborán and Aegean Seas and in the Minorca Rise,[54] highlighting a regional climatic framework of great archaeological relevance.

Speleothem isotope data from the Carburangeli cave (northern Sicily) record high-resolution rainfall signal between *c.* 9.0–6.0 ka BP (Figures 3A, B).[55] The $\delta^{18}O$ speleothem profiles of the Carburangeli and Wintimdouine caves, recovered from the north-western coast of Morocco,[56] have been correlated to refine the age model of the first by incorporating the numerous U/Th dates from the second (Figures 3A, B). The alignment of the two records (Figures 3C, D) shows

---

[34] Scicchitano et al. 2007; Barbano et al. 2010.
[35] De Martini et al. 2010.
[36] Antonioli et al. 2006a.
[37] Ferranti et al. 2006.
[38] Basso et al. 2008.
[39] Antonioli et al. 2002.
[40] Antonioli et al. 1996; 1999.
[41] Basso et al. 2008.
[42] Andersen et al. 2004; Rasmussen et al. 2014.
[43] Cronin 2010.
[44] Lang 1994.
[45] Sbaffi et al. 2001; Sprovieri et al. 2003; Di Stefano and Incarbona 2004; Essallami et al. 2007; Di Stefano et al. 2015.
[46] Marchal et al. 2002; Martrat et al. 2004; 2014.
[47] Sbaffi et al. 2001; Incarbona et al. 2008.
[48] Bond et al. 2001.
[49] Frigola et al. 2007; Incarbona et al. 2008; Bazzicalupo et al. 2020.
[50] Silenzi et al. 2004; Incarbona et al. 2010a.
[51] Piervitali and Colacino 2001.
[52] Calò et al. 2013.
[53] Margaritelli et al. 2020.
[54] Rodrigo-Gámiz et al. 2014; Cisneros et al. 2016; Gogou et al. 2016; Kontakiotis 2016.
[55] Frisia et al. 2006.
[56] Sha et al. 2019.

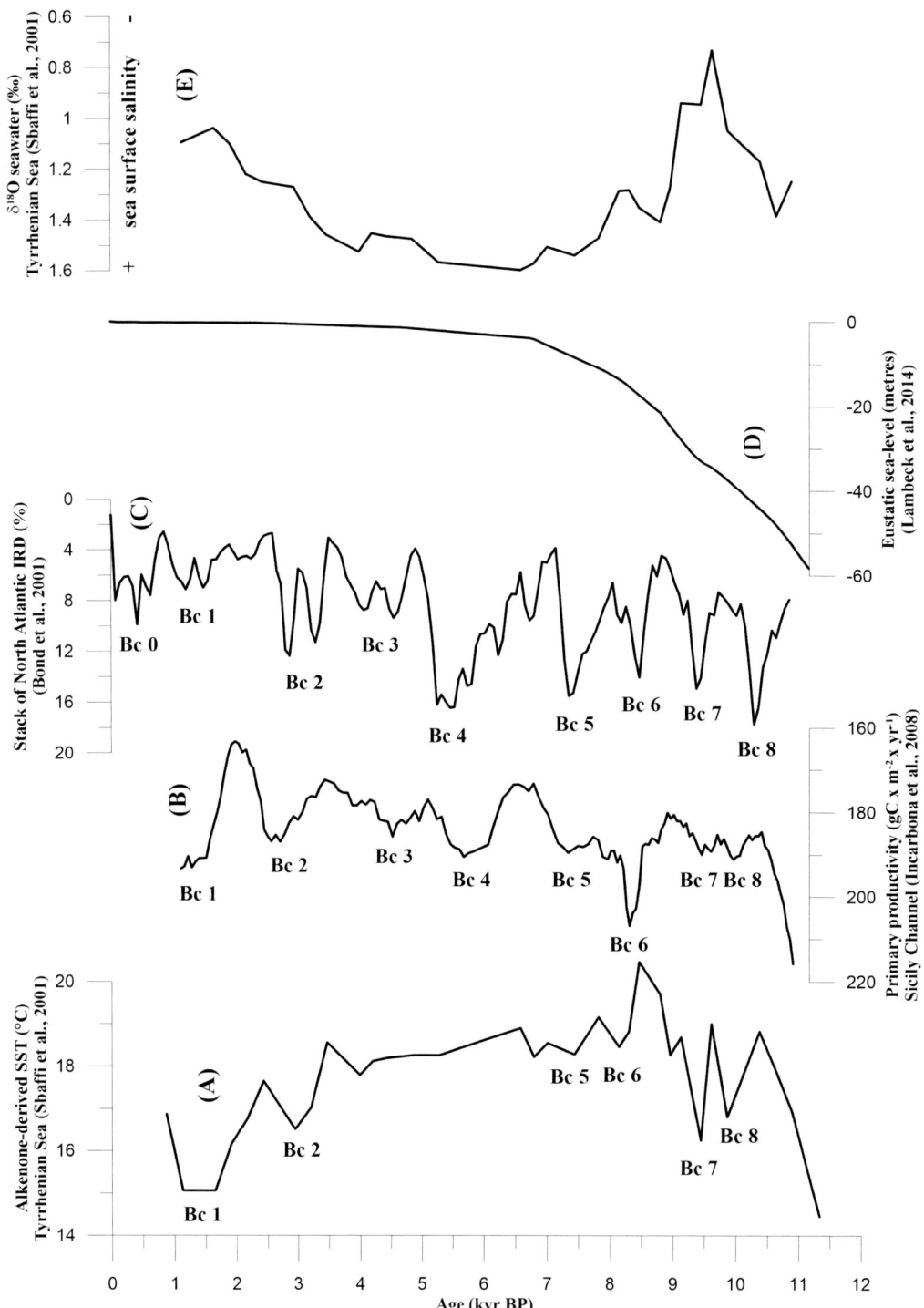

Figure 2: Downcore variations of marine geochemical and micropalaeontological records from the seas around Sicily, in comparison with a standard for Holocene climatic suborbital fluctuations (North Atlantic IRD) and the eustatic sea-level:
A: alkenone-derived estimates of SST in °C for core B79-38 (southern Tyrrhenian Sea, Sbaffi et al. 2001);
B: net primary productivity estimates (gC × m^{-2} × yr^{-1}), derived from percentage values of the coccolith species *Florisphaera profunda*, expressed as a 5-pt running average in the ODP 963 Site (Strait of Sicily) (Incarbona et al. 2008);
C: the stacked IRD record (%) from the North Atlantic Ocean (Bond et al. 2001). Bc (Bond cycles) refers to Holocene pulses of IRD that are assumed as a standard for RCC. Bc are also indicated in the SST and productivity records of the Tyrrhenian Sea and Strait of Sicily;
D: eustatic sea-level variations (metres) estimated by Lambeck et al. (2014);
E: Core B79-38 $\delta^{18}O$ seawater for core B79-38 (southern Tyrrhenian Sea) calculated from *Globigerina bulloides* $\delta^{18}O$ and alkenone-derived estimates of SST (Sbaffi et al. 2001).

similarities regarding the general lightening trend (arrows) and the millennial-scale oscillations, which in turn seem to be well correlated to North Atlantic IRD deposition (Figure 3E). The δ^{18}O data from Carburangeli and Wintimdouine speleothems were resampled to 100-yr window (circles in Figures 3C, D) to estimate their positive correlation, whose result is r = 0.592, n = 22. This preliminary remark suggests that isotope data from Sicily and Morocco may have responded to the same millennial-scale solicitation over this limited 9.0–6.0 ka BP time interval.

The Wintimdouine cave δ^{18}O record documents that the influence of the West African monsoon rainfall expanded to 31° N during the African Humid Period (11.0–5.0 ka BP in north-western Africa), coincident with the Green Sahara,[57] due to the northward shift of the ITCZ.[58] The occurrence of monsoonal precipitation in Sicily has to be excluded from many proxy data, first of all the vegetation pattern.[59] The speleothem δ^{18}O records from Soreq (Israel) and Jeita (Lebanon) caves have been associated to sapropel formations and to North African monsoon precipitation,[60] but their source is from Mediterranean Sea evaporation, even though they may have partially captured the monsoon signal from freshwater discharge into the Levantine Sea.[61] Equally, it is possible to assume that the Carburangeli speleothem δ^{18}O record may have captured the monsoon signal from freshwater discharge into the Atlantic Ocean and the western-central Mediterranean Sea, from north-western African rivers.[62] Thus, the Carburangeli speleothem δ^{18}O signal may not represent a reliable proxy for local precipitation during the Early/Middle Holocene.

Calcite speleothem δ^{13}C may reflect different processes, among others the δ^{13}C value of soil CO_2 and of the host bedrock prior to precipitation, variable CO_2 degassing due to different cave ventilation, vegetation composition/cover, and microbiological activity above the cave.[63] Frisia et al. 2006 interpreted the significant decrease in δ^{13}C observed in the Carburangeli speleothem at c. 7.8 ka BP as related to the establishment of C3 plant communities and a permanent vegetation cover above the cave (Figure 3F). However, this explanation is in conflict with δ^{18}O data from the same speleothem (Figure 3A) and with pollen data from Sicily records that, at the date of the Frisia et al. 2006 publication, was limited to Pergusa Lake.[64] As discussed above, the Carburangeli speleothem δ^{18}O signal may be affected by freshwater discharge from North Africa related to monsoonal activity and therefore would not be conclusive evidence of local rainfall. Further data concerning the Holocene vegetation patterns and dynamics from mountain and inland areas of Sicily show an earlier development of deciduous and evergreen forests[65] than in coastal areas, where their onset occurred after the 8.2 ka BP event, with subsequent full establishment at c. 7.0 ka BP.[66]

The increased moisture and the development of evergreen broadleaved forests in southern Sicily after the 8.2 ka BP event were explained by a decline in monsoon activity and the shift of the Hadley circulation cell, causing a weakening of the North Atlantic anticyclone, favouring moist, eastwards airflow towards the Mediterranean.[67] The high level of the Preola Lake between 10.5–7.0 ka BP, associated with no coastal evergreen forests in the same area, was explained with a marked seasonal precipitation contrast, and more specifically with insufficient summer moisture availability,[68] which is in agreement with recent quantitative salinity reconstructions from the same Sicilian coastal sites of Lago di Preola and Gorgo Basso, suggesting dry conditions during the Early Holocene and subhumid to humid conditions during the Middle/Late Holocene.[69] The two hypotheses were recently combined to explain increased precipitation in southern Spain since 7.8 ka BP, as indicated by the Cueva Victoria speleothem δ^{13}C record (Figure 3G).[70] Comparing the Spain and Sicily speleothem δ^{13}C records, it is likely that severe summer drought affected a large portion of the western-central Mediterranean Basin, due to the northward shift of ITCZ during the 10.0–7.8 ka BP (Figures 3F, G) phase of the Green Sahara,[71] while intense winter rainfall was occurring, as documented by high-level lakes, light δ^{18}O seawater in the Mediterranean Sea (Figure 2E), and increased river discharge in the Mediterranean northern borderlands.[72] Atmospheric models support the establishment of a stable, high-pressure cell in the Mediterranean area in summer, at the time of the northernmost ITCZ location.[73] Alternatively, Finné et al. 2019 state that there is no general feature common to the whole Mediterranean area, but quite sharp differences in terms of rainfall regimes between the South and East versus the North and West Mediterranean. They also point out a drought crisis affecting North Africa and

---

[57] Ehrmann et al. 2017; Tierney et al. 2017; Sha et al. 2019.
[58] Rohling et al. 2002; 2015.
[59] Sadori and Narcisi 2001; Tinner et al. 2009; Magny et al. 2011.
[60] Bar-Matthews et al. 2000; Cheng et al. 2015; Rohling et al. 2019.
[61] Wang et al. 2014; Rohling et al. 2015.
[62] Drake et al. 2011; Larrasoaña et al. 2013.
[63] Bar-Matthews et al. 1997; 1999; 2003; Drysdale et al. 2004; Spötl et al. 2005; Frisia et al. 2006; Budsky et al. 2019.
[64] Sadori and Narcisi 2001.
[65] Sadori and Narcisi 2001; Bisculm et al. 2012.
[66] Noti et al. 2009; Tinner et al. 2009; Calò et al. 2012.
[67] Tinner et al. 2009.
[68] Magny et al. 2011.
[69] Curry et al. 2016.
[70] Budsky et al. 2019.
[71] Budsky et al. 2019.
[72] Magny et al. 2011; Toucanne et al. 2015; Di Donato et al. 2018; Filippidi and De Lange 2019; Maiorano et al. 2019; Sha et al. 2019; Wagner et al. 2019.
[73] Pausata et al. 2016.

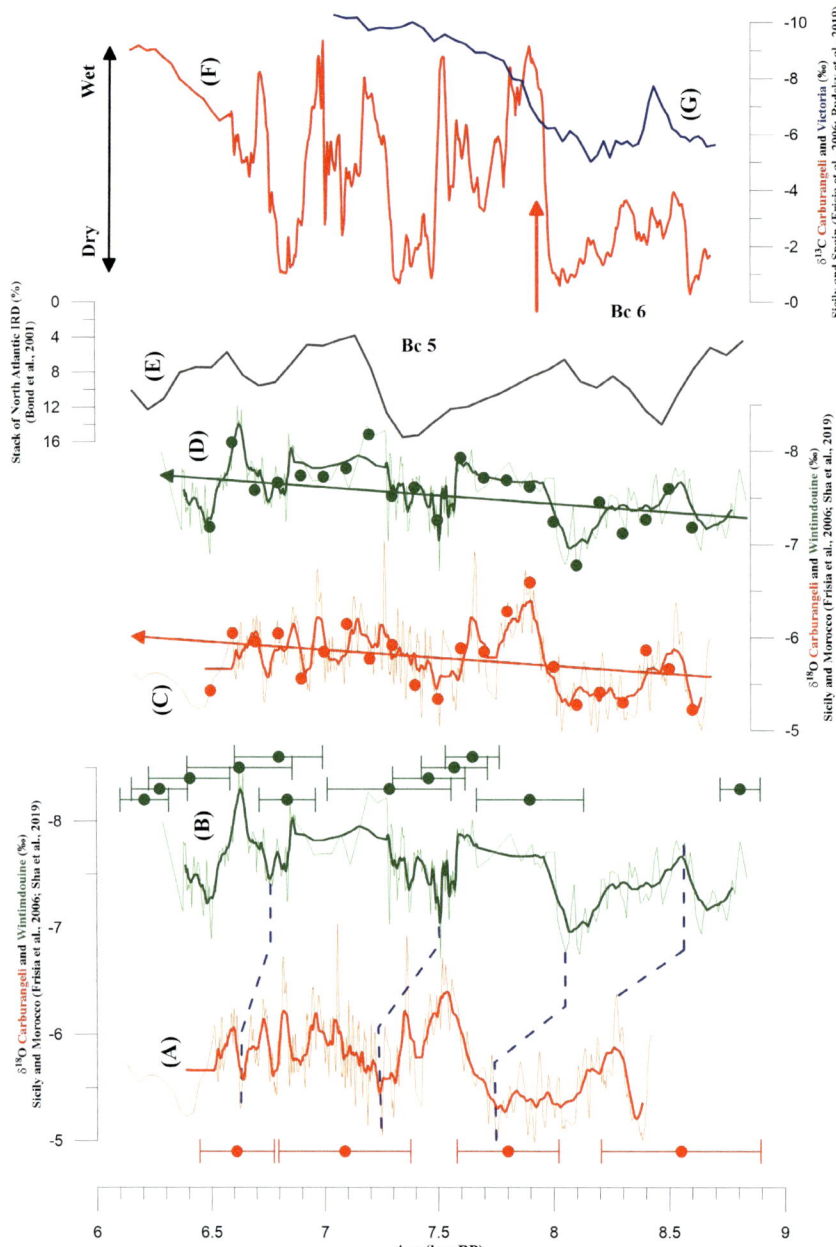

Figure 3: Geochemical profiles of stable isotopes from calcite speleothems and the IRD record from the North Atlantic Ocean, focused on the interval between about 9.0 and 6.0 ka BP:
A: the $\delta^{18}O$ record (‰) from Carburangeli Cave, plotted on the original chronology by Frisia *et al.* (2006). The orange thin line represents raw data. The red thick line is a 19-pt running average. The red circles are the U/Th datings and the vertical red bars are the 2-σ errors.
B: the $\delta^{18}O$ record (‰) from Wintimdouine Cave (Sha *et al.* 2019). The light green thin line represents the raw data. The green thick line is a 9-pt running average. The green circles are the U/Th datings and the vertical green bars are the 2-σ errors. The dashed blue lines are the tie-points used for the correlation and used to assess the new Carburangeli speleothem age model.
C: the $\delta^{18}O$ record (‰) from Carburangeli Cave, plotted on the new chronology after the correlation with the Wintimdouine speleothem. The orange thin line represents raw data. The red thick line is a 19-pt running average. The red circles are samples taken with a 100-yr pace. The red arrow represents the linear fit, whose equation is Y = 0.1716152701 * X - 7.068023323, n = 630.
D: the $\delta^{18}O$ record (‰) from Wintimdouine Cave (Sha *et al.* 2019). The light green thin line represents the raw data. The green thick line is a 9-pt running average. The green circles are samples taken with a 100-yr pace. The green arrow represents the linear fit, whose equation is Y = 0.1785826222 * X - 8.86427158, n = 388.
E: the stacked IRD record (%) from the North Atlantic Ocean (Bond *et al.* 2001). Bc (Bond cycles) refers to Holocene pulses of IRD that are assumed as a standard for RCC.
F: the $\delta^{13}C$ record (‰) from Carburangeli Cave (Frisia *et al.* 2006), plotted on the new chronology after the correlation with the Wintimdouine speleothem. The red thick line represents raw data. The red arrow marks the switch to wet conditions in Sicily after the 8.2 ka BP event. G: the $\delta^{13}C$ record (‰) from Victoria Cave, southern Spain (Budsky *et al.* 2019). The blue thick line represents raw data.

the Eastern Mediterranean *c.* 3.0 ± 0.3 kyr BP, and agree with generally wetter conditions between 8.5–6.1 ka BP.

In conclusion, the comparison between δ^{18}O of Carburangeli and Wintimdouine speleothems suggests that the source of the isotopic composition of the former speleothem may be from the Mediterranean Sea evaporation. More specifically, the speleothem of Carburangeli may have captured the monsoon signal from the freshwater discharge of north-western African rivers into the Atlantic Ocean and the western-central Mediterranean. This new interpretation corroborates the hypothesis that Sicily was prone to prolonged and severe droughts between *c.* 10.0–7.8 ka BP, which hampered the development of coastal evergreen forests, while forests were established in mountain and inland areas under the influence of more favourable orographic conditions.

Main volcanic eruptions and earthquakes affecting the history of Sicilian settlements (S. Rotolo, V. Forgia and S. Pasta)

On the one hand, the mineral-rich and fertile soils of the volcanic areas have long been considered a precious resource already by early Sicilian farmers. Not surprisingly, the foothills of Mt Etna, as well as most of the circum-Sicilian volcanic islands (Pantelleria, Linosa, Ustica, and the whole Aeolian Archipelago) were cultivated since prehistoric times[74] and have been and still are renowned for the quality of their agricultural products.[75] On the other hand, some active volcanic areas were avoided for a long time because they were considered dangerous due to frequent eruptions or poisonous gas emissions, e.g. Lago di Venere at Pantelleria Island,[76] or the foothills of the 'La Fossa' crater on Vulcano Island,[77] where even a brief overnight stay during wind-free periods may be lethal for humans and even migrating birds (Table 1).

However, on several volcanic islands eruptions ceased long before humans arrived. This is the case at Linosa, in the Sicily Channel, whose subaerial activity ended *c.* 500 ka BP, Ustica, whose last eruptive phase dates back to 132 ka BP,[78] Filicudi, Alicudi, and Salina.[79] The north-eastern Sicilian hunter-gatherer communities may have witnessed from afar the volcanic activity of Stromboli Island,[80] and the last Mesolithic and Neolithic groups the final eruptions of Panarea[81] and Pantelleria.[82] Moreover, the frequency, intensity and impact of submarine eruptions on human activities is still poorly understood; an increasing amount of evidence proves their occurrence until recent times close to apparently extinct volcanoes like Linosa,[83] Pantelleria,[84] Salina[85] and Panarea, where fumarolic activity is still present in the locality of Calcara, and underwater near its eastern offshore satellite islets.[86] Repeated localised gas emissions and ephemeral pyroclastic eruptions occurred in the Sicily Channel between the 19th and 20th centuries as well.[87]

The three most powerful eruptions of the last 20 ka BP undoubtedly occurred on Mt Etna. Specifically, there was a plinian eruption in 122 BCE and two earlier sub-plinian eruptions, dated 3.1 and 3.9 ka BP, respectively.[88] According to the Greek historian Diodorus Siculus, the 3.1 ka BP eruption probably caused the westward migration of Sicanian communities from their original settlements located north of Etna. Nevertheless, the most dangerous events linked to volcanic activity were (and very likely still are) not eruptions on their own. This is the case of Etna and Stromboli Island: here several tsunamis that caused many casualties during medieval times,[89] and probably even earlier (Tab. 1), were triggered by the massive landslides affecting the highly unstable and steep slopes of Sciara del Fuoco.

As for the Aeolian islands of Vulcano, Lipari and Stromboli, their almost continuous volcanic activity since prehistoric times had a strong effect on the alternated phases of economic health and decline of Aeolian communities. For instance, three demographic crises which have long been attributed to unfavourable socio-economic situations or to wars/aggressions, may have instead been triggered by volcanic events that disturbed the daily lives of local people and deeply compromised their economic activities and the undertaking of navigation and trade from mainland Sicily to the islands and back.[90] A first crisis occurred in the 4th millennium BCE and involved all the Aeolian Islands, in particular Contrada Diana. This coastal settlement, the main Neolithic site of Lipari Island, experienced a well-documented decline, and local people preferred to settle in areas protected from volcanic activity, such as Rocca del Castello. This demographic process may have depended on one or more tsunamis caused by Sciara del Fuoco and the

---

[74] E.g. Martinelli *et al.* 2021; Speciale *et al.* 2021.
[75] E.g. Laghetti *et al.* 1996; 1998.
[76] Favara *et al.* 2000.
[77] Carapezza *et al.* 2011.
[78] De Vita *et al.* 1998.
[79] 29, 28 and 15.6 ka BP, respectively: see Lucchi *et al.* 2013, and references therein.
[80] Lava flows of Vancori Superiori, *c.* 13.0±1.9 ka BP: Francalanci *et al.* 2013.
[81] Approx. 8.7 ka BP: Lucchi *et al.* 2013.
[82] Approx. 6 ka BP: Jordan *et al.* 2018.
[83] Romagnoli *et al.* 2020.
[84] Submarine eruption in 1891: Conte *et al.* 2014.
[85] Subaerial and submarine hotsprings: Cocchi *et al.* 2019.
[86] Lucchi *et al.* 2013; Esposito *et al.* 2018.
[87] Rotolo *et al.* 2006; Coltelli *et al.* 2016.
[88] Coltelli *et al.* 1998; 2000.
[89] Rosi *et al.* 2019.
[90] Manni *et al.* 2019.

Table 1: Overview of the main events related to Sicilian volcanoes and their consequences on local human communities.

Date/Age	Location	Impact	Eruptive typology	Reference(s)
1991–1993 CE	Etna Mt, fissure from W flank of Valle del Bove	A voluminous lava flow threatened the village of Zafferana Etnea	Flank effusive	Calvari *et al.* 1994
1983 and 1985 CE	Etna Mt, near Rifugio Sapienza	A lava flow destroyed several touristic installations	Flank effusive	Branca and Del Carlo 2004
1981 CE	Etna Mt, N flank fissure (2600 m a.s.l.)	A fast-moving lava flow threatened the village of Randazzo cutting the railway and a main road	Flank effusive, with a very high lava flow rate (up to 640 m3/sec)	Coltelli *et al.* 2014
1930 CE	Stromboli Island	Six casualties and mass migration of local people to Australia and USA	Vulcanian associated with small pyroclastic flows	Rosi *et al.* 2019
1928 CE	Etna Mt, fissure on the E flank (Ripa della Naca, 1200 m a.s.l.)	Lava flow destroyed the village of Mascali	Flank effusive	Duncan *et al.* 1996
1919 CE	Stromboli Island	1 ton ballistic ejecta reached Stromboli village; four casualties	Violent Strombolian	Ponte 1919
1891 CE	Pantelleria Island, Foerstner submarine volcano	Earthquakes and bradyseism damaged several houses at Pantelleria	Submarine eruption 4 km west offshore	Kelly *et al.* 2014
1888–1890 CE	Vulcano Island, locality La Fossa	Fallout of m3-sized lithic blocks over the village of Vulcano Porto; destruction of mining machines at Porto Levante	Vulcanian	Mercalli and Silvestri 1891; De Astis *et al.* 2013
1863 CE	Etna Mt, Central crater	Ash fallout reached Malta Island, *c.* 200 km SW	Sub-Plinian	Branca and Del Carlo 2004
1831 CE	Sicily Strait, Graham Bank	Earthquakes over the entire SW Sicilian sector; ashes reached the town of Sciacca *c.* 40 km NE	Submarine and subaerial (Surtseyan)	Cavallaro and Coltelli 2019
1763 CE	Etna Mt, locality Montagnola	2–3 km high fire fountains; a 10 cm-thick ash layer covered Catania	Hawaiian	Branca and Del Carlo 2004
1669 CE	Etna Mt, locality Monti Rossi	A fast-moving lava flow destroyed nine villages and part of the city of Catania (18 km from the vent)	Flank eruption, initially Strombolian, then effusive, with a very high lava flow rate (300–500 m3/sec)	Mulas *et al.* 2016
1456 CE	Stromboli Island	A tsunami triggered by a major landslide at Sciara del Fuoco reached the Campanian coast	Unrelated to a specific eruption	Rosi *et al.* 2019
1343 CE	Stromboli island	A tsunami originated from Stromboli reached the Campanian coast, 200 km N, and caused hundreds of casualties at Naples harbour and surroundings. Inhabitants left the island	Unrelated to a specific eruption	Rosi *et al.* 2019
1230 CE	Lipari Island, locality Rocche Rosse	Much damage to roads and houses	Strombolian and lava flow	Arrighi *et al.* 2006
1000–1200 CE	Vulcano Island, Vulcanello cone	Local damage. (N.B.: Plinius and Strabo, cited by Mercalli and Silvestri 1891 and by De Fiore 1922, report ages 126 BC and 183 BC, respectively)	Lava flows and Strombolian	Mercalli and Silvestri 1891; De Fiore 1922; Arrighi *et al.* (2006)
776 CE	Lipari Island, locality Mt Pilato	Ash fallout reached Vulcano Island and the Sicilian inland	Violent Strombolian and lava flow	Keller 2002
44 BCE	Etna Mt	Few cm-large lapilli and ash fallout reached Catania	Sub-Plinian	Coltelli *et al.* 2014
122 BCE	Etna Mt ('FG' eruption)	Fallout of a 10 to 20 cm-thick layer of lapilli covered Catania and the nearby villages along the Ionian coast	Plinian	Coltelli *et al.* 2014

Date/Age	Location	Impact	Eruptive typology	Reference(s)
3.1 ka BP	Etna Mt ('FL' eruption)	Migration of Sicani population (referred by Diodorus Siculus)	Sub-Plinian	Coltelli et al. 2014
3.9 ka BP	Etna Mt ('FS' eruption)	Fallout of lapilli covered the entire E slope of Etna and reached Catania	Sub-Plinian	Coltelli et al. 2014
7.0 ka BP	Stromboli Island, locality Secche di Lazzaro	Eruption triggered by the collapse of Sciara del Fuoco, which probably also caused a tsunami	Phreatomagmatic	Giordano et al. 2008
8.4 ka BP	Etna Mt (Valle del Bove collapse)	One or more tsunamis may have destroyed the Near East and Libyan settlements	Sector collapse and generation of a large (25 km^3) debris flow to the Ionian sea	Calvari et al. 1998; Pareschi et al. 2006
10.0 ka BP	Pantelleria Island, locality Fastuca	Pumice fallout over a 20 km^2- wide area	Violent Strombolian	Rotolo et al. 2007; Scaillet et al. 2011

collapse of Stromboli (Tab. 1), and by the explosive eruptions of Gran Cratere at Vulcano. A second crisis occurred between the 9th – 6th centuries BCE, and once again could have derived from the combined effect of an increase in the eruptive activity of Vulcano around 2.9 ka BP and a tsunami triggered by another collapse of Sciara del Fuoco. A third crisis struck the Aeolian communities between the 6th – 11th centuries CE, with a peak in 776 CE during the eruption of Monte Pilato (north-eastern sector of Lipari).

Volcanoes also provide obsidian, one the main raw materials exploited in prehistoric times. The timing of obsidian exploitation and circulation is strictly connected with location and the age of the natural sources. Soon after the Early Neolithic, obsidian spread throughout the entire Sicilian territory, rapidly prevailing over other more easily available raw materials, such as flint. The extraordinary success of this volcanic glass is based on its technical characteristics. As far as we know, the only available obsidian outcrops of the western Mediterranean were located at Mt Arci (Sardinia), Palmarola (Ponziane Archipelago, Latium), Lipari, and Pantelleria. From this latter island came c. 40% of the obsidian artifacts of north-western Sicily (where chemical analyses were performed), while all obsidian found in eastern Sicily came from Lipari,[91] and only one specimen coming from Palmarola has been found at Ustica.[92] Obsidian reached Sicily from the earliest stage of the Neolithic. At Grotta d'Oriente,[93] layer 5, good quality flint from local sources has been found, with the presence of obsidian artifacts, all in primary position. At Grotta dell'Uzzo, obsidian is present even in pre-Stentinello spits, and increases significantly in the next phase of Stentinello, but flint is still the most common raw material, even in the Stentinello phase,[94] whilst the people living on the Tyrrhenian coasts of eastern Sicily, hence not far from Lipari, very soon preferred the volcanic glass, such as those living in the Neolithic settlement (Stentinello II and Diana horizons) of San Martino near Spadafora, where the pre-formed cores of obsidian from Lipari represent >98% of the raw materials.[95]

Around the mid Neolithic (*Ceramiche tricromiche* horizon) the obsidian from Lipari gradually prevails over the other raw materials in the hilly and mountainous sites of Sicily, i.e. Le Rocche near Roccapalumba[96] and Vallone Inferno, where obsidian represents 87.5% and flint only 12.5% of the artefacts.[97]

Over recent decades, increasing attention has also been paid to the impact of earthquakes and tsunamis on human history.[98] Additionally, the increasing accuracy of seismic inventories and dating[99] have allowed us to better evaluate the social, economic, and demographic consequences of major seismic events on Mediterranean people. For instance, recent studies[100] have pointed out that several tsunamis, like the one initiated by the Thera-Santorini eruption c. 3.6 ka BP, may have damaged or even wiped out Sicilian coastal settlements.

Regarding historical seismic activity in Sicily, the most affected areas have been – and still are – the Peloritani mountains,[101] the Etna region, the Hyblaean Plateau,[102] and the inner sector of south-western Sicily. Combining historical documents and scientific data, we know that several wealthy Greek cities were shaken by seismic events, such as Selinunte in the 4th century BCE and c. 330–340 CE, whilst Catania was probably damaged

---

[91] Tykot 2017; 2019.
[92] Foresta Martin et al. 2017.
[93] Lo Vetro et al. 2016.
[94] Collina 2012.
[95] Quero et al. 2019.
[96] Mannino 2012; Italiano et al. 2018.
[97] Natali and Forgia 2018.
[98] Galadini et al. 2006; Bottari 2016; Stiros 2019.
[99] E.g. Locati et al. 2014; Guidoboni et al. 2007–2013.
[100] Smedile et al. 2011; Martini et al. 2012.
[101] Barbano et al. 2014.
[102] De Martini et al. 2010; Smedile et al. 2011.

by a poorly documented earthquake preceding the 122 BCE eruption (Tab. 1). In the early 1st century CE at least one earthquake and a tsunami struck the coastal areas of southern Calabria and north-eastern Sicily, where their combined effects caused the destruction of several cities, e.g. Abakainon and Tyndaris.[103] The latter was entirely rebuilt and flourished again, although a later earthquake in the 4th/5th century CE ruined the city forever. Agrigentum was also heavily damaged by an earthquake shortly after 360 CE, while the area of Capo d'Orlando was hit *c.* the mid 7th century CE.[104] In 365 CE, an earthquake – whose underwater epicentre was located not far from Crete – caused a tsunami that devastated the eastern and southern shores of the whole Mediterranean, destroying most of the North African settlements (even large cities such as Alexandria) and probably also caused major damage along the eastern and southern coasts of Sicily.[105]

According to historical references,[106] several strong earthquakes also hit the western and central sectors of the Aeolian Archipelago (e.g. the Mw = 6:2 in March 1786, or the recent Mw = 6:1 in April 1978), causing severe damage and casualties in the surrounding localities.[107]

Looking for recent colonisation paths for plants and animals by means of a phylogenetic lens (S. Pasta)

Many articles[108] have pointed out the extremely high genetic originality of the Sicilian populations of several widespread woody species, testifying the important role played by Sicily in terms of central and southern European plants that colonised the island through the Strait of Messina during Quaternary glacial events. With no doubt the most favourable (humid) microhabitats of the island (e.g. thalwegs and the north-facing slopes of the mountain ranges near the northern coast) allowed the survival and the *in situ* evolution of many species which disappeared elsewhere in the Mediterranean Basin. This explains the relatively high number of woody species and woody endemics living on the island.[109] For the same reasons, Sicily acted as a reservoir of genetic diversity, and, as soon as a new interglacial started, forest tree species surviving on the island were able to make their way back by crossing the Strait of Messina and contribute to the re-colonisation of the whole Italian Peninsula and Europe migrating northwards.[110]

Recent genetic analyses show that some species may have followed different paths to reach Sicily. In fact, an increasing number of publications on perennial herbs,[111] woody plants (*Arbutus unedo*),[112] butterflies (*Melanargia* spp.),[113] and frogs,[114] suggests that during the Pleistocene – or even in more recent times – both plants and animals had several opportunities to also reach Sicily through the Sicily Channel, and from there to southern Europe.

Even if some vascular plants and animals may have exploited the close vicinity of the North African and Sicilian coasts during the last glacial maxima, most of the vertebrates and humans were unable to do the same (see next paragraphs). In fact, all the available literature on Pleistocene faunistic assemblages[115] agrees in reporting that the vertebrates which colonised Sicily came from the Italian Peninsula, presumably exploiting passable land-bridges during the glacial maxima, particularly the last one.

Late Pleistocene and Holocene vertebrate faunas (S. Pasta and L. Sineo)

Although the first humans may have arrived in Sicily during the repeated low stand phases of the Late Middle Pleistocene (between *c.* 370 and 75 ka BP) in close conjunction with the dispersal of the animal assemblage referred to the *Palaeoloxodon* (*Elephas*) *mnaidriensis* Faunal Complex, a growing bulk of evidence testifies that they achieved a much more intense and stable colonisation of the island only afterwards.[116] In fact, the scarcity of vertebrate findings suggests that Sicilian habitats must have been much less attractive than those of continental Italy for both migrating herbivores and hunter-gatherer populations until the Late Pleistocene.

The turnover of Sicilian Quaternary mammalian assemblages is arranged in several distinct faunal complexes,[117] whose differences in terms of composition and degree of endemism are rather sharp, probably as a result of the varying frequency and ease of dispersal mechanisms from Africa and Europe. In fact the palaeogeographic evolution of the island was driven by tectonics and glacial and eustatic marine cycles, which in turn controlled the Middle and Late Pleistocene vertebrate faunal dispersal from adjacent landmasses towards Sicily. More specifically, one or more temporary connections between Sicily and Calabria, through the

---

[103] Bottari 2016.
[104] Bottari *et al.* 2008.
[105] Stiros 2001; Gerardi *et al.* 2012.
[106] Rovida *et al.* 2015.
[107] Cultrera *et al.* 2017.
[108] E.g. Michaud *et al.* 1995; Vendramin *et al.* 1998; Lumaret *et al.* 2002; Vettori *et al.* 2004; De Castro and Maugeri 2006.
[109] Médail and Diadema 2009; Guarino and Pasta 2018; Médail *et al.* 2019.
[110] Demesure *et al.* 1996; Dumoulin-Lapègue *et al.* 1997; Petit *et al.* 2002.

[111] E.g. *Linaria* sect. *Versicolores*, Fernández-Mazuecos and Vargas 2011; *Ambrosina bassii*, Troia *et al.* 2012; *Helminthotheca* spp., Tremetsberger *et al.* 2016.
[112] Santiso *et al.* 2016.
[113] Habel *et al.* 2018.
[114] Stock *et al.* 2008.
[115] E.g. Garilli *et al.* 2020, and references therein.
[116] Sineo *et al.* 2015.
[117] Bonfiglio *et al.* 2002.

Strait of Messina and between southern Calabria and Italian Peninsula through the Isthmus of Catanzaro, seem to have significantly affected the timing and modalities of the most recent vertebrate colonisation of Sicily and the Maltese Archipelago.

The two most recent faunal complexes (called 'Grotta di San Teodoro-Contrada Pianetti' and 'Castello'), dated to the last glacial cycle until LGM (c. 75 to 23–21 ka BP) and just after the LGM (c. 23–21 to 11.6 ka BP) respectively,[118] prove that vertebrate dispersal from Europe became more frequent. In fact, these faunal complexes appear increasingly similar to those of southern Italy, albeit impoverished.[119] The faunal history of this period is dominated by extinction events (large predators, e.g. lions; fallow deer; endemic dormice and shrews), and by the dispersal of 'continental' small mammals, e.g. *Microtus* (*Terricola*) gr. *savii*, *Crocidura* cf. *sicula*, *Apodemus* cf. *sylvaticus* and *Erinaceus europaeus*, and large mammals such as *Equus hydruntinus*, *Cervus elaphus*, and *Bos primigenius*. The extinction of the endemic small mammals that survived for the whole Middle Pleistocene is probably due to the severe climatic deterioration during the Last Glacial cycle, combined with the arrival of small-sized terrestrial predators, e.g. *Mustela* and *Vulpes* from Italy.[120] Interestingly, the dispersal to Sicily of *Microtus savii*, a vole with a fossorial habit, and of *Equus hydruntinus*, a wild ass unable to swim, implies that, at least once, a fully exposed connection (probably a temporary land-bridge related to eustatic low stand) had formed. This could have happened during the LGM, whose relatively arid conditions certainly favoured *E. hydruntinus*, which preferred open spaces provided by steppe-like grasslands, as did its other genetically similar Asian relatives.[121] The replacement of the Middle Pleistocenic vertebrate fauna probably took place stepwise. In fact, the Grotta di San Teodoro-Contrada Pianetti Faunal Complex still includes hyenas, several endemic animals and a high number of medium-sized herbivores (*Equus hydruntinus*, *Paleoloxodon mnaidriensis*, *Cervus elaphus siciliae*, *Bison priscus siciliae*, *Bos primigenius siciliae*, and *Sus scrofa*), whilst in the Castello Faunal Complex the turnover is complete, with no endemics, and the main preys are only four, i.e. the red deer,[122] the wild ass (*Equus hydruntinus*), the auroch (*Bos primigenius*), and the wild boar (*Sus scrofa*), as shown by the most recent papers on the Late Pleistocene/Early Holocene Sicilian vertebrate faunas.[123] The distribution range of *E. hydruntinus* became highly fragmented into discrete subpopulations during the Holocene;[124] in Sicily it may have survived until the Mesolithic.[125]

The faunistic analysis at the Late Pleistocene-Holocene transition in Sicily sheds light on the role of local ecological constraints (mainly climate and palaeogeography) and human activity on the Holocene faunal turnover of the island. Apparently, the last endemic herbivores (*Elephas mnaidriensis*, *Cervus elaphus siciliae*, *Bos primigenius siciliae*, and *Bison priscus siciliae*), as well as the large predators (the spotted hyena and the cave bear), living on the island since the Late Middle Pleistocene, became extinct during the Pleniglacial-Late Glacial interval, i.e. when humans probably did not yet occur on the island. Later on, a heterogeneous cohort of vertebrates (lynxes, aurochs, roe and red deer, martens, weasels, hares, wild cats), colonised the island, together with humans; among them, only steppe asses and lynxes were probably affected by human hunting activity, becoming extinct at the transition with Holocene. By contrast, the overall pattern of faunistic turnover in Sicily appears to have been mostly driven by climatic fluctuations and geodynamic events (modulating the phases of the connection and isolation with the mainland), Instead, anthropogenic impact seems to have been very low until the recent Holocene, when humans influenced the island's faunal composition with massive hunting, and mostly with the active and passive introduction of alien animal species.[126]

The human legacy in ancient Sicily (G. D'Amore and L. Sineo)

The archaeological evidence for Pleistocene human presence in Sicily has been a disputed topic for more than a century.[127] Summarising, the lower Pleistocene peopling hypothesis, mostly based on casual, random and badly documented findings of lithic artifacts in the Trapani and Agrigento provinces, has been progressively abandoned, at least waiting for clear[128] and less circumstantial evidence.[129]

According to the present state of our knowledge, humans arrived very late in Sicily, and their migration was probably favoured by a temporary connection of the island with Calabria during the LGM. Following this interpretation, the first settlers arrived within a time lapse of 20.0–15.0 cal ka BP.[130] In fact, the earliest unambiguous archaeological record belongs to the Late Upper Paleolithic period, represented by stone

---

[118] Garilli *et al.* 2020.
[119] Bonfiglio *et al.* 2007.
[120] Bonfiglio *et al.* 2007.
[121] Catalano *et al.* 2020b.
[122] *Cervus elaphus*, the most important protein source for the Sicilian upper Palaeolithic hunter-gatherer communities according to Mannino *et al.* 2011a.
[123] Cassoli and Tagliacozzo 1982; Tagliacozzo 1993; Villari 1995; Burgio *et al.* 2005; Mangano 2005; Mannino and Thomas 2010.
[124] Crees and Turvey 2014.
[125] Salari and Masseti 2016; Catalano *et al.* 2020b.
[126] Petruso and Sineo 2012.
[127] See Galland *et al.* 2019, and references therein.
[128] Bonfiglio *et al.* 2000; Garilli *et al.* 2020.
[129] Di Maida 2020.
[130] Incarbona *et al.* 2010b; Mannino *et al.* 2011a.

artifacts related to the Late Epigravettian culture, found sometimes in association with human skeletal remains. The paucity of these finds does not allow us to infer much in terms of the physical aspects and genetic relationships of the first inhabitants of Sicily, and highlights the importance of the most significant Late Pleistocene Sicilian site, the cave of San Teodoro (Acquedolci, Messina). Here, several skeletal remains and archaeological finds of human frequentation[131] were found in association with faunal remains of the so-called 'Castello Faunal Complex'. An AMS-^{14}C date obtained from the 'San Teodoro 1' (ST1) human specimen gave a conventional radiocarbon age of 12.580±0.130 yr BP, thus confirming a Late Glacial, Late Epigravettian age of the site;[132] moreover, a conservative estimate of 15.0–11.0 ka BP remains the most probable date also for the San Teodoro individuals 2-4. D'Amore et al. 2009 restudied the whole cranial sample of San Teodoro and compared it with a wide array of both prehistoric and recent samples through different multivariate techniques. As a result, the morphometric pattern of San Teodoro people appears to be very close to western European groups of similar antiquity, in particular to those from central and southern Italy, corroborating the hypothesis that these first settlers probably came from peninsular Italy by sea during the Late Pleistocene. According to an alternative (and weaker) hypothesis, the people from San Teodoro originated from immigrants who arrived by land during a low sea-level episode corresponding to the LGM regression, c. 18.0 Uncal. Ka BP.

Early Holocene Mesolithic hunter-gatherers of Sicily could be the result of the localised evolution of Paleolithic individuals who migrated to the island during the Late Pleistocene, or issue from the overlapping of different and more recent migrations. According to the first hypothesis, Paleolithic gene flow from the Italian Peninsula could have been the primer of human distribution on the island, producing a quite homogeneous local pool, as a result of a common response to rather uniform ecological constraints. Paleolithic hunter-gatherers could represent the effective genetic base of ancient Sicily, while Mesolithic ones might represent a cultural transition. Following the second scenario, a rather intense gene flow during the Mesolithic could have involved different people, whose evolution inside the island would subsequently have been under common ecological pressures.[133]

To test the hypothesis proposed by D'Amore et al. 2009, the same authors carried out a multivariate analysis[134] focused on the linear traits of the whole Mesolithic adult cranial sample available from Sicily, and pointed out a general similarity among western/central European Late Paleolithic and Mesolithic groups. These results underline the major role played by gene flow among European hunter-gatherer populations. This gene flow gradually decreased during the transition from the Late Paleolithic to the Mesolithic, and even more for the following main cultural transitions, could depend on the starting of cultural segregation and genetic isolation.[135]

If a Mesolithic transition took place – at least in Sicily – through a meaningful cultural/biological shift in local populations, it might be considered as a consequence of the climatic and environmental changes that occurred during the transition from the Pleistocene to the Holocene, which may have induced major changes in both the ecological constraints and the resources available to humans.[136] Indeed, the final part of the Pleistocene and the transition to the Holocene were characterised by increasing climate stability, which created opportunities for hunter-gatherer groups to repeatedly occupy areas in a move towards a mobile-forager/semi-sedentary behaviour.[137] Without doubt, the Sicilian Mesolithic people who lived during the Early Holocene (11.7–8.2 ka BP) experienced high summer temperatures and summer droughts,[138] with deep consequences for local plant and faunal assemblages, and consequently on human foraging opportunities, as it has been well documented in some key sites, such as Uzzo Cave.[139] Until 8000–7000 cal. BP, here and elsewhere in the Sicilian lowlands, *Olea* and evergreen *Quercus* tree stands were rare, and forests may have occurred only higher up in the mountain ranges, where moisture availability was sufficient.[140] Thus, harvesting edible tree fruits may have required too much time and energy for people living close to the coast. Afterwards, forests expanded in response to climate change, improving the resource availability and changing the diet options of Neolithic people.

After the earliest colonisation of Sicily, local human communities probably spread along the coasts and successfully inhabited the whole island. In fact, the number of Sicilian skeletal specimens found within Mesolithic archaeological contexts represents the richest record from the whole of Italy. Well-excavated Mesolithic horizons lie in several sites, mainly caves, such as Grotta della Molara (Palermo), Grotta dell'Uzzo (Trapani) and Grotta d'Oriente (Favignana island, Egadi archipelago), and returned as many as 17 individuals.

---

[131] Graziosi 1947; Mallegni 2005; D'Amore et al. 2009; Galland et al. 2019, and supplementary materials.
[132] Mannino et al. 2011a.
[133] Galland et al. 2019.
[134] D'Amore et al. 2010.
[135] Galland et al. 2019.
[136] Galland et al. 2019.
[137] Sineo et al. 2015.
[138] Zanchetta et al. 2007; Sadori et al. 2008; Incarbona et al. 2010b; Curry et al. 2016.
[139] Piperno 1985.
[140] Tinner et al. 2009; Noti et al. 2009; Calò et al. 2012.

Direct ¹⁴C dates are available for the individuals 'Molara 2' (8.600±100 ¹⁴C yr BP), 'Uzzo 5' (9.270±100 ¹⁴C yr BP), and 'Oriente B' (9.377±25 ¹⁴C yr BP).[141]

Multidisciplinary research has revealed many bio-anthropological aspects related to the lifestyle, dietary habits, health conditions, and the subsistence strategies of Sicilian Mesolithic human groups, especially from the specimen-rich site of Uzzo. For instance, dental wear/microwear and trace elements analysis suggested good nutritional and health conditions, with regular consumption of animal protein, also of marine origin,[142] but with a relatively high sugar intake due to abundant plant consumption.[143]

The only line of indirect and independent evidence that could provide support to the hypothesis of a genetic discontinuity during the transition between Late Epigravettian/Late Glacial and Mesolithic-Early Holocene periods, might be represented by palaeogenomic studies.[144] In the last few years, developments in massive sequencing techniques have allowed the generation of an unprecedented amount of genomic data from past populations.[145] To date, only two Palaeo-Mesolithic Sicilian specimens have produced published data. A first palaeogenetic study on the 'Oriente B' individual (Mesolithic, directly dated by AMS ¹⁴C to 10.683–10.544 yr BP), focused on the mtDNA HVR1 region only, suggested that humans living in Sicily during the Early Holocene could have originated from groups that migrated from the Italian Peninsula around the LGM.[146] Recently, Catalano et al. 2020a generated a whole genome aDNA data set from the 'Oriente C' individual (putative stratigraphic date of 14.210–13.770 ka BP). This individual belonged to haplogroup U2'3'4'7,8,9; Van de Loosdrecht et al. 2020 reported this same haplogroup for two Early Mesolithic specimens from Uzzo, and other haplogroups of the U clade, very common among Upper Paleolithic and Mesolithic hunter-gatherers of Europe, for further nine individuals from Uzzo. These results broadly suggest a strong genetic relationship of local communities with western European Palaeo-Mesolithic hunter-gatherers, a genetically homogeneous population which likely spread from the Atlantic seaboard of Europe in the west, to Sicily in the south, to the Balkan Peninsula in the southeast;[147] in more depth, these data indicate that the Early Mesolithic Sicilian hunter-gatherers were a highly drifted sister lineage to Early Holocene western European hunter-gatherers, with further substantial gene flow from south-eastern Europe between the Early and Late Mesolithic.[148] In the light of these findings, Catalano et al. 2020a pointed out that the Oriente B individual needs to be processed by a genome-wide approach, looking for continuity or discontinuity with the Oriente C results. At present, only our morphometric results seem to depict a discontinuity scenario.[149]

The Mesolithic-Neolithic transition and the dawn of productive economy in Sicily (end of the 7th/beginning of the 5th millennium BCE) (V. Forgia and S. Pasta)

The transition between the Mesolithic and the Neolithic, related to the change of subsistence strategies, has been interpreted as the response of last communities of hunter-gatherers to the ongoing changes of natural landscapes and linked to the interaction of these communities with the first farmers.[150] Thus, we must briefly introduce the environmental and economic framework before this transition occurred.

According to Mannino et al. 2011a, evidence from Sicilian territories, occupied since the Late Pleistocene, suggests that the diet of local people was increasingly based on animal proteins, mainly originating from red deer. At the beginning of the Holocene, local human groups show an increasingly sedentary behaviour, accompanied by an intensification and diversification of resource exploitation.[151] Moreover, during the transition between the Late Pleistocene and Early Holocene, archaeological data show that diet progressively adapted to the environmental constraints, especially in those sites affected by the retreat of the coastline, caused by sea-level rise.[152]

More specifically, Late Pleistocene hunter-gatherers spent the coldest months of the year close to the coast, moving inland in late spring and summer.[153] In the Early Holocene, by contrast, marine molluscs were exploited all year, suggesting a change in seasonal mobility strategies and, probably, frequent moves within more restricted territories. Marine foods were marginal in the diets of both Late Pleistocene and Early Holocene hunter-gatherers,[154] although an increase of marine resources has been recorded at the beginning of the Holocene (c. 9.0 ka BP), perhaps due to the growth of Mediterranean Sea productivity.[155] The karst areas of north-western (including the Egadi Islands) and south-eastern Sicily were by far the most densely inhabited

---

[141] Lo Vetro and Martini 2012; Sineo et al. 2015.
[142] Including whales and dolphins: Mannino et al. 2015; Colonese et al. 2018.
[143] Borgognini Tarli et al. 1993.
[144] Galland et al. 2019.
[145] E.g. Posth et al. 2016; Mathieson et al. 2018.
[146] Mannino et al. 2012.
[147] Mathieson et al. 2018.
[148] van de Loosdrecht et al. 2020.
[149] Galland et al. 2019.
[150] Tiné and Tusa 2012.
[151] Mannino et al. 2011b.
[152] Lo Vetro et al. 2016.
[153] Mannino et al. 2011b.
[154] Colonese et al. 2011; Mannino et al. 2011a; 2011b; Sparacello et al. 2020.
[155] Colonese et al. 2018.

territories of Sicily during the Late Pleistocene,[156] when they were probably exploited from the coastal plains to the upland areas, but the home range of local human groups became more restricted in the Early Holocene.

A deeper insight into the major environmental and landscape modifications that occurred in Sicily between the Late Pleistocene and the Early Holocene helps us understand the spread of early farming activities at the end of 7th, beginning of the 6th millennium BCE. In the area of Trapani and Favignana, on the western part of the island, ecological indicators show the presence of open lands during the Late Pleistocene.[157] The oxygen isotopic composition of land-snail shells[158] suggests a shift towards a wetter climate during the Early Holocene. As a consequence, gradual afforestation affected the Sicilian coastal areas, yet more slowly, and with some delay with respect to other Mediterranean countries.[159]

Underwater archaeological explorations, in combination with submarine geomorphology and, in general, studies about sea-level changes,[160] are shedding new light on early maritime voyages[161] and providing new clues as to the environmental and landscape changes which triggered the economic transition towards the Mesolithic and Early Neolithic of Sicily. In some cases, the coastline was actually quite far from current positions, and rather rapid sea-level changes deeply modified or even submerged the coastal territories important for Sicilian Palaeolithic communities. Isolidda di Macari (San Vito Lo Capo, Trapani) was near the coastline, even in the Pleistocene,[162] like several other major western Sicilian Epigravettian and Mesolithic sites, e.g. Grotta d'Oriente[163] and Grotta Racchio.[164] Looking towards the east, the coastline of the Gulf of Termini Imerese (Palermo) was much more advanced when the sea-level was between 120–80 m lower than the current position, and rose quickly after the Younger Dryas,[165] when Riparo del Castello starts to be used by human groups, as attested by its articulated Epigravettian and Mesolithic deposits.[166]

One of the most lively debated topics is how Early Neolithic economic strategies interacted with the local Mesolithic substratum.[167] Unfortunately, several sites of the Mediterranean and Balkan area, as well as in the Near East, show a hiatus in the coincidence of this transitional period.[168] The lack of archaeological layers corresponding to this transition has been linked to climatic changes that occurred between 8.5–8.0 ka BP.[169] In particular, the 8.2 ka BP climatic event was a period of broad climatic change[170] which strongly affected the environmental patterns of both the Mediterranean Basin and central Europe, and, as a consequence, economic subsistence activity and the settlement strategies of human societies.[171] This event also had a striking taphonomic impact on the last Mesolithic and first Neolithic sites located in valley bottoms and in karst areas, hampering the formation and preservation of relevant deposits.[172] Caves such as Grotta dell'Uzzo (San Vito Lo Capo, Trapani), on the contrary, do preserve archaeological layers dating back to the transitional phase,[173] showing no gaps between the two horizons. Evidence from the transitional layers suggests the acceptance of newcomers and the gradual adaptation of earlier settlers to the external technical and cultural inputs.

The earliest archaeological presence of a fully Neolithic horizon in Sicily is linked to the so-called 'Impressed Ware' culture (also named 'Archaic' or 'ceramiche impresse arcaiche' culture), that spread into the island from southern Italy. This kind of pottery has been found, and absolutely dated, in spit 10 of trench F at Grotta dell'Uzzo (Figure 4) (6390–6220 BCE) and in layers 16–17 of trench B at Grotta del Kronio (Figure 4) (5990–5740 BCE)[174] (Tab. 2).

Among the novelties introduced with the Neolithic 'package' in Sicily, it is worth mentioning the use of obsidian as a raw material to produce lithic artifacts. The use of this volcanic glass is linked to a prehistoric eruption that occurred on Lipari Island.[175] One of the oldest records of obsidian artifacts in Sicily occurs in a context (6020–5790 BCE) that lies between the end of the Mesolithic period and the early Sicilian Neolithic, at Grotta d'Oriente (Figure 4) on Favignana Island.[176] Here, the co-occurrence of new raw materials and a few remains of domesticated vertebrates over a cultural Mesolithic substratum without pottery probably testify to one of the first contacts between Mesolithic and Neolithic communities in western Sicily.[177]

---

[156] Tusa 1999.
[157] Incarbona et al. 2010b; Martini et al. 2012.
[158] Lo Vetro and Martini 2016.
[159] Tinner et al. 2009; Mercuri and Sadori 2014.
[160] Benjamin et al. 2017.
[161] Lo Presti et al. 2019.
[162] Lo Vetro and Martini 2016.
[163] D'Amore et al. 2010; Mannino et al. 2012; Colonese et al. 2014; Lo Vetro and Martini 2016; Catalano et al. 2020a.
[164] Baglioni et al. 2012; Lo Vetro et al. 2016.
[165] Caruso et al. 2011.
[166] Nicoletti and Tusa 2012b.
[167] Lo Vetro and Martini 2012; Tiné and Tusa 2012.
[168] Berger and Guilaine 2009.
[169] Rohling and Pälike 2005.
[170] At some sites in Europe it brought more moisture and thus increased biomass and biodiversity, in Sicily likely dry; see, e.g., Tinner et al. 2009.
[171] Berger and Guilaine 2009.
[172] Berger and Guilaine 2009.
[173] Lo Vetro and Martini 2012; Tiné and Tusa 2012; Mannino et al. 2015.
[174] Tiné and Tusa 2012.
[175] Tykot 2019.
[176] Lo Vetro and Martini 2012.
[177] Lo Vetro and Martini 2016.

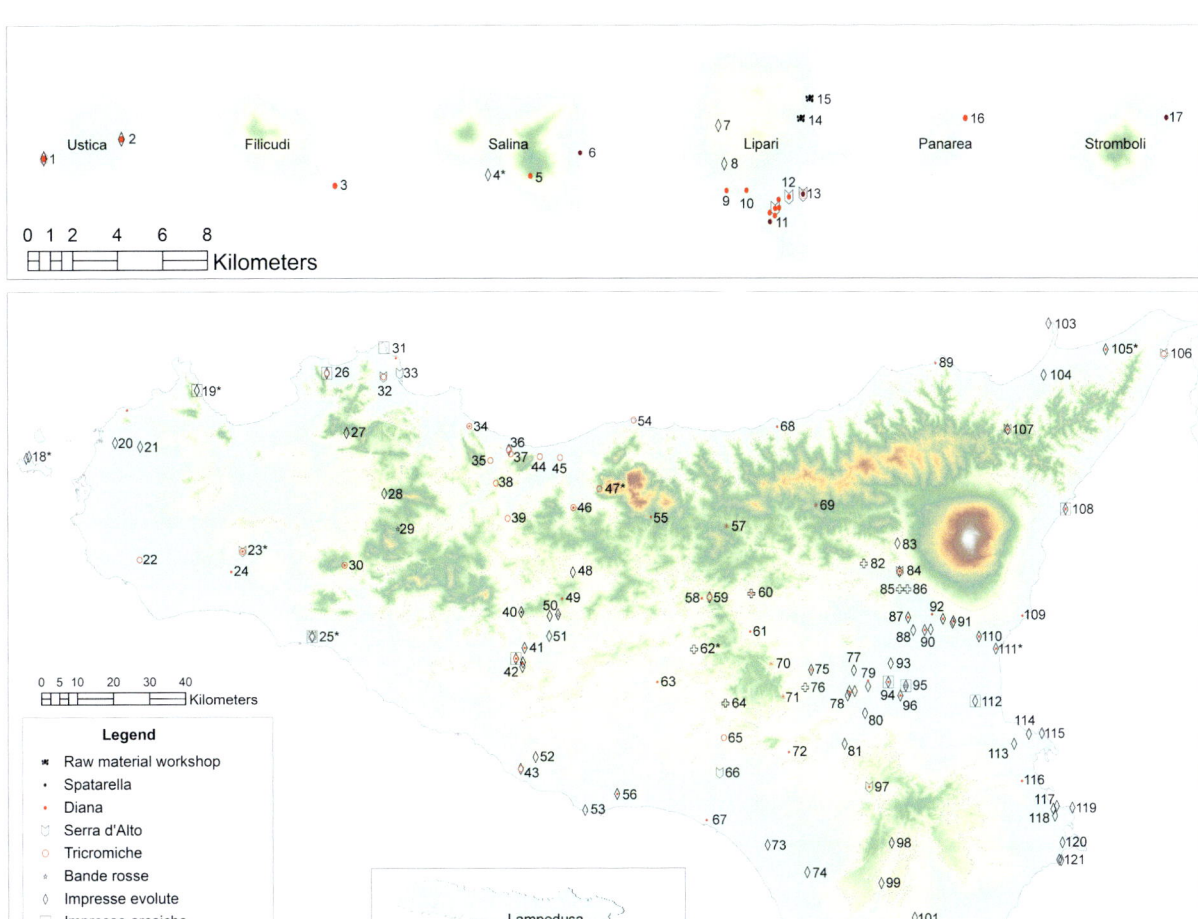

Figure 4: Distribution map of Sicilian Neolithic sites (* = Early to Middle Neolithic sites with radiocarbon dates):
1. Punta Spalmatore; 2. Piano Cardoni; 3. Capo Graziano; 4. Rinicedda; 5. Monte Fossa delle Felci; 6. Brigadiere; 7. Castellaro vecchio; 8. Cicerata; 9. Fumarolo; 10. Mercorella; 11. Predio Megna, Piano Greca, San Bartolo, Fossa, Mulino a Vento, Spatarella; 12. Diana; 13. Castello, Acropoli; 14. Lami; 15. Papesca; 16. Calcara; 17. San Vincenzo; 18. Grotte Oriente* e Uccerie*; 19. Grotta dell'Uzzo*; 20. Maiorana; 21. Chiappera; 22. Castedduzzo; 23. Partanna-Stretto*; 24. Castello della Pietra; 25. Kronio*; 26. Grotta dei Puntali; 27. Serre Mirabella; 28. Cicio; 29. Spolentino; 30. Favarotti; 31. Grotta Regina, Grotta Antro Nero; 32. Grotta Molara; 33. Monte Pellegrino; 34. Grotta Mazzamuto; 35. Ciacca; 36. Grotta Puleri; 37. Grotta Geraci; 38. Grotta Grande; 39. Le Rocche; 40. Corvo; 41. San Paolino; 42. Fontanazza, Serra del Palco, Masseria Zellante; 43. Piano Vento; 44. Mura Pregne; 45. Rocca del Drago; 46. Grotta di Bommartino; 47. Vallone Inferno*; 48. Centosalme; 49. Polizzello; 50. Grotte, Bragamè; 51. Fra' Gaetano; 52. Grotta Infame Diavolo; 53. Caduta; 54. Grotta delle Giumente; 55. Grotta del Vecchiuzzo; 56. Casalicchio Agnone; 57. Monte Barbagiano; 58. Case Bastione; 59. Realmesi; 60. Casa Recifori; 61. C.da Marcato; 62. Borgo Cascino*; 63. Tornambè; 64. Monte Navone; 65. Rocca di Maio; 66. Disueri; 67. Collina di Gela; 68. Caronia; 69. Lago Ancipa; 70. C.da Bosco; 71. Gelso; 72. Poggio delle Pille; 73. Tatappì; 74. Vittoria; 75. Giresi; 76. Crunici; 77. Impennate; 78. Torricella, Vannuto, Cozzo Santa Maria, La Montagna; 79. Poggio delle Forche, Feccia di Vino; 80. Rocchicella-Palikè; 81. Serravalle; 82. Pizzo Tamburino; 83. Barbaro; 84. Adrano; 85. Cavaleri; 86. Picone-Riparo Cassataro; 87. Poira; 88. Poggio Bianco; 89. Tono; 90. Poggio Monaco, Masseria Cafaro; 91. Valcorrente, Fontana di Pepe; 92. Salinelle San Marco, Giaconia; 93. Castellito; 94. Masseria Scavo; 95. Perriere Sottano; 96. Sciccaria; 97. Via Capuana, Licodia Eubea; 98. Calaforno; 99. Ragusa Ibla; 100. San Francischiello; 101. Cava Lazzaro; 102. Bruca; 103. Punta Messinese; 104. Barcellona; 105. San Martino*; 106. Torrente Boccetta; 107. Riparo della Sperlinga di San Basilio; 108. Schisò; 109. Gelso; 110 Pezzamandra; 111. Catania-Acropoli*; 112. Bonvicino; 113. Petraro; 114. Gisira; 115. Campolato; 116. Palombara; 117. Targia, Stentinello; 118. Fondo Ferrante; 119. Santa Panagia; 120. Milocca; 121. Isola di Ognina; 122. Pachino; 123. Grotta di Calafarina, Vulpiglia; 124. Cala Pisana (data sources: Fugazzola et al. 2004; AA.VV. 2012; Giannitrapani 2017; Quero et al. 2019; Forgia 2019).

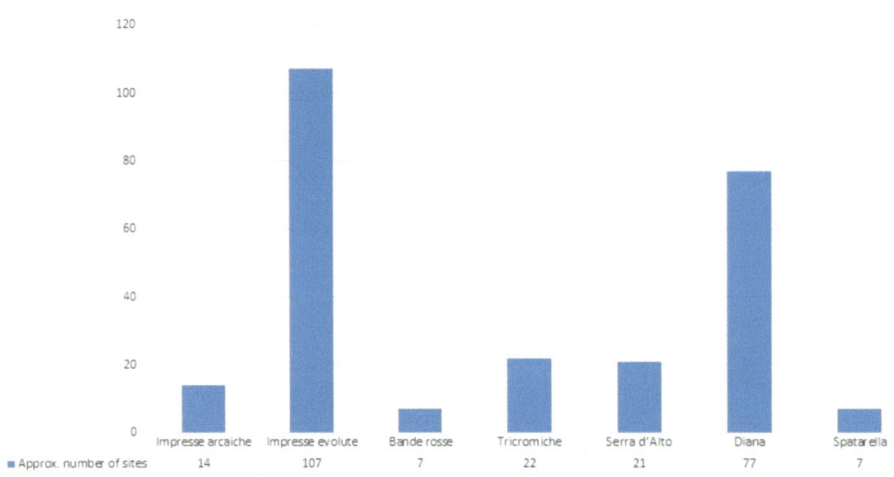

Figure 5: Approximate number of Neolithic sites in Sicily, separated by chrono-cultural horizons. (data sources: Fugazzola Delpino *et al.* 2004; AA.VV. 2012; Giannitrapani 2017; Forgia 2019).

Three more horizons characterise the development of the 6th millennium BCE: 1) the so-called advanced Impressed Ware or 'Impresse evolute' of *Stentinello I* pottery style; 2) the advanced Impressed Ware or 'Impresse evolute of *Stentinello II* pottery style' (after 5500 BCE), in association or not with painted ware; and 3) trichrome ware or 'Ceramiche tricromiche'.[178]

The *Stentinello I* pottery style is identified by impressed decoration, with common motifs spread over the eastern part of Sicily, and a wider spectrum of decorative options and techniques (in the form of incisions rather than impressions) throughout the western part of the island, where it takes the name of 'Kronio pottery style', from the homonymous cave near Sciacca (Figure 4). In sites of *Stentinello II*, the presence of painted pottery represents a clear hint of cultural contact with groups characterised exclusively by a painted-ware pottery style. The 'Capri-Lipari-Scaloria Alta' or 'tricromiche' painted pottery, quite common in Southern Italy,[179] has been clearly recognised at Contrada Stretto (Partanna), Le Rocche/Fiaccati near Roccapalumba,[180] Vallone Inferno in the Madonie mountains,[181] and at Lipari (Figure 5).[182]

According to the seasonal use identified from the archaeological surveys carried out on the relevant sites, early farming and herding activities in Sicily developed along coastal sites and lowlands (e.g. Uzzo, Favignana, Mt Kronio), where Neolithic communities preferred to occupy partially open landscapes. At a later stage, more articulated production activities and changes in herd management[183] led some groups to start the exploitation of the hilly hinterland (e.g. Le Rocche) and uplands (e.g. Vallone Inferno), forcing the opening of already afforested uplands. Based on some paleoenvironmental proxies, an increase of anthropogenic fire impact has been hypothesised to have occurred along the northern mountain ranges of the island *c.* 7.0 ka BP.[184]

Thanks to its almost continuous archaeological record, Grotta dell'Uzzo represents a key site for understanding the changes in human subsistence strategies in the coastal areas of Sicily. Several investigations carried out there[185] document a constant frequentation of the cave, all year round, during the Mesolithic/Neolithic transition. Being rich in habitats suitable for different wild edible plants[186] and animal preys, the patchy and partially open landscape surrounding the site at that time surely favoured human settlement by the last hunters-gatherers. Mannino *et al.* 2015 documented a huge increase in fish remains and cetaceans and the decrease in micromammalians and birds. The prevalence of red deer over wild boar among the mammals is also well documented. During the Early Neolithic (*Ceramiche impresse* and *Stentinello I*), hunting and fishing are still documented, as well as the collection of shellfish between autumn and spring, but not during the summer.

Additionally, five new species of domesticated animals are documented at Uzzo Cave, i.e. *Bos taurus*, *Ovis aries*

---

[178] Pessina and Tinè 2008.
[179] Pessina and Tinè 2008.
[180] Tiné and Tusa 2012.
[181] Forgia *et al.* 2013.
[182] Bernabò Brea and Cavalier 1980.
[183] Natali and Forgia 2018; Forgia 2019.
[184] Bisculm *et al.* 2012.
[185] e.g. Tagliacozzo 1993; Mannino and Thomas 2012.
[186] Pasta *et al.* 2020.

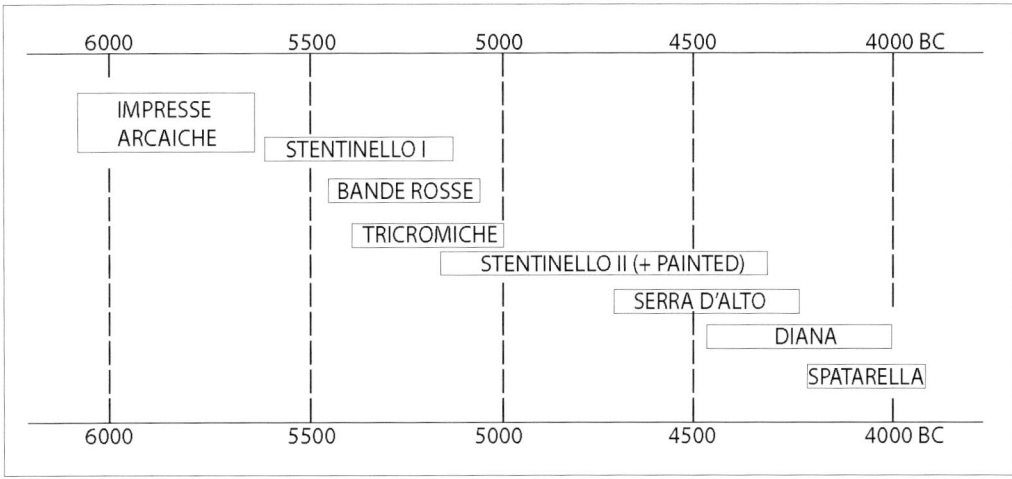

Figure 6: Diachronic overview of the Sicilian Neolithic cultural facies (modified after Pessina *et al.* 2008).

and *Capra hircus*, clearly introduced from abroad, pigs, and dogs, likely locally domesticated. During the second phase of the Neolithic (*Stentinello II* and *Ceramiche tricromiche*), the rarefaction of the faunal remains inside the cave suggests a shift in the use of the site, exploited as a shelter for herders and their flocks.[187]

Much more data could be eventually obtained by analysing the data of currently submerged Neolithic sites. Nonetheless, it is possible to outline the following preliminary framework for the Early and Middle Neolithic of Sicily. Most of the Early Neolithic sites seem to be concentrated near the coast, with the exception of a few dubious cases lacking absolute dates, where early presences have often been recognised only at surface level. To date, Uzzo and Kronio are the only Sicilian Early Neolithic sites provided with radiometric dates. Archaeozoological remains, as well as the proxies of anthropogenic fire impact found in several palaeoenvironmental research projects, suggest that pastoral activities affected the Sicilian uplands since the Middle Neolithic. A broad count of the sites covering the whole range of the Neolithic period in Sicily (Figure 5) allows us to highlight the high proportion of those referred to as *Impresse evolute* (both *Stentinello I* and *Stentinello II* or Kronio pottery styles). This result is likely dependent on the long duration of these stylistic aspects, spanning over a millennium (Figure 5), while the comparable high number of sites belonging to the *Diana* pottery style (see the following section) could be the consequence of a period of demographic growth.[188] Apart from some localities in the Madonie mountain range, almost no record is currently available for the hilltops and mountain ranges, while an outstandingly high number of sites has been found on the Simeto Plain, as well as within the Aeolian Archipelago. This apparently sharp unbalance may be an artefact, because the inlands and uplands of Sicily are still very poorly investigated, while the long-lasting and in-depth research carried out over the last fifty years in the Aeolian Archipelago has provided very detailed information on the demographic, socioeconomic, and cultural features of local communities.[189]

The Late Neolithic and the Metal Ages in Sicily (S. Pasta, V. Forgia and C. Speciale)

Subsequent Neolithic horizons are characterised by the *Serra d'Alto* and *Diana* wares (Figure 6). While in Sicily the Serra d'Alto pottery style overlaps with *Stentinello II* (5000–4500 BCE), the *Diana* style (4500–4000 BCE) coincides with a broad cultural province involving southern Italy and its last phase (*Diana D*), also known as *Spatarella*, representing the transition to the Early Copper Age (4000–3800 BCE).[190]

Without doubt, the human groups living in Sicily at the transition between the Neolithic and the Metal Ages formed an eastern pool which strongly influenced the genetic pattern of the human groups who settled the island during the following millennia. The data issuing from pollen analyses carried out in the permanent ponds of the mountain ranges of northern Sicily[191] show a remarkable agreement in pointing out a complex alternation of forest and open land phases throughout the Metal Ages. According to these studies, synchronous phases of grassland establishment occurred at the Late Copper/Early Bronze Age transition (*c.* 4500 cal BP, 2550 BCE), during the Early Bronze Age (4200–3800 cal BP, 2250–1850 BCE) and the Early Iron age (2800–2600 cal BP, 850–650 BCE). Considering that human density remained low during most of the Metal Ages, and that

---

[187] Tagliacozzo 1997.
[188] Tusa 1999.
[189] Leighton 1999; Martinelli and Lo Cascio 2018.
[190] Pessina and Tinè 2008.
[191] Bisculm *et al.* 2012; Tinner *et al.* 2016.

for a long time lithic technology still provided the main tools required to modify local natural environments, climate and anthropogenic activities (e.g. slash-and-burn) might have affected synergistically these oscillations in tree cover.

During the Copper Age (4000–2300 BCE), a wide variety of historical trajectories and environmental constraints led to the development of rather distinctive cultures.[192] Like elsewhere in the Mediterranean Basin,[193] local human communities underwent a deep reorganisation in terms of identity, location, and density in order to adapt to local climatic patterns and resource availability. Agriculture and pastoralism continued to provide the economic subsistence base for local communities. Well-protected places with permanent water and near the main transhumant tracks[194] represented from this period ideal sites for the establishment of human communities, eventually driven by cyclic and repeated arid events.

Within the Late Copper Age (LCA), Pacciarelli *et al.* 2015 has proposed an 'early', rather stable phase (between 2800–2750 and 2600–2550 BCE), characterised by the *Malpasso* stylistic aspect (sometimes associated with the so-called *Sant'Ippolito* painted style), an 'intermediate' phase (between 2600–2550 and 2350–2300 BCE), signalled by the arrival of the Bell Beaker complex, and a 'final' phase (between 2350–2300 to 2200–2150 BCE), where the *Naro-Partanna* pottery style may be considered an evolution of the previous *Malpasso-Sant'Ippolito* style, and a precursor of the *Castelluccio* painted pottery of the Early Bronze Age.

The transition from LCA to the Early Bronze Age (*c.* 2200 BCE) was shaped by various events and by the emergence of new forms of social structures based, probably, on the integration of indigenous communities with groups coming from Eastern Mediterranean countries.[195] During this time frame, ongoing changes in terms of geographic pattern of occupation, hierarchy of settlements, forms of resource exploitation and management, and funerary rituals also triggered a profound re-shaping of local cultural landscapes. A diachronic overview based on the available knowledge on the distribution of Bronze Age settlements in western Sicily confirms this really dynamic picture, with earlier settlements mostly concentrated on the hilltops of Palermo and Trapani provinces and the Belice area, and a predominantly coastal location (including some circum-Sicilian islands such as Favignana, Pantelleria and Ustica) for the sites referred to the *Castelluccio* and *Thapsos* chrono-cultural horizons.[196]

The Bronze Age was also a period of complex commercial and cultural relationships, with distant countries using the Sicilian coastlands as important landing points and hubs for their trade routes. In this context, not only the main island, but also the Aeolian Islands and Malta played major roles in the international metal trade network, involving numerous north-western European (Cornwall), west Mediterranean (e.g. the Balearic Islands, Sardinia, southern France, southern Spain, Tuscany) and the East Mediterranean (e.g. Cyprus, Greece, Aegean islands, Lebanon).[197]

The Early Bronze Age (2200–1700 BCE) is characterised by regionally long-lasting cultural aspects, e.g. *Capo Graziano* on the Aeolian Islands, or *Castelluccio* on the main island, spreading throughout the main island (but to date apparently poorly represented in the north-eastern sector), and overlapped with *Rodì-Vallelunga-Tindari*, mostly concentrated over the northern and western parts of the island.[198] The north-western part of the island is also characterised by the so-called *Moarda* style, which shares many common traits with LCA horizons and is subject to western (mainly Iberian and Sardinian) influences.[199] These data match rather well with recent findings issuing from human genetics, suggesting that steppe pastoralist ancestry colonised the island by ~2200 BCE, in part from Iberia.[200]

Despite the socio-economic and demographic growth, the *Castelluccio* complex (strictly connected with the Maltese people and showing affinities with Apulian, western Balkan and Helladic communities) still preserves several archaic features, similar to those of LCA societies. Probably the most remarkable of these features is the very rare occurrence of metal objects, and the lithic industry, which was still intensively exploiting flint mines, concentrated in the Hyblaean Plateau.[201] Demographic trends show a continuity between the Late Copper Age and Early Bronze Age, especially in central Sicily.[202]

During the Middle Bronze Age (1700–1300 BCE), cultural and commercial exchange with civilisations from the eastern and central Mediterranean regions (Mycenae, Aegean islands, Cyprus and Malta)[203] further increased and predominated, as testified by the style of the manufacts of the cultural sphere of *Thapsos*

---

[192] McConnell 1997; Leighton 1999; Cazzella and Maniscalco 2012; Pacciarelli *et al.* 2015; Giannitrapani and Iannì, *in press*.
[193] E.g. Lillios *et al.* 2019.
[194] Forgia 2019; 2019b.
[195] Pacciarelli *et al.* 2015.
[196] Nicoletti and Tusa 2012a.
[197] Fokkens and Harding 2013.
[198] Ardesia 2014.
[199] E.g. Bell Beaker; Tusa 1999.
[200] Olalde *et al.* 2018; Fernandes *et al.* 2020, and references therein.
[201] Copat 2020.
[202] Giannitrapani *et al. in press*.
[203] Raneri *et al.* 2015, Russell 2017; Tanasi 2010.

(south-eastern Sicily), a huge emporium, and probably one of the first pro-urban settlements of the western Mediterranean, and of *Punta Milazzese* in the Aeolian Islands.[204] Circulation of metals and prestige goods, in parallel with an increase in social complexity, are testified by funerary rituals.[205] The Middle Bronze Age is also a phase of increasing conflict in the Mediterranean.[206] Settlements tend to be concentrated only in some specific defensive places and their numbers decrease when compared to the EBA. Central Sicily is less populated, while the most important settlements are concentrated along the coasts,[207] and fortified villages become increasingly common.[208] Genetic research provides some clues as to the possible causes of rapid changes and ethnic unrest, suggesting that during this timeframe Iranian-related ancestry arrived during the second half of the 2nd millennium BCE, contemporary with its spread in the Aegean area.[209]

The Late Bronze Age (*Bronzo Recente* and *Finale*, 1250–950 BCE) is a period of major instability, probably exacerbated by climatic variance,[210] and without doubt Sicily was also affected by large-scale population replacement after the Bronze Age.[211] Around the mid 13th century BCE, many small coastal settlements were probably attacked and destroyed, no doubt either temporarily or partially abandoned. People went inwards and upwards and built the first large villages (e.g. Pantalica, Dessueri, Polizzello) with hundreds of inhabitants, while the Ausonian groups – strictly linked to the Italian Peninsula – spread throughout north-eastern Sicily and around the Aeolian Archipelago. These changes point out the presence of two distinct cultural districts, differentiating the 'Pantalica Nord' archaeological horizon, with local roots in the Early to Middle Bronze Age cultural traditions of the island – constantly under the Aegean sphere of influence – from the allochthonous Ausonian cultural complex, originating from Southern Italy.[212] Of the three different groups traditionally reported by classic literature to live in Sicily before Greek colonisation, i.e. Sicanians, Elymians and Siculi/Sikels – refined from possible political contingent interests of Greek colonists affecting the reliability of the sources – Sicanians have been identified with the Pantalica Nord cultural aspect, while the long interaction between local and foreign (peninsular) traditions possibly led to the Sikel cultural identity.[213]

Among the most fragile and crucial geographical contexts, the Aeolian Islands represent one of the key laboratories for the study of past human-environment relationships. The archipelago was colonised during the Neolithic, but the most relevant and best investigated sites are the villages from the Bronze Age (4.2–3.0 ka BP).[214] According to the analyses of macrobotanical remains, the techniques adopted to build huts were similar over the entire archipelago, but the raw materials used varied from island to island according to the locally available timber resources. For instance, the use of oak poles was limited to the Late Bronze Age on Lipari, while on Filicudi Ericaceae represented at least 50% of each archaeological unit, and oak remains are seldom found.[215] Analysing local agricultural choices, striking differences between Filicudi and Lipari were also highlighted: *Hordeum vulgare* was widely cultivated on Filicudi, while people from Lipari probably exploited a wider spectrum of crops (including free-threshing wheats).[216] Further analyses are needed to assess if such differences depend on the differences in land suitability among islands or issue from an abrupt change in subsistence systems that occurred between the Early and Late Bronze Age.

The pattern of occupation and the subsistence strategies were adapted to local and regional climatic dynamics, as highlighted by the isotopic analyses carried out on wood remains and seeds. The occupation by people coming from the Aegean area during the Early Bronze Age was probably triggered by favourable climatic conditions. On the other hand, the increased aridity at the end of this phase (~1950–1700 BCE), and another dry phase at the end of the Middle Bronze Age (~1500–1250 BC) did not affect the distribution of local villages, whose economies were probably supported by food resources coming from extra-insular contacts.[217]

During the Iron Age[218] the three different ethnic-cultural groups were already occupying their core territorial areas: the Sicanians were located in the south-western part of the island (Sant'Angelo Muxaro/Polizzello), the Elymians in western and north-western Sicily, and the Siculi/Sikels in central and eastern Sicily (Pantalica Sud); between the 8th and 7th centuries BCE the Phoenicians and Greeks[219] settled its coasts. Actually, those civilisations had a close commercial and cultural relationship with Sicily before the colonisation started;[220] mainland Greek cities and traders from Cyprus, the Aegean islands and Lebanon

---

[204] Alberti 2013.
[205] Tanasi and Veca 2019.
[206] Van Dommelen and Knapp 2014; Bietti Sestieri 2018.
[207] Albore Livadie *et al.* 2003.
[208] Vanzetti *et al.* 2013; Cazzella and Recchia 2014.
[209] Fernandes *et al.* 2020.
[210] E.g. Kaniewski *et al.* 2013.
[211] Fernandes *et al.* 2020.
[212] Bietti Sestieri 1997.
[213] Palermo and Tanasi 2005; Procelli 2016.

[214] Martinelli and Lo Cascio 2018.
[215] Speciale *et al.* 2016; Speciale *2021*.
[216] Speciale *et al.* 2016.
[217] Caracuta *et al.* 2012; Fiorentino and Recchia 2015; Speciale *et al.* 2016.
[218] 950 to 735 BCE according to Leighton 1999.
[219] Morris 1996; Niemeyer 2006.
[220] Crielaard 1998; Nijboer 2006.

had frequent contacts with Sicilians already, many centuries before the first emporia (trade-harbours) and colonies were founded. During the transition between the Iron Age and the archaic Greek colonial period the Sicilian communities seem to have experienced a demographic crisis, inducing the abandonment of the inner, hilly territories, but the apparent gap in the archaeological record could also be ascribed to a bias in the interpretation of the results issuing from surface surveys.[221] Starting from the Archaic period, local communities continued or restarted settling on well-defended hilltops along the main rivers, enhancing their traditional vocation for agro-pastoral economy.[222] Their strategies of territorial exploitation differ from those adopted by Greek colonists and are peculiar, diverse and articulated at the regional scale, responding to an increasing number of inner and external socio-political constraints. For instance, in the territory of Himera, the indigenous communities exploited a hierarchic settlement system, with networks based on eminent sites located on the hilltops and the smaller rural sites surrounding them.[223]

The combined effect of land use intensification of agro-pastoral activities by local people, and forest clearing and crop cultivation to satisfy food requirements soon after the arrival of the first generation of Greek colonists, was responsible for the sudden thinning of tree and shrub cover throughout the Sicilian territory, as recorded by pollen analyses, c. 800 BCE.[224] Additionally, Phoenician/Punic arrivals probably induced major changes in the landscape near their main colonies, clearing woodlands and maquis to make room for crop fields and orchards,[225] and shaping many river mouths and coastal ecosystems of western Sicily. In fact, as they needed huge amounts of salt to support the massive trade of the fish-products they collected in the shallow waters of the western Sicilian coasts and the Sicily Channel, they started very soon to create artificial salt pans, reshaping the shores between Trapani and the Marsala Lagoon.[226]

Holocene dynamics of Sicilian vegetation inferred from pollen, plant macro-remains and charcoal data(S. Pasta, C. Speciale, C. Morales-Molino, G. Di Pasquale and W. Tinner)

All the sea and air temperature reconstructions from Sicily point to remarkably severe climatic conditions during the LGM.[227] At that time most of the island was likely covered by steppe-like discontinuous vegetation.[228] Small populations of mesophilous species (e.g. *Abies nebrodensis*, *Zelkova sicula*) survived in favourable (warmer and more humid) microhabitats. After the LGM some of the temperate and Mediterranean species that had survived the LGM locally, contributed to the re-colonisation of the central Mediterranean area when the climate ameliorated, i.e. between 19.0–11.7 ka BP,[229] with warming steps around 19, 16, 14.7 and 11.7 ka BP, as illustrated by novel quantitative temperature reconstructions from Italy and the Alps covering the past 24,000 years.[230] Just after the LGM, cool and arid climatic conditions were still prevalent in Sicily, whose emerged surface was c. 30% larger than today (Figure 1). Steppe-like vegetation conditions remained unchanged throughout the Heinrich Stadial 1 (also known as Oldest Dryas) until c. 14.7 cal. ka BP. The following period of transition towards the onset of the Holocene was punctuated by at least two abrupt suborbital climatic fluctuations, the Bølling-Allerød (an overall warm phase between 14.7–12.9 cal. ka BP) and the Younger Dryas (a cold phase between 12.9–11.7 cal. ka BP), as pointed out by marine sediments.[231]

Several recent studies investigated the connections between environmental changes and socio-cultural dynamics in prehistoric and historical times in Sicily.[232] Archaeobotanical studies aimed at reconstructing past environments are now common in most archaeological excavations, although in Sicily the picture is still rather incomplete.[233] Plant macro-remains in archaeological contexts (on-site record) are mainly represented by charred wood fragments, but also by seeds and fruits, which may provide information about local and regional environmental conditions[234] and trace the intensity, timing and speed of major environmental and land use changes.

In general, the information provided by pollen and charcoal diagrams coming from archaeological sites is quite different from that obtained by coring in wetlands (lakes and mires). As the archaeobotanical on-site records from archaeological sites are more effective in reconstructing the local agricultural activities during a certain settlement time, the role of these data for the reconstruction of ancient landscapes is widely acknowledged,[235] as: 1) they inform us about palaeoeconomics as they directly result from human selection; 2) apart from some rare cases of long-

---

[221] Burgio 2002.
[222] Spatafora 2002.
[223] Belvedere et al. 2002.
[224] E.g. Noti et al. 2009; Tinner et al. 2009, 2016; Bisculm et al. 2012; Calò et al. 2012.
[225] Moricca et al. 2021.
[226] Carusi 2008.
[227] Incarbona et al. 2010b.
[228] Lang 1994.
[229] Sadori et al. 2008; Incarbona et al. 2010b.
[230] Finsinger et al. 2019.
[231] Incarbona et al. 2010b.
[232] Martini et al. 2009; Mercuri et al. 2013; Zanchetta et al. 2013; Kouli et al. 2015; Pacciarelli 2015; Izdebski et al. 2016; Sadori et al. 2016; Marignani et al. 2017.
[233] Mariotti Lippi et al. 2018.
[234] Araus et al. 2014; Hrisova et al. 2016.
[235] Asouti and Kabukcu 2014; Roberts et al. 2017.

distance transport, they mirror tree occurrences within a few km-radius; 3) they are easy to date thanks to the strict link of the sample with specific cultural phases; and 4) they supply information where water bodies are lacking, a common feature in Mediterranean countries.

Conversely, quite often the data issuing from coring campaigns are not supported by archaeological data collected in the surroundings to which they can be correlated only by the chronology. On the other hand, the palaeoecological records coming from off-site mires and lakes are better suited to reconstruct the uninterrupted course of human impact (including arable and pastoral activities) over long periods. Moreover, if compared to independent evidence of climate change and human impact, off-site records are particularly suited to assess the overall local to regional vegetation dynamics in response to both human and climatic forcing.[236]

To help readers visualise the different vegetation patterns at different altitudes, two synthetic diachronic schemes of the evolution of Sicilian plant assemblages are provided, based on the pollen diagrams obtained from a lowland/coastal locality, Gorgo Basso (Figure 7a) and a hill/mountain locality, Urio Quattrocchi (Figure 7b).

With the exception of the pioneering research carried out by Bertolani Marchetti *et al.* 1984 on the Madonie mountains and the extensive pollen analysis of Lago di Pergusa in central Sicily,[237] no historical data on the Holocene dynamics of Sicilian plant communities were available until fifteen years ago, when several field surveys were carried out with the aim of disentangling the impact of the climatic patterns from anthropogenic disturbance in shaping the island's Holocene landscapes. The smallest investigated permanent water bodies (<1 ha) provided very valuable information on local plant communities, whilst the largest ones (i.e. Biviere di Gela and Preola lakes, *c.* 120 and 25 ha, respectively) provided extra-local vegetation reconstructions for larger portions of the Sicilian territory.

The coring campaigns carried out in the brackish coastal lagoons of Bivicre di Gela[238] and the wetland system of Lago Preola and Gorghi Tondi near the shore of south-western Sicily[239] provided paramount information on the main driving factors and the timing of the disruption of island's coastal woody communities over almost the entire Holocene (10.5 ka BP in the case of Gorgo Basso, Figure 7a). While maquis represented the most mature plant community before 8–7 ka BP, evergreen broadleaved forest was the dominant vegetation type during the Middle Holocene. The interplay between fire frequency, forest cover and human disturbance remains lively debated and some questions seem unsolved. For instance, Noti *et al.* 2009, Tinner *et al.* 2009, and Calò *et al.* 2012 pointed out that fire was usually associated with natural maquis conditions or anthropogenic forest disruption in the coastal landscapes, while Sadori *et al.* 2015 argued that fire was more frequent when forest cover was high at Lago di Pergusa in inland Sicily. However, at other inland sites such as Urio Quattrocchi (Figure 7b), Gorgo Lungo, Gorgo Tondo, and Urgo Pietra Giordano fire increase was usually related to forest openings as deriving from slash-and-burning activities.[240] Most rainfall reconstructions used to put forward a fire increase in response to a general Late Holocene aridification are based on pollen data, so that the drier conditions inferred from this proxy may well be a reflection of human-induced deforestation and ruderalisation that would induce a seeming xerophytisation.[241] Indeed, novel pollen-independent quantitative palaeoclimatic evidence from the warmest and driest environments of Sicily shows that the Late Holocene was not particularly dry, instead it was comparable to the subhumid to humid Middle Holocene and significantly wetter than the Early Holocene.[242] This new evidence is corroborated by recent process-based modelling results suggesting that in the absence of human impact, even the warmest and driest areas of Sicily would be forested today.[243]

Post-glacial afforestation, triggered by increasing humidity, probably started around 11.0 ka BP and/or was faster in inland and upland territories while in the lowlands it took place only *c.* 3000–4000 years later at the Early to Middle Holocene transition. This delay in afforestation is explained by the dry conditions that impeded forest growth in the very warm lowlands until ecological requirements in terms of moisture availability were met during the Middle Holocene.[244] The mountain ranges of northern Sicily were occupied by species-rich closed forests, dominated by *Quercus pubescens, Q. cerris, Fagus sylvatica, Ulmus* sp. and *Ilex aquifolium*, that reached their maximal cover between 9.7–6.8 ka cal BP before Neolithic impact became significant.[245] Open pasturelands established already at *c.* 7–6 ka BP, when fire activity increased for forest disruption. In general, growing fire activity appears to be strictly linked to local land use during the Middle and Late Holocene.[246] Palynological evidence (pollens and NPP) suggests that pastoral activities became increasingly important

---

[236] E.g. Tinner *et al.* 2009; Gassner *et al.* 2020.
[237] Sadori and Narcisi 2001; Sadori *et al.* 2013.
[238] Noti *et al.* 2009.
[239] Tinner *et al.* 2009; Calò *et al.* 2012.
[240] Bisculm *et al.* 2012; Tinner *et al.* 2016.
[241] Tinner *et al.* 2016.
[242] Curry *et al.* 2016.
[243] Henne *et al.* 2015.
[244] Curry *et al.* 2016.
[245] Bisculm *et al.* 2012; Tinner *et al.* 2016.
[246] Bisculm *et al.* 2012; Tinner *et al.* 2016.

A Gorgo Basso (Sicily, 6 m a.s.l.)

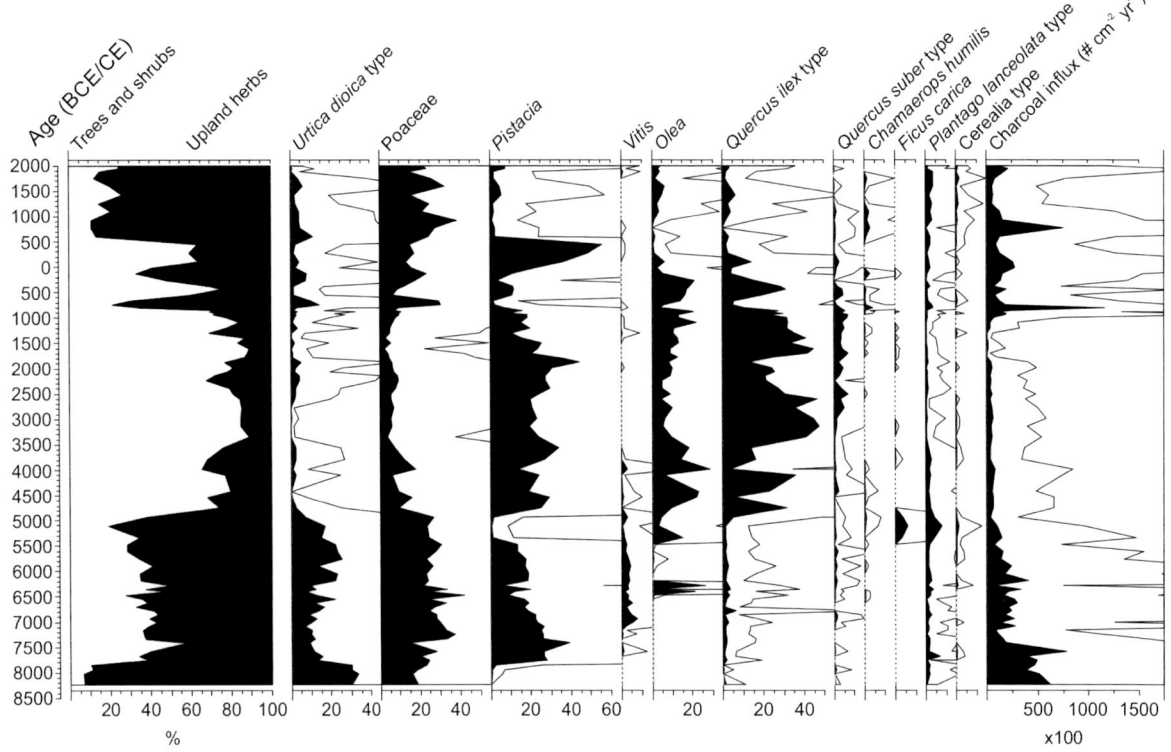

B Urio Quattrocchi (Sicily, 1044 m a.s.l.)

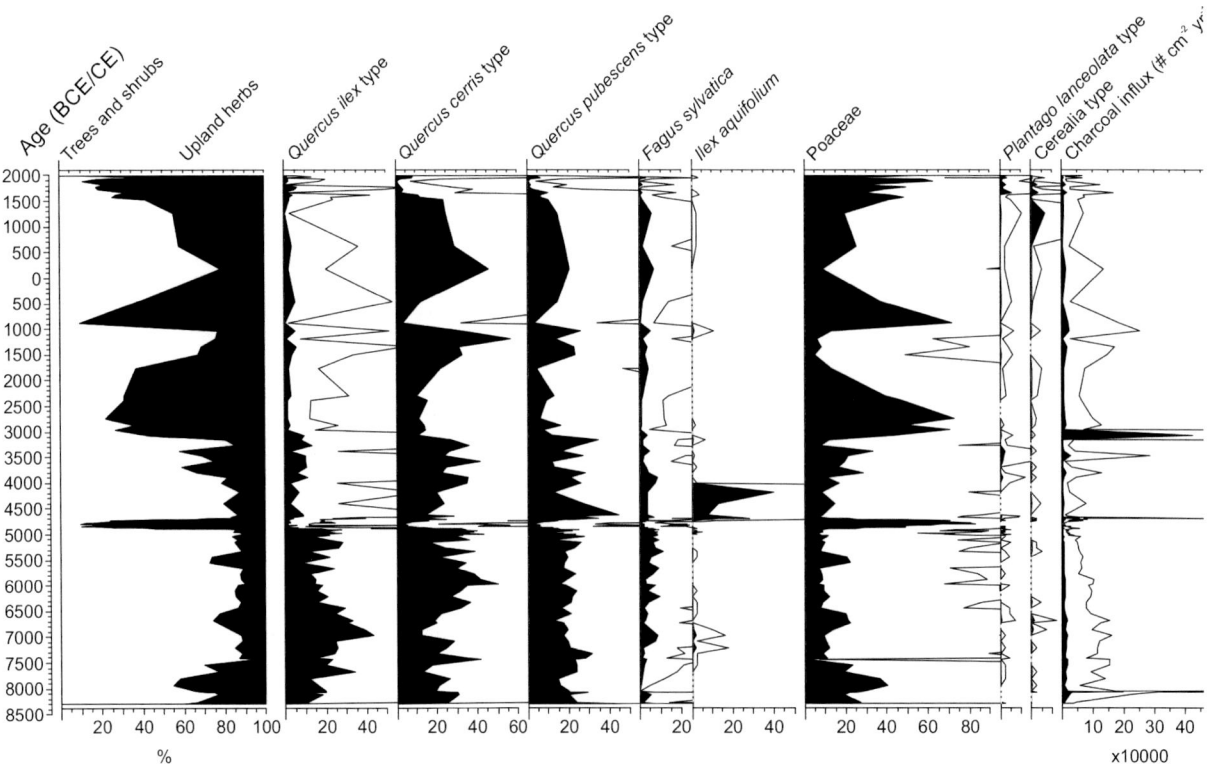

Figure 7: Pollen diagrams for Gorgo Basso (a) and Urio Quattrocchi (b).

in the mountain sites of northern Sicily, whilst cereal production was not as prominent as in the lowlands. In the Madonie mountain range, the first evidence for land use and slash-and-burn activities at *c.* 7–6 ka BP chronologically coincides with the *Ceramiche tricromiche* horizon at the Vallone Inferno rock-shelter,[247] used as sheepfold during spring.

One of the oldest Sicilian study areas is Grotta dell'Uzzo (Trapani), located on the north-western coast of the island, a key archaeological site for the transition from Mesolithic to Neolithic (~10.0–6.7 ka BP) in the central Mediterranean. Here, the high proportion of evergreen sclerophyllous woody taxa, such as evergreen oaks, *Erica* spp. and *Phillyrea* spp. in the local archaeobotanical assemblage, mirrors the regional establishment and prevalence of a natural sclerophyllic maquis during the Early Holocene up to *c.* 7 ka BP.[248] Finds of seeds and charcoal fragments attest to the early exploitation (*c.* 8–7.5 ka BP) of olive trees[249] and grapevines, and probably involves the selection of local cereal crop varieties,[250] whilst palaeoecological evidence from coastal southern Sicily points to a peculiar fig and cereal-based agriculture type during the Early Neolithic, *c.* 7.5–7 ka BP.[251]

Preliminary anthracological analyses on the charcoals from two caves of Favignana Island, in the Egadi archipelago,[252] are broadly consistent with other local and regional palaeoenvironmental data sets. For instance, the Mesolithic (12.2–12.0 uncal. ka BP)[253] charcoal data from Grotta d'Oriente do not show variations in plant selection: *Pistacia* cf. *lentiscus* is the most represented taxon, followed by Rosaceae subfam. Maloideae (probably *Crataegus* and/or *Pyrus* spp.). Other taxa typical of Mediterranean shrublands, e.g. *Erica* sp., *Myrtus communis*, *Cistus* spp., *Arbutus unedo*, *Juniperus* spp., *Rhamnus/Phillyrea*, and *Prunus* spp., are recorded in low numbers. A monocotyledon specimen has been attributed to the dwarf palm, *Chamaerops humilis*. The taxa recorded and their relative abundances do not display significant differences in the levels dated to the Early Holocene (approximatively between 10.2–9.5 ka BP). In fact, the preliminary results from Grotta delle Uccerie (13.3–12.8 uncal. ka BP)[254] are consistent with those from Grotta d'Oriente, showing no substantial differences both in terms of identity (*Pistacia lentiscus*, *Juniperus*, *Arbutus unedo*, *Rhamnus/Phillyrea* where found there,[255] and relative abundance of the taxa represented in the anthracological diagram.

The exploitation of wood resources on the island of Ustica (Palermo) during prehistoric times was studied through the first archaeobotanical analyses from the Neolithic and Bronze Age.[256] This interesting case study provided some interesting clues about the effect of a prehistoric community on a limited resource, i.e. the pristine vegetation of a tiny island (*c.* 8 km^2). Local climatic and topographic patterns, and the current floristic composition of last nuclei of maquis, suggest that the unspoiled woody vegetation was dominated by high evergreen maquis (or even forest) communities. Neolithic human communities and their agro-pastoral activities affected the local vegetation, triggering the disappearance of native pine trees and holm oaks. Olive trees were probably introduced only in the Middle Bronze Age as a crop, forming near-natural evergreen woodland nuclei during the phases of abandonment of local cultivated lands. The interpretation of historical maps allows us to quantify land use changes across the last three centuries, and proved to be a very useful tool to disentangle the complicated role played by climatic factors and intermittent human disturbance on the appearance/disappearance of woody species.

More recently, during the Copper Age, the archaeobotanical record from Phase 5 (*c.* 4.3–4.2 ka BP) of Hut 5 at Case Bastione, near Enna in central Sicily,[257] shows that the construction of huts inside the village relied on the intense exploitation of timber resources around the site, mostly deciduous oaks. Some data suggest that drier conditions related to the so-called '4.2 ka BP event, a 300-year-long intense drought period at the transition from the Middle to Late Holocene, with different effects and timing over the whole Mediterranean area,[258] may have affected local human communities. However, in the available Sicilian records referring to this period there is no evidence for major production crises, such as a collapse of cereals and/or weeds.[259] On the contrary, at some sites this period is characterised by higher abundances of cultural indicators and the expansion of open land, e.g. at Biviere di Gela or at Lago di Preola,[260] which does not exclude synergistic detrimental effects of increasing land use and drought on forest cover. The intensification of subsistence activities in central Sicily, increasingly focused on cereal cultivation, accompanied by a more rational use of target species (barley or different species of *Triticum* in different areas),[261] and a highly specialised pastoralism,[262] can be interpreted as a consequence of local demographic growth, and

---

[247] Scillato, Palermo; Forgia *et al.* 2013.
[248] Tinner *et al.* 2009.
[249] Schicchi *et al.* 2021.
[250] Costantini 1989; Zohary *et al.* 2012.
[251] Tinner *et al.* 2009.
[252] Poggiali *et al.* 2012.
[253] Martini *et al.* 2007.
[254] Martini *et al.* 2007.
[255] F. Poggiali, pers. comm.
[256] Speciale *et al.* 2021.
[257] Speciale *et al.* 2020.
[258] Weiss 2016; Bini *et al.* 2019; Di Rita *et al.* 2019.
[259] E.g. Noti *et al.* 2009; Tinner *et al.* 2009; 2016; Bisculm *et al.* 2012; Calò *et al.* 2012.
[260] Noti *et al.* 2009; Calò *et al.* 2012.
[261] Speciale *et al.* 2020.
[262] Giannitrapani *et al.* 2014.

resulted in a rising impact on vegetation, especially in the Early Bronze Age.[263]

After a general reduction of vegetation cover *c.* 2200 BCE at the onset of the Bronze Age, perhaps related to supra-regional climatic trends referred to as the '4.2 ka BP event',[264] woodlands seem to slightly recover *c.* 1800 BCE, only to decline again *c.* 1500–1200 BCE.[265] At that time, domesticated mammals (mainly goats and sheep, but also cows and pigs) represented the main protein source. The foraging activity of these animals (especially over browsing by goats and the mechanical damages to forest undergrowth and recruitment by pigs) may have caused a major reduction/disruption of local forest and shrubland ecosystems; this could also explain the rather low amounts of wild preys (red deer and wild boar) found at the two sites of inner Sicily.[266]

According to EBA anthracological data from the southeastern site of Castelluccio, local vegetation was mainly dominated by *Olea europaea*, *Quercus ilex*, *Pistacia lentiscus*, and deciduous oaks.[267] The most intense periods of forest clearance were caused by increasingly frequent grazing and burning disturbance, as elsewhere in the Mediterranean.[268] Indeed, burning induced the expansion of sclerophyllous maquis and xerophilous shrubland across the Sicilian lowlands, given that all the palynological data from close to the coast underline that the final demise of natural evergreen broadleaved forest ecosystems took place between 900 BCE and 200 CE, starting in the Iron Age and reaching the most intense phase during Greek colonisation and Roman times.[269]

The phase of early contact between local communities and Greek colonisers (7th/6th centuries BCE) provides an interesting case study in western Sicily. In his comparison between Selinunte (a coastal city) and Monte Polizzo (a defended hilltop), Stika *et al.* 2008 identified different subsistence strategies due not only to cultural choices but also complying with local environmental resources: free-threshing wheats were more suitable for the Monte Polizzo area, and probably non-Greek communities traded their crops with Selinunte. Although the number of charcoal fragments is rather low to provide a clear picture of the vegetation near Monte Polizzo, the roof beams used there were from forest trees (mainly deciduous oaks), suggesting the existence of abundant tree cover in the surroundings at that time.

In approximately the same period, the anthracological record from the Greek site of Himera (648–409 BCE)[270] allows us to reconstruct the coastal forest landscape around the mouth of the Imera river, on the northern coast of Sicily. The examined charcoal samples come from the charred remains of fuelwood used in funeral pyres dating back to the 6th/5th centuries BCE, and point to the dominance of cork oak (*Quercus suber*) in the local woodlands. These data suggest that (either due to natural factors or to human selection) an almost monospecific cork oak woodland covered the coastal plain around the site. Interestingly, the results from charcoal analyses from Thamusida (central Moroccan coast) between the second half of the 1st century and the 3th century CE,[271] are very similar, also supporting some recent studies that show that cork oak may occur on soils with high water content.[272] Hence, the occurrence of mono-specific cork oak forests in ancient times may have been linked to peculiar environmental features and might not only be a result of human influence, as postulated by Carrión *et al.* 2000 for the Iberian Peninsula.

Pollen analysis from the archaeological layers of Villa del Casale (Piazza Armerina, Enna) provided a diachronic interpretation of the environment and the economy of local community between the 1st and 6th centuries CE.[273] During the first centuries of the Empire, the villa was already surrounded by rather degraded vegetation, probably as a consequence of the spread of crop cultivation, with a peak of cereal pollen record during 4th century CE, supporting the historical reports of the vast and long-lasting exploitation of Sicily as granary for the Roman Empire. The Pergusa record suggests that several cooling phases, notably between 450–750 CE,[274] may have favoured the spread of farming activities during Late Antiquity[275] by increasing moisture availability.

During early Medieval times (700–800 CE), pollen records suggest land use intensification over the whole Mediterranean Basin.[276] This general pattern was confirmed by the palaeoecological investigations carried out on Pantelleria Island;[277] more specifically, the data collected from the core at Lago di Venere shed light on the profound changes to local vegetation cover from 800 CE up to the present day, and highlighted the past occurrence of several forest tree species, such as deciduous oaks, which no longer occur on the island. Interestingly, here the cold and humid conditions

---

[263] Sadori *et al.* 2013.
[264] Di Rita and Magri 2019.
[265] Tinner *et al.* 2009.
[266] De Simone *et al.* 2019.
[267] Crispino 2018.
[268] Moreno *et al.* 1998; Colombaroli *et al.* 2008; Vannière *et al.* 2011.
[269] Noti *et al.* 2009; Tinner *et al.* 2009; Calò *et al.* 2012.

[270] Carniato 2011.
[271] Allevato *et al.* 2017.
[272] Petroselli *et al.* 2013.
[273] Montecchi and Mercuri 2016.
[274] Sadori *et al.* 2013.
[275] Sadori *et al.* 2016; Mercuri *et al.* 2019.
[276] E.g. Atherden and Hall 1999; Colombaroli *et al.* 2007; Morales-Molino *et al.* 2019.
[277] Calò *et al.* 2013.

Table 2: Radiocarbon dates from the Early to Middle Neolithic sites of Sicily (end 7th/beginning 5th millennium BCE, modified after Natali and Forgia 2018) (B: bone; C: charcoal, MS: marine shell). All dates have been calibrated with the OxCal 4.3 software (Bronk Ramsey 2009) using the IntCal13 curve (Reimer et al. 2013). n.a = not available/assessed.

Site	Context	Lab Code	Age BP±	Age cal BCE	Material	Horizon	References
Uzzo	F, 12	OxA-13662	7744±33	6650–6490	MS	Castelnovian or Impresse arcaiche	Lo Vetro and Martini 2012; Tinè and Tusa 2012
Uzzo	F, 12	OxA-V-2364-43	7753±36	6650–6490	B	n.a.	Mannino et al. 2015
Uzzo	F, 12	MAMS-16238	7957±25	7040–6700	B	n.a.	Mannino et al. 2015
Uzzo	F, 11	KIA-36032	7175±45	6210–5920	B	n.a.	Mannino et al. 2015
Oriente	5	LTL877A	7040±55	6020–5790	C	Castelnovian-Early Neolithic?	Lo Vetro and Martini 2012; 2016
Uzzo	F, 10	OxA-X-2071-31	7413±39	6390–6220	n.a.	Impresse arcaiche	Tinè and Tusa 2012
Uzzo	F, 10	OxA-13661	7410±32	6380–6220	MS	Impresse arcaiche	Tinè and Tusa 2012
Kronio	B, 16-17	LTL-211A	6991±60	5990–5740	n.a.	Impresse arcaiche	Tinè and Tusa 2012
Uzzo	F, 9	OxA-V-2364-40	7006±34	5990–5800	B	n.a.	Mannino et al. 2015
Uzzo	F, 9	KIA-36033	7020±45	6010–5790	B	n.a.	Mannino et al. 2015
Uzzo	F, 7-9	P-2733	6750±70	5770–5530	C	Impresse evolute	Tinè and Tusa 2012
Uzzo	F, 7	OxA-13808 7640-7510	7173±37	6110–5980	MS	n.a.	Mannino et al. 2015
Uzzo	W, 15	UD-165	6720±80	5750–5480	n.a.	Impresse evolute	Tinè and Tusa 2012
Uzzo	F, 7	OxA-13808	7137±37	6080–5920	MS	Impresse evolute	Tinè and Tusa 2012
Uzzo	F, 2	KIA-36035	6985±40	5990–5750	B	n.a.	Mannino et al. 2015
Stretto Partanna	US46	R-291	6630±120	5750–5330	-	Tricromiche (Stentinello?)	Tinè and Tusa 2012
Catania Acropoli	US144	LTL15581A	6476±45	5520–5340	B	Impresse evolute	Nicoletti 2015
Vallone Inferno	3.4.N H22-1	Beta-408858	6340±30	5460–5220	C	Ceramiche tricromiche	Forgia 2019
Stretto Partanna	US76	R-292	6260±110	5480–4960	n.a.	Ceramiche tricromiche	Tinè and Tusa 2012
Uccerie	Layer 1	n.a.	5921±50	4940–4690	C	Impresse evolute, Kronio	Martini et al. 2012
San Martino	Layer 20	LTL3521A	5757±50	4720–4490	C	Impresse evolute, Stentinello II	Quero et al. 2019
San Martino	Layer 23	LTL3519A	5853±60	4850–4550	C	Impresse evolute, Stentinello II	Quero et al. 2019
Torcicoda, Borgo Cascino	Neolithic hearth	Beta-134.71	5690±120	4826–4331	C	n.a.	Giannitrapani 2017

promoted *Pinus pinaster* and *Erica arborea* at the expense of *Q. ilex* and *Olea*, while crop production showed no significant trend before or during the LIA. Only after the LIA, c. 1880 AD, did cereal cultivation increase significantly, likely as a consequence of agricultural intensification more than climate warming.

As Castrorao Barba (2017) pointed out, the distribution and the demography of Sicilian farms and rural villages between Late Antiquity and the Middle Ages did not follow a unique pattern: many sites experienced a strong crisis during the 3rd century CE, after which they flourished one hundred years later, before gradually being abandoned between the 5th and 6th centuries, sometimes recovering or being re-colonised between the 7th and 8th centuries, and being inhabited until the 10th – 11th centuries CE.

The site of Contrada Castro near Corleone, Palermo[278] covers a period from the 7th to the 11th century CE. Local vegetation, reconstructed on the basis of both wood charcoals and seeds found *in situ*, fits well with the pattern of other Sicilian sites from the early Middle Age known to date. Results indicate an intense exploitation of local woody species, such as *Quercus ilex, Q. pubescens, Fraxinus ornus, Acer campestre, Populus nigra, Ulmus canescens, Ostrya carpinifolia*, and *Pistacia terebinthus*. The unexpected absence of olive trees, usually very common in rural areas of inner Sicily colonised by Berberian farmers, and increasing in the pollen records[279] remains unexplained. Probably local people preferred animal breeding instead of investing in the time-demanding practices required to maintain olive grove cultivation.

Intensive cork oak forest clearing during the Middle Ages (12th – 13th centuries CE) is well-documented by the charcoal record from the Medieval settlement of Segesta (north-western Sicily), where this tree species was heavily exploited for building purposes until the 18th century CE.[280] As a result of woodland overexploitation, nowadays only a few, small and scattered nuclei testify to the previous occurrence of vast cork oak stands in the nearby territories. At this same site, small, charred branches of *Abies* were recovered in contexts dated to the 12th/13th centuries CE.[281] Although wood anatomy does not allow us to classify at species level, these charcoals could be referred to *A. nebrodensis*, nowadays only occurring on the Madonie mountains. between 1350 and 1700 m a.s.l. As these remains were probably functionally related to the furnaces, and the only known Sicilian population of *Abies* is located more than 200 km from Segesta, firs might have occurred on nearer mountain ranges.[282]

Between overexploitation and abandonment: lessons learned from Sicilian landscape dynamics over the last three millennia (S. Pasta and C. Speciale)

To disentangle the impact of climatic change and human activities on ecosystems, land suitability, soil erosion, and geological hazard is challenging. A multidisciplinary approach like the one proposed here may allow the scientific community to accomplish this. In fact, the analysis of the diachronic data from specific archeological sites, like the ones briefly discussed in the previous paragraphs, offers the chance to take stock of concluded socio-ecological cycles and draw general conclusions about the carrying capacity of given territories within given timeframes.

Nevertheless, many issues are still pending. For instance, it is difficult to evaluate the real impact of the introduction of agriculture during the Neolithic without a reliable estimation of the population density at that time. In the end, a better reconstruction of past demographic fluxes and socio-economic structures enables a better understanding of landscape dynamics. As a matter of fact, the size and the number of Neolithic settlements was higher than previously expected, as well as their impact on natural resources.[283] Indeed, early agro-pastoral activities, albeit somehow discontinuous or intermittent in the uplands,[284] probably started to change forever the composition, structure and distribution of many local ecosystems, usually through excessive disturbance, resulting, e.g., from slash-and-burn activities or (over)browsing by domestic animals.[285]

The subsistence economy of the human communities grew in complexity during the Metal ages, and was mostly based on cereal cultivation and livestock husbandry, hence local communities must have already intensely shaped the landscape surrounding their settlements. Along with demographic growth, there was a substantial change of settlement density and distribution, economical systems, and introduced (cereal crop, domesticated herbivores) species.

Later, the impact of the Greek colonies and the Roman Empire on the Sicilian landscape was even more severe – like elsewhere in the Mediterranean Basin.[286] Increasing demographic pressure before and during Greek colonisation caused massive forest clearance to establish more extensive croplands to feed the many densely populated colonies,[287] which also needed wood for urbanisation, irrigation systems, heating, cooking, ship building, and military and mining purposes. The spread of agricultural practices to meet the food requirements of a continuously rising populations[288] eventually caused soil erosion,[289] large landslides, frequent flooding, and, in the long run, noticeable reduction of soil fertility.[290]

Massive forest clearance under the rule of the Greeks and Romans (particularly between the 5th century

---

[278] Castrorao *et al.* 2018; Bazan *et al.* 2020.
[279] Sadori *et al.* 2016.
[280] Castiglioni and Rottoli 1997.
[281] Castiglioni and Rottoli 1997.
[282] Di Pasquale *et al.* 2014; Pasta *et al.* 2019.
[283] Boyle *et al.* 2011.
[284] Tinner *et al.* 2016.
[285] Rius *et al.* 2009; Revelles *et al.* 2015; Braje *et al.* 2017; Bouby *et al.* 2020.
[286] Hughes and Thirgood 1982; Sadori *et al.* 2016.
[287] Forni 1988; De Angelis 2002.
[288] Carter *et al.* 1985; Carter and Costantini 1994; De Angelis 2016.
[289] Neboit 1984; Ayala and French 2005; Heinzel and Kolb 2011.
[290] Tainter 2006.

BCE and 1st century CE) changed forever the Sicilian landscapes and caused the rarefaction or even extinction of vulnerable woody plants that once dominated entire ecosystems. This was probably the fate of many conifers – unable to recover after the recurrent wildfires and heavy grazing/browsing related to forest clearance process. For instance, the dense fir forests (probably *A. nebrodensis*) – once covering the tops of the Nebrodi mountains – were completely destroyed by the end of the 1st century CE.[291] Deforestation had also permanent consequences on the vegetation of the inner part of the island, devoted to crop cultivation, and with no, or very few, native woodlands for two thousand years.[292] In fact, the extreme scarcity and isolation of the remnant forest fragments represented an insurmountable obstacle for the encroachment of wood species and for forest recover. More in-depth research focused on sedimentology and soil micromorphology is needed to assess the impact of forest clearance on both soil erosion and vegetation degradation; without doubt deforestation triggered massive topsoil erosion and probably induced the spread of badlands (locally called 'calanchi') in the warmest and driest sectors of southern and central Sicily.

Along with deforestation and soil erosion, the increasing demand of water for urban and agricultural needs[293] caused profound and permanent modifications in the regional water network. In fact, overexploitation of water resources changed forever the course, shape, water balance, and sedimentation regime of the main Sicilian rivers, i.e. Simeto, Salso, Fiume Grande or Imera, Platani, and Belìce. The water tables of these rivers, once navigable, currently show a strong seasonal oscillation, and these watercourses host only a few, scattered and impoverished nuclei of riparian forests, woodlands, and thickets.[294] Similar changes also affected smaller watercourses located near several important Greek settlements, such as Selinunte,[295] Morgantina,[296] and Camarina, where the salinisation of local rivers, the depletion of the water table, and the development of extensive marshes – increasing the probabilities of malaria – likely issued from the heavy deforestation and consequent hydrogeological alterations.[297] In the long run, however, human impact must have initiated a more efficient use of local water resources[298] to face seasonal shortages, and compensate for the loss of previously available springs due to freshwater overexploitation.

Several historical studies[299] provide clues to the Medieval distribution and use of Sicilian woodlands. Yet, a big knowledge gap still needs to be filled to reconstruct the history of land use changes and natural resource exploitation (e.g. extent, timing and modalities of agro-pastoral activities) that occurred within Sicilian territory over the last thousand years. In fact, although the most important phases of deforestation were already far in the past, large, patchy woodlands – such as the so-called 'Genoard' (= Paradise on Earth) or 'Parco', located south-west of Palermo, the fief of 'Birribaida' in south-western Sicily,[300] and the 'Saltus Camarinensis' in south-eastern Sicily – were destroyed during the Middle Ages, or even later, as testified by the three case studies/examples shortly to be discussed below. As a matter of fact, the last synchronous phase of grassland establishment recorded in the northern mountain ranges of Sicily[301] occurred during the Middle Ages (800–700 cal BP, 1150–1250 CE).

Research focused on the village of Brucato (Termini Imerese, Palermo) allowed Beck-Bossard (1981, 1984) to provide diachronic and topographic syntheses of the faunistic and floristic specimens (i.e. cereals, legumes, fruits and nuts, trees, and other plants) found in the village, and to draw some conclusions on the local subsistence economy between the 13th/14th centuries BCE. As far as tree species are concerned, deciduous oaks (*Quercus* spp.) and beeches (*Fagus sylvatica*), probably used for roof beams, were the most common wooden remains, testifying to logging and transport activities from the woods of the Madonie mountains,

Sugar cane cultivation was introduced into Sicily by Arabs *c*. the 9th century CE. It was carried out in the swamps and river mouths of almost all the coastal plains of the islands. The environmental costs of this activity were unsustainable, as sugar cane cultivation required huge amounts of water. Additionally, to sustain the increasing demand of wood and fuel needed for sugar production, most of the remnant forest nuclei near Palermo were cut down between the 12th – 16th centuries CE.[302] During this time lapse, before the Mediterranean sugar market was outcompeted by that of central and outh America, the capital of Sicily represented one of the main centres of sugar production and export worldwide.[303]

Analyses of the wood charcoal remains of the furnaces from Burgio (Agrigento), used for pottery and bell production, and dating to the 16th/17th centuries CE, have identifies *Fagus* sylvatica, *Quercus* cf. *pubescens*,

---

[291] Pasta *et al.* 2019.
[292] Sadori *et al.* 2013.
[293] Crouch 1993; De Angelis 2003, 2016; Evans 2016.
[294] Brullo and Spampinato 1991.
[295] Rabbel *et al.* 2014.
[296] Crouch 1984; Bruno *et al.* 2015.
[297] Vacante 2016.
[298] Crouch 1984; Collin Bouffier 2009.

[299] E.g. Lombard 1959; Higounet 1966; Falkenhausen 1980; Bresc 1983; Corrao 1988; Dentici Buccellato 1994; Quattrocchi 2017.
[300] La Rosa *et al.* 2021.
[301] Tinner *et al.* 2016.
[302] Trasselli 1955.
[303] Galloway 1989.

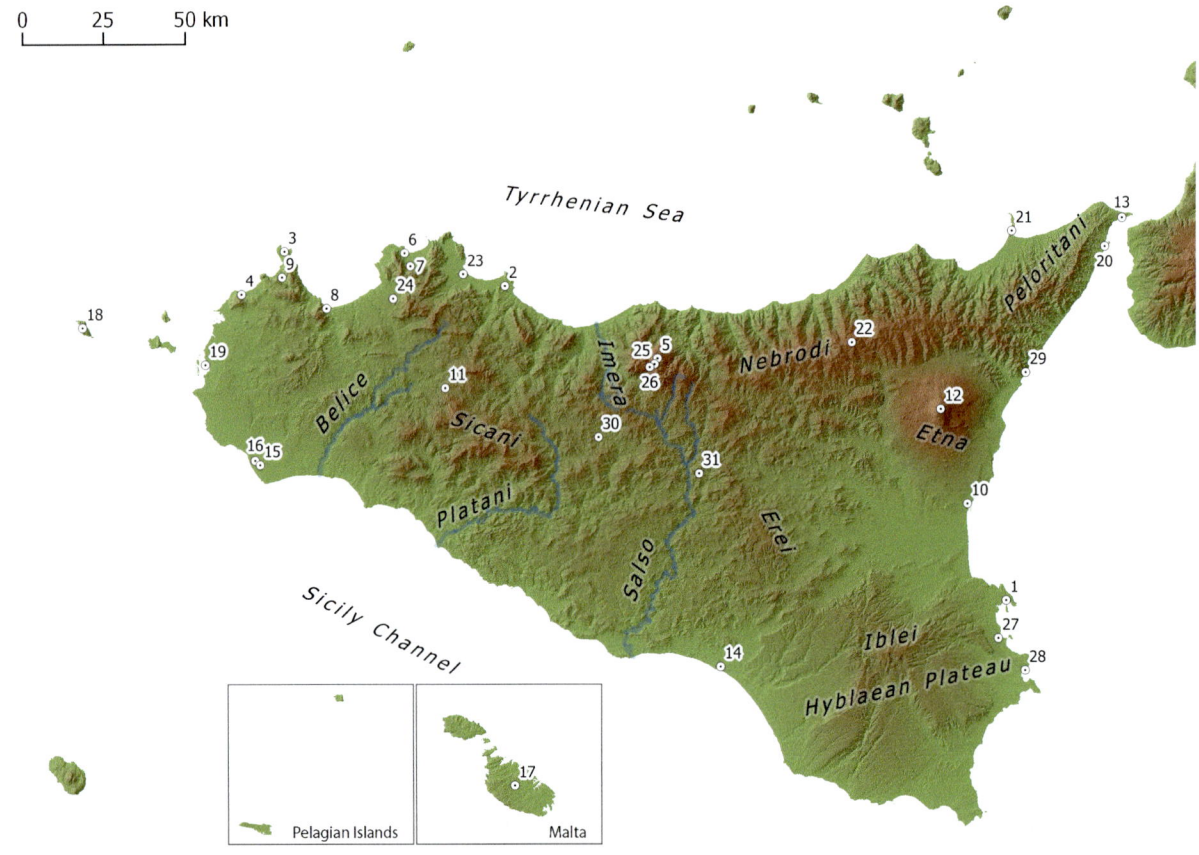

Figure 8: Localities in the paragraphs focused on the geological, morphological and climatic evolution of Sicily: Augusta (1); Bagheria (2); Capo San Vito (3); Bonagia (4); Carbonara Massif (5); Carburangeli (6); Carini (7); Castellammare del Golfo (8); Castelluzzo (9); Catania (10); Corleone (11); Etna Mts. (12); Ganzirri (13); Gela (14); Gorgo Basso (15); Lago Preola (16); Malta (17); Marettimo (18); Marsala Lagoon (19); Messina (20); Milazzo (21); Monte Soro (22); Palermo (23); Partinico (24); Piano Zucchi (Madonie Mts.) (25); Portella Colla (Madonie Mts.) (26); Priolo (27); Siracusa (28); Taormina (29); Vallelunga Pratameno (30); Villarosa (31).

Quercus ilex, *Olea europaea*, *Prunus* sp., Rosaceae subfam. Maloideae, Corylaceae, *Pinus* spp., and, probably, *Vitis vinifera*. Notwithstanding the small sample size and probable overrepresentation due to the selection of wood of higher calorific value, the presence of *Fagus sylvatica* is a remarkable feature. In fact, considering the distance to the current nearest beechwoods (c. 100 kms), the species may have occurred on the tops of the Sicani mountains, whose highest peaks exceed 1300 m a.s.l. In actual fact, the last regular, although very rare pollen occurrences of *F. sylvatica* from western Sicily, date back to the period between the 2nd – 8th centuries CE.[304] During the last two centuries, mixed evergreen/semideciduous oak forests survived only on the top of the Sicani mountains, whilst the surroundings of Burgio are currently characterised by a mostly open and degraded landscape.[305] The data issuing from the pollen analyses carried out on the permanent ponds in the forest of Ficuzza[306] seem to confirm those obtained from charcoal analyses, pointing to a first major collapse of forest cover c. 1000 CE, and a second between the 15th/16th centuries CE.

Sicilian landscape history between the past and future: perspectives, challenges and wishes (S. Pasta)

Further collaborations with experts from other disciplines may provide new points of view and enable deeper insights into the already available interpretations concerning Sicilian landscape history and dynamics. For example, using geo-statistical tools to reconstruct species and vegetation distribution patterns,[307] and applying dynamic, process-based models, may enable us to better understand the

---

[304] Tinner *et al.* 2016.
[305] Fiorentino and D'Oronzo 2007.
[306] Tinner *et al.* 2016.
[307] E.g. Kaplan *et al.* 2009; Kleinen *et al.* 2011; Fyfe *et al.* 2015; Marquer *et al.* 2017; Roberts *et al.* 2018; 2019; Zanon *et al.* 2018; Goedecke *et al.* 2020; Racimo *et al.* 2020.

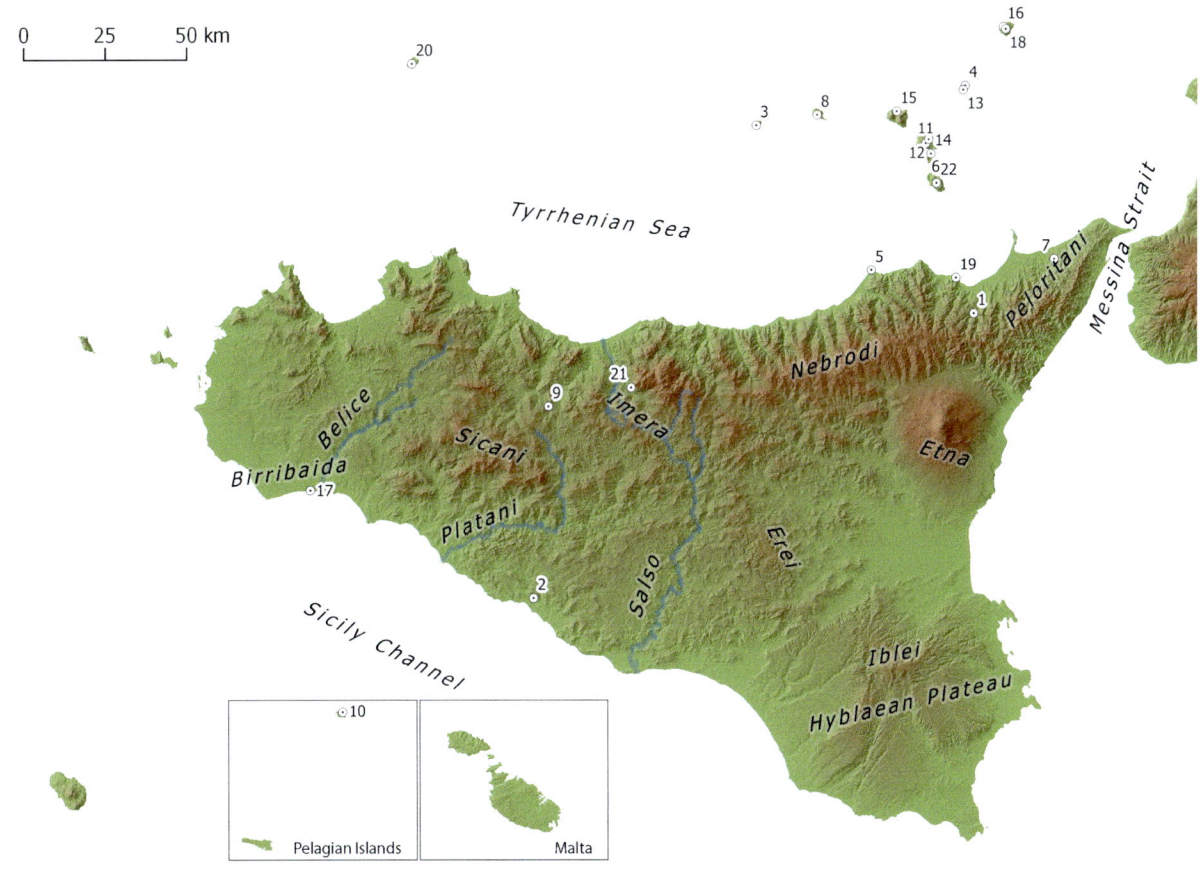

Figure 9: Localities in the paragraph relating to volcanic and seismic activity:
Abakainon (1); Agrigentum (2); Alicudi Island (3); Calcara (Panarea Island) (4); Capo d'Orlando (5); Contrada Diana (Lipari Island) (6); Contrada San Martino (Spadafora) (7); Filicudi Island (8); Le Rocche (Roccapalumba) (9); Linosa Island (10); Lipari Island (11); Monte Pelato (Lipari Island) (12); Panarea Island (13); Rocca di Castello (Lipari Island) (14); Salina Island (15); Sciara del Fuoco (Stromboli Island) (16); Selinunte (17); Stromboli Island (18); Tyndaris (19); Ustica Island (20); Vallone Inferno (Scillato) (21); Vulcano Island (22).

factors shaping past and future vegetation dynamics,[308] and the historical impacts of varying land use and anthropogenic pressure.[309]

To better evaluate the past patterns of species occurrences, when interpreting pollen diagrams it is important to consider that plants and plant communities may respond quickly not only to land use intensification but also to abandonment. In fact, in many areas natural or anthropogenic disasters (e.g. large wildfires, wars, epidemics, tsunamis, volcanic eruptions, landslides, floods, extreme drought events) may have triggered the temporary cessation of agro-pastoral activities over large territories for decades, or even centuries. Plant ecologists and vegetation scientists have pointed out that after abandonment the speed and the result of progressive succession processes after land abandonment depend not only on local environmental features (climate, soil, disturbance regime), but also on the structure of the mosaic-like patchwork formed by the semi-natural and cultural components of the landscape. For instance, recent research carried out in Sicily[310] shows that if woody species and forest nuclei are widespread and close enough to the abandoned agro-pastoral sites, the full recover of forests and an extensive re-shaping of the whole landscape may happen within a few decades.

Population genetics represent another promising tool for the historical reconstruction of the cultural and socio-economic fluxes throughout the Mediterranean, and may lead us to rethink our paradigms on the speed and path of (and the role played by Sicily in) the Neolithic revolution. One such example is the recent paper[311] who shed new light on the history of early cultivation and domestication of *Vitis vinifera* on the island. Additionally, recently published pollen spectra represent a basic tool to document the gradual or sudden increase of pollen grains belonging to

---

[308] Henne *et al.* 2015.
[309] Romano *et al.* 2021.
[310] E.g. Rühl *et al.* 2005; 2006; La Mela Veca *et al.* 2016.
[311] De Michele *et al.* 2019.

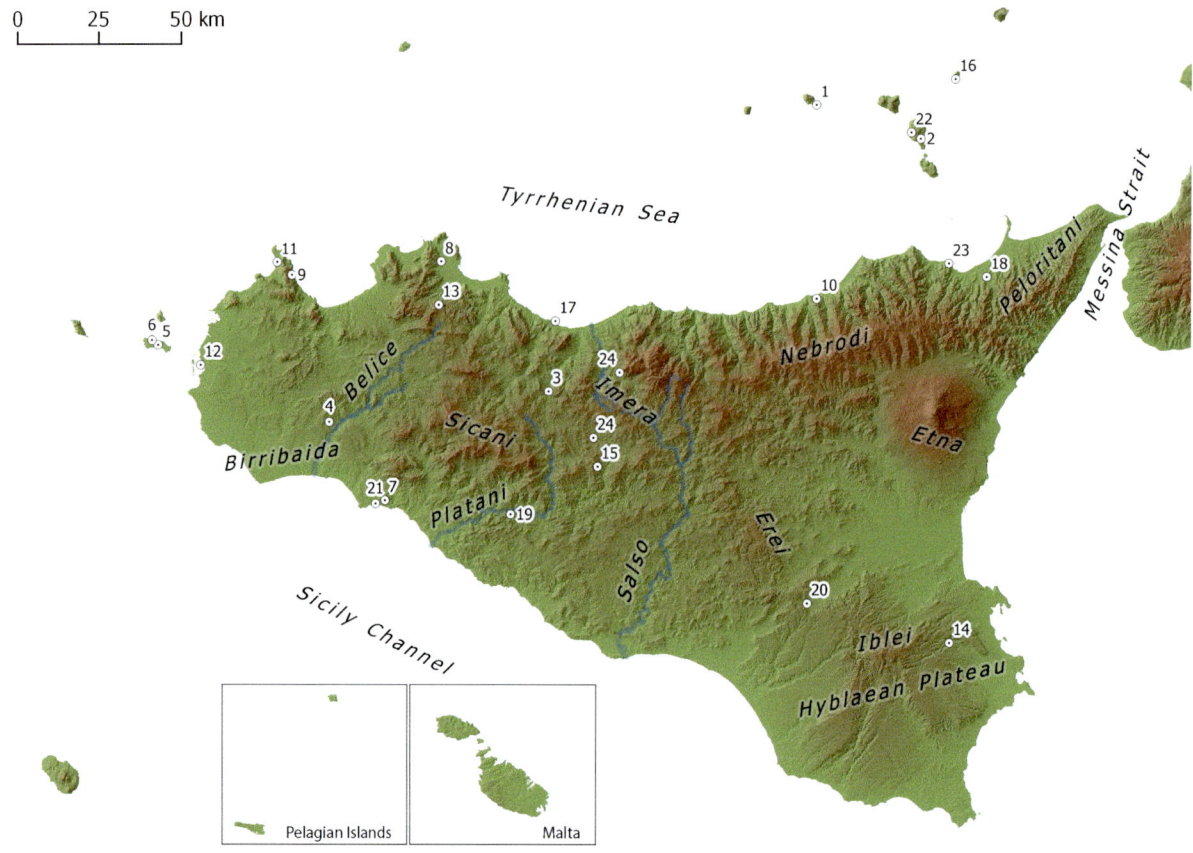

Figure 10: Other localities from the paragraph on proto-history:
Capo Graziano (Filicudi Island) (1); Contrada Diana (Lipari Island) (2); Contrada Le Rocche/Fiaccati (Roccapalumba) (3); Contrada Stretto (Partanna) (4); Favignana Island (5); Grotta d'Oriente (Favignana Island) (6); Grotta del Kronio (Sciacca) (7); Grotta della Molara (8); Grotta dell'Uzzo (9); Grotta di San Teodoro (10); Isolidda di Macari (11); Marsala Lagoon (12); Moarda (13); Pantalica (14); Polizzello (15); Punta Milazzese (Panarea Island) (16); Riparo del Castello (Termini Imerese) (17); Rodì Milici (18); Sant'Angelo Muxaro (19); Sciacca (21); Sant'Ippolito (20); Serra d'Alto (Lipari Island) (22); Tyndaris (23); Vallone Inferno (Scillato) (24); Vallelunga Pratameno (25).

important cultivated species (e.g. *Olea europaea* and *Vitis vinifera*), and to clarify whether and when woody plants of silvicultural and/or agronomical interest (e.g. *Castanea sativa* and *Corylus avellana*) were introduced by humans during the Holocene or if they should be considered native.

Increasing importance must be given to human sciences,[312] promoting a more intense dialogue between archaeologists, who may point out the cultural and demographic patterns of local population dynamics, life and earth scientists, who may assess the agro-sylvo-pastoral carrying capacity at the landscape level, linguists who may help with toponym interpretations to detect past occurrences and recent extinctions, and historians, who should be able provide a huge amount of information extracted from travel reports, chronicles, private and public archives, deeds, demographic censuses, reports on epidemics, cadastres, and other official documents.

Contrasting processes (abandonment of traditional agro-forestry practices, increasing grazing/browsing pressure by introduced wild ungulates, the encroachment of woody plants) are currently driving the vegetation dynamics of Sicilian hill systems and mountain ranges.[313] Many patches of semi-natural landscape near the coast, in the plains, and on the low hills of inner Sicily have been destroyed over last fifty years to provide land for intensive agriculture (e.g. greenhouses for tomatoes and strawberries, artichoke plantations, vineyards, olive groves, and citrus and peach orchards). At the same time, several industrial clusters were built, obliterating forever some of the most beautiful and productive coastal plains of Sicily (e.g. Milazzo, Termini Imerese, Augusta, Gela, and Porto Empedocle), whose fertile soils, mild climate and freshwater resources have attracted and sustained human communities over millennia.

[312] Izdebski *et al.* 2016.

[313] E.g. Lasanta-Martínez *et al.* 2005; Bianchetto *et al.* 2015.

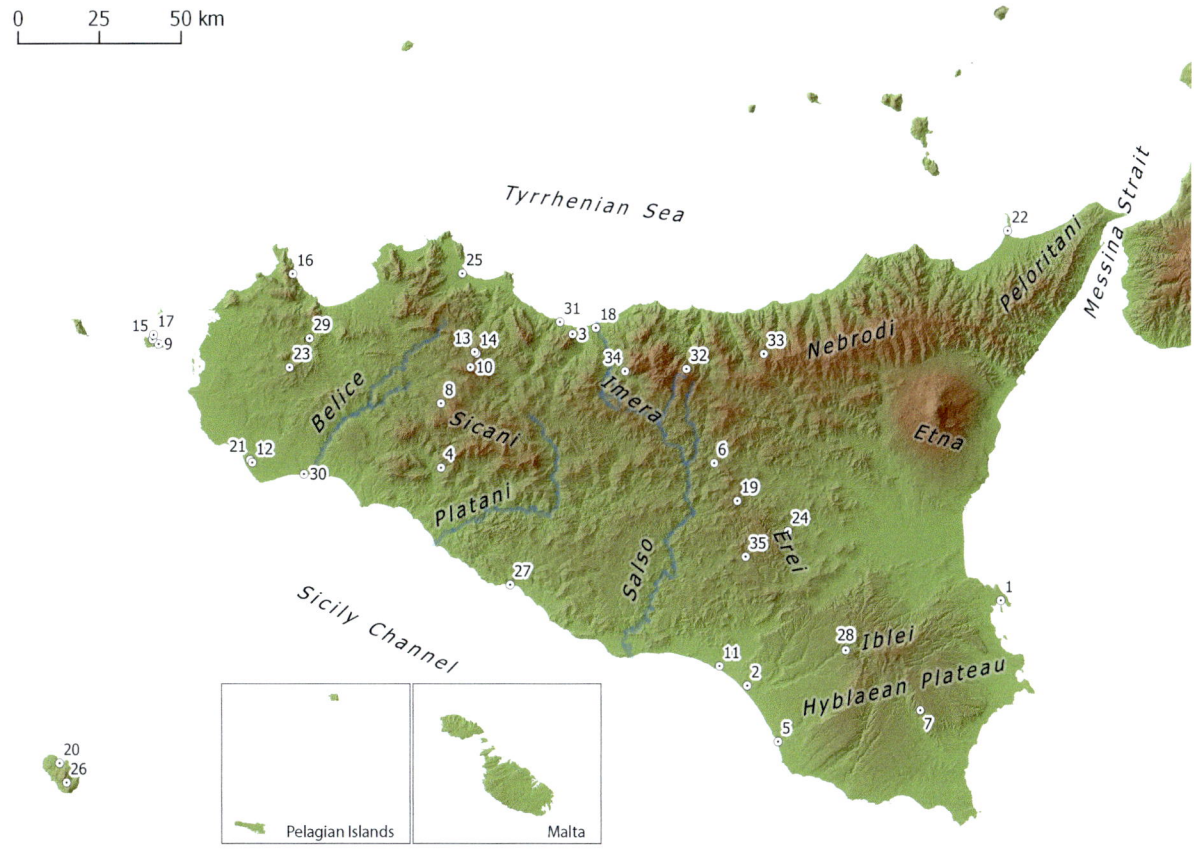

Figure 11: Localities cited in the paragraph dealing with palaeo- and archaeobotany:
Augusta (1); Biviere di Gela (2); Brucato (3); Burgio (4); Camarina (5); Case Bastione (Enna) (6); Castelluccio (7); Contrada Castro (Corleone) (8); Favignana Island (9); Ficuzza (10); Gela (11); Gorghi Tondi (12); Gorgo Lungo (Ficuzza) (13); Gorgo Tondo (Ficuzza) (14); Grotta d'Oriente (Favignana Island) (15); Grotta dell'Uzzo (16); Grotta delle Uccerie (17); Himera (18); Lago di Pergusa (19); Lago di Venere (Pantelleria Island) (20); Lago Preola (21); Milazzo (22); Monte Polizzo (23); Morgantina (24); Palermo (25); Pantelleria Island (26); Porto Empedocle (27); Saltus Camarinensis (28); Segesta (29); Selinunte (30); Termini Imerese (31); Urgo Pietra Giordano (32); Urio Quattrocchi (33); Vallone Inferno (Scillato) (34); Villa del Casale (Piazza Armerina (35).

Historia magistra vitae: indeed, we must share our understanding about past landscape evolution to raise awareness of the short-, mid- and long-term consequences of errors in managing our territory. But, of course, human history is not only a tale of mistakes and destruction: in fact, in many cases local people have been able to perform less impactful ways of exploiting local resources, and/or have left unspoiled very large portions of their environments. For example, we can highlight the growing interest of the scientific community in our very old 'monumental' trees[314] and old-growth forests, corresponding, as they do, to places of special, 'sacred' significance.[315] Understanding the reasons behind the distribution pattern of these living landmarks may help us in reviving past, sustainable management strategies and policies.[316]

Acknowledgements

We are grateful to Federico Poggiali (Department of Archaeology, Classics and Egyptology, University of Liverpool, UK) for sharing some unpublished data on charcoal analysis carried out in some Sicilian Mesolithic caves; to Tommaso La Mantia (Dept. SAAF, University of Palermo) for his essential support of the fieldwork carried out by the research group of Bern; to Riccardo Guarino (Dept. STEBICEF, University of Palermo, Italy), Giuseppe Garfi (IBBR-CNR, Unit of Palermo), and Pietro Lo Cascio (Nesos, Lipari), whose critical advice and helpful suggestions improved the quality of this contribution. We are also indebted to Giuseppe Vendramin (IBBR-CNR, Unit of Florence, Italy) for sharing information on the possible applications of forest plant genetics.

---

[314] Schicchi and Raimondo 2007.
[315] Verschuuren *et al.* 2010; Frascaroli *et al.* 2015.
[316] Spies *et al.* 2016; Badalamenti *et al.* 2018.

Bibliography

Agate, M., Basilone, L., Di Maggio, C., Contino, A., Pierini, S. and Catalano, R. 2017. Quaternary marine and continental unconformity-bounded stratigraphic units of the NW Sicily coastal belt. *Journal of Maps* 13(2): 425–437.

Alberti, G. 2013. Issues in the absolute chronology of the Early-Middle Bronze Age transition in Sicily and southern Italy: a Bayesian radiocarbon view. *Journal of Quaternary Sciences* 28: 630–640.

Albore Livadie, C., Cazzella, A., Marzocchella, A. and Pacciarelli, M. 2003. La struttura degli abitati del Bronzo antico e medio nelle Eolie e nell'Italia meridionale, in Istituto Italiano di Preistoria e Protostoria (ed.) *Atti della XXXV Riunione Scientifica in memoria di Luigi Bernabò Brea 'Le comunità della preistoria italiana. Studi e ricerche sul Neolitico e le Età dei Metalli'* (Lipari, 2–3 giugno 2000): 113–142. Firenze: Istituto Italiano di Preistoria e Protostoria.

Allevato, E., Buonincontri, G.M.P., Pecci, A., D'Auria, A., Papi, E., Saracino, A. and Di Pasquale, G. 2017. Wood exploitation and food supply at the border of the Roman Empire: the case of the vicus of Thamusida-Sidi Ali ben Ahmed (Morocco). *Environmental Archaeology* 22(2): 200–217.

Amorosi, A., Bini, M., Giacomelli, S., Pappalardo, M., Ribecai, C., Rossi, V., Sammartino, I. and Sarti, G. 2013. Middle to late Holocene environmental evolution of the Pisa Plain (Tuscany, Italy) and early human settlements. *Quaternary International* 303: 93–106.

Andersen, K.K., Azuma, N., Barnola, J.-M., Bigler, M., Biscaye, P., Caillon, N., Chappellaz, J., Clausen, H.B., Dahl-Jensen, D., Fischer, H., Flückiger, J., Fritzsche, D., Fujii, Y., Goto-Azuma, K., Grønvold, K., Gundestrup, N.S., Hansson, M., Huber, C., Hvidberg, C.S., Johnsen, S.J., Jonsell, U., Jouzel, J., Kipfstuhl, S., Landais, A., Leuenberger, M., Lorrain, R., Masson-Delmotte, V., Miller, H., Motoyama, H., Narita, H., Popp, T., Rasmussen, S.O., Raynaud, D., Rothlisberger, R., Ruth, U., Samyn, D., Schwander, J., Shoji, H., Siggard-Andersen, M.-L., Steffensen, J.P., Stocker, T., Sveinbjörnsdóttir, A.E., Svensson, A.M., Takata, M., Tison, J.-L., Thorsteinsson, T., Watanabe, O., Wilhelms, F. and White, J.W.C. 2004. High-resolution record of Northern Hemisphere climate extending into the last interglacial period. *Nature* 431: 147–151.

Antonioli, F., Chemello, R., Improta, S. and Riggio, S. 1999. *Dendropoma* lower intertidal reef formations and their palaeoclimatological significance, NW Sicily. Marine Geology 161(2–4): 155–170.

Antonioli, F., Cremona, G., Immordino, F., Puglisi, C., Romagnoli, C., Silenzi, S., Valpreda E. and Verrubbi, V. 2002. New data on the Holocenic sea-level rise in NW Sicily (Central Mediterranean Sea). *Global and Planetary Change* 34(1–2): 121–140.

Antonioli, F., Ferranti, L., Lambeck, K., Kershaw, S., Verrubbi, V. and Dai Pra, G. 2006b. Late Pleistocene to Holocene record of changing uplift rates in southern Calabria and northeastern Sicily (southern Italy, Central Mediterranean Sea). *Tectonophysics* 422: 23–40.

Antonioli, F., Kershaw, S., Renda, P., Rust, D., Belluomini, G., Cerasoli, M., Radtke, U. and Silenzi, S. 2006a. Elevation of the last interglacial highstand in Sicily (Italy): A benchmark of coastal tectonics. *Quaternary International* 145–146: 3–18.

Antonioli, F., Lo Presti, V., Morticelli, M.G., Bonfiglio, L., Mannino, M.A., Palombo, M.R., Sannino, G., Ferranti, L., Furlani, S., Lambeck, K., Canese, S., Catalano, R., Chiocci, F.L., Mangano, G., Scicchitano, G. and Tonielli, R. 2014. Timing of the emergence of the Europe-Sicily bridge (40–17 cal ka BP) and its implications for the spread of modern humans, in J. Harff, G. Bailey and F. Lüth (eds) *Submerged Landscapes of the Continental Shelf.* Geological Society, London, Special Publications 411(1): 111–144.

Araus, J., Ferrio, J. and Voltas, J. 2014. Agronomic conditions and crop evolution in ancient Near East agriculture. *Nature Communications* 5: 3953.

Ardesia, V. 2014. La cultura Rodì-Tindari-Vallelunga in Sicilia: Origini, diffusione e cronologia alla luce dei recenti studi. Parte 2. *IpoTESI di Preistoria* 6: 99–170.

Arrighi, S., Tanguy, J.C. and Rosi, M. 2006. Eruptions of the last 2200 years at Vulcano and Vulcanello (Aeolian Islands, Italy) dated by high-accuracy archeomagnetism. *Physics of the Earth and Planetary Interiors* 159: 225–233.

Asouti, E. and Kabukcu, C. 2014. Holocene semi-arid oak woodlands in the Irano-Anatolian region of Southwest Asia: natural or anthropogenic? *Quaternary Science Reviews* 90: 158–182.

Atherden, M. and Hall, J. 1999. Human impact on vegetation in the White Mountains of Crete since AD 500. *The Holocene* 9: 183–193.

Ayala, G. and French, C. 2005. Erosion modelling of past land use practices in the Fiume di Sotto di Troina river valley, North-Central Sicily. *Geoarchaeology* 20(2): 149–167.

Badalamenti, E., Pasta, S., La Mantia, T. and La Mela Veca, D.S. 2018. Criteria to identify old-growth forests in the Mediterranean: A case study from Sicily based on literature review and some management proposals. *Feddes Repertorium* 129: 25–37.

Baglioni, L., Lo Vetro, D. and Martini, F. 2012. L'arte parietale del gruppo dell'Isolidda (S. Vito lo Capo, Trapani). Atti XLII Riunione Scientifica dell'Istituto italiano di Preistoria e Protostoria, 'L'arte preistorica in Italia' (Trento - Riva del Garda - Val Camonica, 2007). *Preistoria Alpina* 46(1): 109–111.

Baldanza, B. and Triscari, M. 1987. *Le miniere dei Monti Peloritani.* Biblioteca dell'Archivio Storico Siciliano, 8: 283 + LII pp.

Bar-Matthews, M., Ayalon, A., Gilmour, M., Matthews, A. and Hawkesworth, C.J. 2003. Sea-land oxygen isotopic relationships from planktonic foraminifera and speleothems in the Eastern Mediterranean region and their implication for paleorainfall during interglacial intervals. *Geochimica Cosmochimica Acta* 67: 3181–3199.

Bar-Matthews, M., Ayalon, A. and Kaufman, A. 2000. Timing and hydrological conditions of Sapropel events in the Eastern Mediterranean, as evident from speleothems, Soreq cave, Israel. *Chemical Geology* 169: 145–156.

Bar-Matthews, M., Ayalon, A. and Kaufman, A. 1997. Late Quaternary Paleoclimate in the Eastern Mediterranean Region from Stable Isotope Analysis of Speleothems at Soreq Cave, Israel. *Quaternary Research* 47: 155–168.

Bar-Matthews, M., Ayalon, A., Kaufman, A. and Wasserburg, G.J. 1999. The Eastern Mediterranean paleoclimate as a reflection of regional events: Soreq cave, Israel. *Earth and Planetary Science Letters* 16: 85–95.

Barbano, M.S., Castelli, V., Pantosti, D. and Pirrotta, C. 2014. Integration of historical, archaeoseismic and paleoseismological data for the reconstruction of the early seismic history in Messina Strait (south Italy): The 1st and 4th centuries AD earthquakes. *Annals of Geophysics* 57(1): S0192.

Barbano, M.S., Pirrotta, C., and Gerardi, F. 2010. Large boulders along the South-Eastern Ionian coast of Sicily: storm or tsunami deposits? *Marine Geology* 275(1–4): 140–154.

Barbera, G., Cullotta, S., Rossi-Doria, I., Rühl, J. and Rossi-Doria, B. 2010. *I paesaggi a terrazze in Sicilia: Metodologie per l'analisi, la tutela e la valorizzazione*. Palermo: ARPA Studi e Ricerche vol. 7: 529.

Basso, D., Bernasconi, M.P., Robba, E. and Marozzo, S. 2008. Environmental evolution of the Marsala sound, Sicily, during the last 6000 years. *Journal of Coastal Research* 24: 177–197.

Bazan, G., Speciale, C., Castrorao Barba, A., Cambria, S., Miccichè, R. and Marino, P. 2020. Historical suitability and sustainability of Sicani Mountains landscape (Western Sicily): An integrated approach of phytosociology and archaeobotany. *Sustainability* 12: 3201.

Bazzicalupo, P., Maiorano, P., Girone, A., Marino, M., Combourieu-Nebout, N., Pelosi, N., Salgueiro, E. and Incarbona, A. 2020. Holocene climate variability of the Western Mediterranean: Surface water dynamics inferred from calcareous plankton assemblages. *The Holocene*, 30(5): 691–708.

Beck-Bossard, C. 1981. L'alimentazione in un villaggio siciliano del XIV secolo, sulla scorta delle fonti archeologiche. *Archeologia Medievale* 8: 311–320.

Beck-Bossard, C. 1984. Le mobilier ostéologique et botanique, in J.M. Pesez (ed.) *Brucato. Histoire et archéologie d'un habitat médiéval en Sicile*, vol. 2: 615–671. Rome: Collection de l'Ecole française 78.

Belvedere, O., Bertini, A., Boschian, G., Burgio, A., Contino, A., Cucco, R.M. and Lauro, D. 2002. *Himera III.2*. Roma: l'Erma di Bretschneider.

Benallack, K., Green, A.N., Humphries, M.S., Cooper, J.A.G., Dladla, N.N. and Finch, J.M. 2016. The stratigraphic evolution of a large back-barrier lagoon system with a non-migrating barrier. *Marine Geology* 379: 64–77.

Benjamin, J., Rovere, A., Fontana, A., Furlani, S., Vacchi, M., Inglis, R. H., Galili, E., Antonioli, F., Divan, D., Miko S., Mourtzas, N., Felja, I., Meredith-Williams, M., Goodman-Tchernov, B., Kolaiti, E. and Gehrels, R. 2017. Late Quaternary sea-level changes and early human societies in the central and eastern Mediterranean Basin: An interdisciplinary review. *Quaternary International* 449: 29–57.

Berger, J.F. and Guilaine, J. 2009. The 8200 cal BP abrupt environmental change and the Neolithic transition: A Mediterranean perspective. *Quaternary International* 220(1–2): 31–49.

Bernabò Brea, L. and Cavalier, M. 1980. *Meligunis Lipara*. Palermo: S.F. Flaccovio.

Bertolani Marchetti, D., Accorsi, C.A., Arobba, D., Bandini Mazzanti, M., Bertolani, M., Biondi, E., Braggipo, G., Ciuffi, C., De Cunzo, T., Della Ragione, S., Forlani, L., Guido, A.M., Lolli, F., Montanari, C., Paoli, P., Raimondo, F.M., Rossitto, M. and Trevisan Grandi, M. 1984. Recherches géobotaniques sur les Monts Madonie (Sicile du Nord). *Webbia* 38(1): 329–348.

Bianchetto, E., Buscemi, I., Corona, P., Giardina, G., La Mantia, T. and Pasta, S. 2015. Fitting the stocking rate with pastoral resources to manage and preserve Mediterranean forestlands: a case study. *Sustainability* 7: 7232–7244.

Bietti Sestieri, A.M. 1997. Sviluppi culturali e socio-politici differenziati nella tarda età del Bronzo della Sicilia, in Tusa S. (ed.), *Catalogo della Mostra 'Prima Sicilia: Alle Origini della Società Siciliana'* (Palermo, 18 October – 22 December 1997): 473-491. Siracusa: Ediprint.

Bietti Sestieri, A.M. 2018. *L'Italia nell'età del Bronzo e del Ferro. Dalle palafitte a Romolo (2200-700 a. C.)*: 405. Roma: Carocci Editore.

Bini, M., Zanchetta, G., Perşoiu, A., Cartier, R., Català, A., Cacho, I., Dean, J.R., Di Rita, F., Drysdale, R.N., Finnè, M., Isola, I., Jalali, B., Lirer, F., Magri, D., Masi, A., Marks, L., Mercuri, A.M., Peyron, O., Sadori, L., Sicre, M.A., Welc, F., Zielhofer, C. and Brisset, E. 2019. The 4.2 ka BP event in the Mediterranean region: An overview. *Climate of the Past* 15(2): 555–577.

Bisculm, M., Colombaroli, D., Vescovi, E., van Leeuwen, J.F.N., Henne, P.D., Rothen, J., Procacci, G., Pasta, S., La Mantia, T. and Tinner, W. 2012. Holocene vegetation and fire dynamics in the supra-mediterranean belt of the Nebrodi Mountains (Sicily, Italy). *Journal of Quaternary Science* 27(7): 687–698.

Bonasera, F. 1963. *L'isola di Ustica*. Palermo: Annali della Facoltà di Economia e Commercio, Università di Palermo, n. 2.

Bonfiglio, L., Esu, D., Mangano, G., Masini, F., Petruso, D., Soligo, M. and Tuccimei, P. 2007. The Late Pleistocene vertebrate-bearing deposits at San Teodoro Cave (North-Eastern Sicily): preliminary data on faunal diversification and chronology. *Quaternary International* 190: 26–37.

Bonfiglio, L., Marra, A.C. and Masini, F. 2000. *The contribution of Quaternary vertebrates to palaeoenvironmental and palaeoclimatological reconstructions in Sicily*. Geological Society, London, Special Publications 181(1): 171–184.

Bonfiglio, L., Mangano, G., Marra, A.C., Masini, F., Pavia, M. and Petruso, D. 2002. Pleistocene Calabrian and Sicilian paleobioprovinces. *Geobios*, Mém. spéc. 24: 29–39.

Bonfiglio, L., Marra, Borgognini Tarli, S.M., Canci, A., Piperno, M. and Repetto, E. 1993. Dati archeologici e antropologici sulle sepolture mesolitiche della Grotta dell'Uzzo (Trapani). *Bollettino di Paletnologia Italiana* 84: 85–179.

Bottari, C. 2016. Archaeoseismology in Sicily: Past earthquakes and effects on ancient society, in S. D'Amico (ed.) *Earthquakes and their impact on Society*: 491–504. Cham, Switzerland: Springer International Publishing.

Bottari, C., Stiros, S.C. and Teramo, A. 2009. Archaeological evidence for destructive earthquake in Sicily between 400 BC and AD 600. *Geoarchaeology* 24: 147–175.

Bouby, L., Marinval, P., Durand, F., Figueiral, I., Briois, F., Martzluff, M., Perrin, T., Valdeyron, N., Vaquer, J., Guilaine, J. and Manen, C. 2020. Early Neolithic (ca. 5850-4500 cal BC) agricultural diffusion in the Western Mediterranean: An update of archaeobotanical data in SW France. PLoS ONE 15 (4): e0230731.

Boyle, J.F., Gaillard, M.-J., Kaplan, J.O. and Dearing, J.A. 2011. Historic land use and carbon budgets: A critical review. *The Holocene* 21(5): 715–722.

Braje, T.J., Leppard, T.P., Fitzpatrick, S.M. and Erlandson, J.M. 2017. Archaeology, historical ecology and anthropogenic island ecosystems. *Environmental Conservation* 44(3): 286–297.

Branca, S. and Del Carlo, P. 2004. Eruptions of Mt. Etna during the past 3,200 years: A revised compilation integrating the historical and stratigraphic records, in A. Bonaccorso, S. Calvari, M. Coltelli, C. Del Negro, S. Falsaperta (eds) *'Mt Etna Volcano Laboratory'*. Geophysical Monograph Series 143: 1–27.

Brandolini, P., Cappadonia, C., Luberti, G. M., Donadio, C., Stamatopoulos, L., Di Maggio, C., Faccini, F., Stanislao, C., Vergari, F., Paliaga, G., Agnesi, V., Alevizos, G. and Del Monte, M. 2019. Geomorphology of the Anthropocene in Mediterranean urban areas. Progress in Physical Geography: *Earth and Environment*; doi.org/10.1177/0309133319881108.

Bresc, H. 1983. 'Disfari et perdiri li fructi e li aglandi': economie e risorse boschive nella Sicilia medievale (XIII–XV secolo). *Quaderni Storici* 54: 941–963.

Brullo, S. and Spampinato, G. 1991. La vegetazione dei corsi d'acqua della Sicilia. *Bollettino dell'Accademia Gioenia di Scienze naturali* (4) 23 (336)[1990]: 119–252.

Bruno, G., Bobbo, L. and Bruno, A.F. 2015. Evidenze di cambiamento climatico desunte da dati idrogeologici e dagli schemi di funzionamento della fontana monumentale di Morgantina (Sicilia), in G. Bruno (ed.) *Atti del Convegno nazionale di Geoarcheologia 'La Geoarcheologia come chiave di lettura per uno sviluppo sostenibile del territorio'* (Aidone, EN, 4–5 luglio 2014). *Geologia dell'Ambiente* 2015-2 (suppl.): 7–18.

Buchner, G. 1949. Ricerche di sui giacimenti in ossidiana e sulle industrie di ossidiana in Italia. *Rivista di Scienze Preistoriche* 4(3–4): 162–185.

Budsky, A., Scholz, D., Wassenburg, J.A., Mertz-Kraus, R., Spötl, C., Riechelmann, D.F.C., Gibert, L., Jochum, K.P. and Andreae, M.O. 2019. Speleothem δ13C record suggests enhanced spring/summer drought in Mediterranean anomaly ? *The Holocene* 29: 1113-1133.

Burgio, A. 2002. Resuttano (IGM 260 III SO). *Forma Italiae* 42. Firenze: Leo S. Olschki Editore.

Burgio, E., Costanza, C., Di Patti, C. and Mannino, G. 2005. Attuali conoscenze sulle faune a vertebrati dei siti preistorici della Sicilia occidentale, in I. Fiore, G. Malerba, S. Chilardi (eds) *Atti del 3° Convegno Nazionale di Archeozoologia* (Siracusa, 3–5 novembre 2000), Studi di Paletnologia II, Collana del Bullettino di Paletnologia: 145–171. Roma: Istituto Poligrafico dello Stato.

Calò, C., Henne, P.D., Curry, B.B., Magny, M., van Leeuwen, J.F.N., Vescovi, E., La Mantia, T., Pasta, S., Vannière, B. and Tinner, W. 2012. Spatio-temporal patterns of Holocene vegetation change in southern Sicily. *Palaeogeography, Palaeoclimatology, Palaeoecology* 323–325: 110–122.

Calò, C., Henne, P.D., Eugster, P., van Leeuwen, J.F.N., Gilli, A., Hamann, Y., La Mantia, T., Pasta, S., Vescovi, E. and Tinner, W. 2013. 1200 yrs of decadal-scale variability of Mediterranean vegetation and climate at Pantelleria Island, Italy. *The Holocene* 23(10): 1477–1486.

Calvari, S., Coltelli, M., Neri, M., Pompilio, M. and Scribano, V. 1994. The 1991–1993 Etna eruption: chronology and lava flow-field evolution. *Acta Vulcanologica* 4: 1–14.

Calvari, S., Tanner, L.H. and Groppelli, G. 1998. Debris-avalanche deposits of the Milo Lahar sequence and the opening of the Valle del Bove on Etna volcano (Italy). *Journal of Volcanological and Geothermal Research* 87: 193–209.

Caracuta, V., Fiorentino, G. and Martinelli, M.C. 2012, Plant remains and AMS: dating climate change in the Aeolian Islands (Northeastern Sicily) during the 2nd Millennium BC. *Radiocarbon* 54: 689–700.

Carapezza, M.L., Barberi, F., Ranaldi, M., Ricci, T., Tarchini, L., Barrancos, J., Fischer, C., Perez, N., Weber, N., Di Piazza, A. and Gattuso, A. 2011. Diffuse CO2 soil degassing and CO2 and H2S concentrations in air and related hazards at Vulcano Island (Aeolian Arc, Italy). *Journal of Volcanological and Geothermal Research* 207(3–4): 130–144.

Carniato, P. 2011. La sughereta come bosco puro naturale: dati archeobotanici dalla necropoli di Imera (500 A.C., Sicilia Settentrionale). Tesi di Laurea Sperimentale, Università degli Studi di Napoli 'Federico II', Facoltà di Agraria, Corso di Laurea in Scienze Forestali e Ambientali.

Carrión, J.S., Parra, I., Navarro, C. and Munuera, M. 2000. Past distribution and ecology of the cork oak (*Quercus suber*) in the Iberian Peninsula: a pollen-analytical approach. *Diversity and Distributions* 6: 29–44.

Carter, J.C. and Costantini, L. 1994. Settlement density, agriculture and the extent of productive land cleared from forest in the time of Roman Empire in Magna Grecia, in B. Frenzel (ed.) *Evaluation of land surface cleared from forests in the Mediterranean region during the time of the Roman Empire*: 101–118. Stuttgart and New York: Gustav Fischer Verlag.

Carter, J.C., Costantini, L., D'Annibale, C., Jones, J.R., Folk, R.L. and Sullivan, D. 1985. Population and agriculture: Magna Grecia in the fourth century BC, in S. Stoddart and C. Malone (eds) *Papers in Italian Archaeology IV, The Cambridge Conference Part I: The Human Landscape*: 281–312. British Archaeological Reports. International Series 243). Oxford: British Archaeological Reports.

Carusi, C. 2008. *Il sale nel mondo greco. Luoghi di produzione, circolazione commerciale, regimi di sfruttamento nel contesto del Mediterraneo antico*. Bari: Edipuglia.

Caruso, A., Cosentino, C., Pierre, C. and Sulli, A. 2011. Sea-level changes during the last 41,000 years in the outer shelf of the southern Tyrrhenian Sea: Evidence from benthic foraminifera and seismostratigraphic analysis. *Quaternary International* 232(1–2): 122–131.

Carveni, P., Neri, M., Benfatto, S., Leonardi, A. and Salleo Puntillo, M. 2016. Importanza paleo-climatica delle morfologie glaciali riscontrate sul versante nord-orientale dell'edificio vulcanico etneo. *Geologia dell'Ambiente* 3(Suppl.): 24–28.

Cassoli, P.F. and Tagliacozzo, A. 1982. La fauna della Grotta di Cala dei Genovesi a Levanzo. *Rivista di Scienze Preistoriche* 37: 48–58.

Castiglioni, E. and Rottoli, M. 1997. I macroresti vegetali, in A. Molinari (ed.) *Segesta II. Il Castello e la Moschea (Scavi 1989-1995)*: 235–257. Palermo: S.F. Flaccovio.

Castrorao Barba, A. 2017. Alcune considerazioni e problematiche sulle dinamiche degli insediamenti rurali in Sicilia tra V e VIII secolo, in P. Arthur and M.L. Imperiale (eds) *VII Congresso nazionale di archeologia medievale*. Pré-tirages (Lecce, 9–12 settembre 2015): 383–387.

Castrorao Barba, A., Miccichè, R., Pisciotta, F., Marino, P., Bazan, G., Aleo Nero, C. and Vassallo, S. 2018. *Archeologia nel territorio dei Monti Sicani (Harvesting Memories project). L'insediamento di lunga durata di Contrada Castro (Corleone, Palermo). Prima campagna di scavo 2017.* FOLD&R 416: 1–21.

Catalano, G., Modi, A., Mangano, G., Sineo, L., Lari, M. and Bonfiglio, L. 2020b. A mitogenome sequence of an *Equus hydruntinus* specimen from the Late Quaternary site of San Teodoro Cave (Sicily, Italy). *Quaternary Science Reviews* 236: 106280.

Catalano, G., Lo Vetro, D., Fabbri, P.F., Mallick, S., Reich, D., Rohland, N., Sineo, L., Mathieson, I. and Martini, F. 2020a. Late Upper Palaeolithic hunter-gatherers in the Central Mediterranean: New archaeological and genetic data from the Late Epigravettian burial Oriente C (Favignana, Sicily). *Quaternary International* 537: 24–32.

Catalano, R., Valenti, V., Albanese, C., Accaino, F., Sulli, A., Tinivella, U., Gasparo Morticelli, M., Zanolla, C. and Giustiniani, M. 2013. Sicily's fold-thrust belt and slab roll-back: The SI.RI.PRO. seismic crustal transect. *Journal of the Geological Society, London* 170(3): 451–464.

Cavallaro, D. and Coltelli, M. 2019. The Graham volcanic field offshore southwestern Sicily (Italy) revealed by high-resolution seafloor mapping and ROV images. *Frontiers in Earth Science* 7: 311.

Cazzella, A. and Maniscalco, L. 2012. L'età del Rame in Sicilia, in R. Giglio and S. Tusa (eds) *Atti della XLI Riunione Scientifica 'Dai Ciclopi agli Ecisti: Società e territorio nella Sicilia preistorica e protostorica'* (San Cipirello, PA, 16-19 novembre 2006): 81–104. Firenze: Istituto italiano di Preistoria e Protostoria.

Cazzella, A. and Recchia, G. 2014. Bronze Age fortified settlements in Southern Italy and Sicily, in G. Bartoloni and L.M. Michetti (eds) Mura di legno, mura di terra, mura di pietra: Fortificazioni nel Mediterraneo antico. *Scienze dell'Antichità* 19(2-3): 45–64.

Cheng, H., Sinha, A., Verheyden, S., Nader, F.H., Li, X.L., Zhang, P.Z., Yin, J.J., Yi, L., Peng, Y.B., Rao, Z.G., Ning, Y.F. and Edwards, R.L. 2015. The climate variability in northern Levant over the past 20,000 years. *Geophysical Research Letters* 42: 8641–8650.

Chew, S.R. 2001. *World ecological degradation: Accumulation, urbanization, and deforestation, 3000 B.C.-A.D. 2000.* Walnut Creek, CA: AltaMira.

Cisneros, M., Cacho, I., Frigola, J., Canals, M., Masqué, P., Martrat, B., Casado, M., Grimalt, J.O., Pena, L.D., Margaritelli, G. and Lirer, F. 2016. Sea surface temperature variability in the central-western Mediterranean Sea during the last 2700 years: A multi-proxy and multi-record approach. *Climate of the Past* 12: 849–869.

Cocchi, L., De Ritis, R., Casalbore, D., Romagnoli, C., Lucchi, F., Tranne, C.A. and Ventura, G. 2019.

Seamount-volcanic island transition and evolution from fissural to central activity inferred by the magnetic modeling of Salina Island (Tyrrhenian Sea). *JGR Solid Earth* 124(5): 4323–4342.

Collin Bouffier, S. 2009. La géstion de l'eau en Sicile Grecque. État de la question. *Pallas* 79: 65–79.

Collina, C. 2012. Sistemi tecnici e chaînes opératoires alla grotta dell'Uzzo (Trapani) : analisi tecnologica delle industrie litiche dai livelli mesolitici e neolitici, in R. Giglio and S. Tusa (eds) *Atti della XLI Riunione Scientifica 'Dai Ciclopi agli Ecisti : Società e territorio nella Sicilia preistorica e protostorica'* (San Cipirello, PA, 16–19 novembre 2006): 447–459. Firenze: Istituto Italiano di Preistoria e Protostoria.

Colombaroli, D., Marchetto, A. and Tinner, W. 2007. Long-term interactions between Mediterranean climate, vegetation and fire regime at Lago di Massaciuccoli (Tuscany, Italy). *Journal of Ecology* 95: 755–770.

Colombaroli, D., Vannière, B., Emmanuel, C., Magny, M. and Tinner, W. 2008. Fire-vegetation interactions during the Mesolithic-Neolithic transition at Lago dell'Accesa, Tuscany, Italy. *The Holocene* 18: 679–692.

Colonese, A.C., Lo Vetro, D. and Martini, F. 2014. Holocene coastal change and intertidal mollusc exploitation in the central Mediterranean: Variations in shell size and morphology at Grotta d'Oriente (Sicily). *Archaeofauna* 23: 181–192.

Colonese, A.C., Lo Vetro, D., Landini, W., Di Giuseppe, Z., Hausmann, N., Demarchi B., D'Angelo, C., Leng, M.J., Incarbona, A., Whitwood, A.C. and Martini, F. 2018. Late Pleistocene-Holocene coastal adaptation in central Mediterranean: Snapshots from Grotta d'Oriente (NW Sicily). *Quaternary International* 493: 114–126.

Colonese, A.C., Mannino, M.A., Bar-Yosef Mayer, D.E., Fa, D.A., Finlayson, J.C., Lubell, D. and Stiner, M.C. 2011. Marine mollusc exploitation in Mediterranean prehistory: An overview. *Quaternary International* 239: 86–103.

Coltelli, M., Cavallaro, D., D'Anna, G., D'Alessandro, A., Grassa, F., Mangano, G., Patanè, D. and Gresta, S. 2016. Exploring the submarine Graham Bank in the Sicily Channel. *Annals of Geophysics* 59 (2): S0208.

Coltelli, M., Del Carlo, P., and Vezzoli, L. 1998. Discovery of a Plinian basaltic eruption of Roman age at Etna volcano, Italy. *Geology* 26(12): 1095–1098.

Coltelli, M., Del Carlo, P. and Vezzoli, L. 2000. Stratigraphic constraints for explosive activity in the past 100 ka at Etna Volcano, Italy. *International Journal of Earth Sciences* 89: 665–677.

Coltelli, M., Marsella, M., Proietti, C. and Scifoni, S. 2014. The case of the 1981 eruption of Mt. Etna: an example of very fast moving lava flows. *Geochemistry, Geophysics, Geosystems* 13(1):Q01004.

Conte, A.M., Martorelli, E., Calarco, M., Sposato, A., Perinelli, C., Coltelli, M. and Chiocci, F.L. 2014. The 1891 submarine eruption offshore Pantelleria Island (Sicily Channel, Italy): Identification of the vent and characterization of products and eruptive style. *Geochemistry Geophysics Geosystems* 15: 2555–2574.

Copat, V. 2020. On pots, people and cultural landscape: The Hyblean Mountains (Sicily) in the Early Bronze Age. *Journal of Archaeological Scientitifc Reports* 30: 102219.

Corrao, P. 1988. Per una storia del bosco e dell'incolto in Sicilia fra XI e XIII secolo, in B. Andreolli and M. Montanari (eds) *Il bosco nel Medioevo*: 351–368. Bologna: CLUEB.

Crees, J. and Turvey, S. 2014. Holocene extinction dynamics of Equus hydruntinus, a late-surviving European megafaunal mammal. *Quaternary Science Reviews* 91: 16–29.

Crielaard, J.P. 1998. Surfing on the Mediterranean web: Cypriot long-distance communications during the eleventh and tenth centuries BC, in V. Karageorghis and N. Stampolidis (eds) *Eastern Mediterranean: Cyprus-Dodecanese-Crete 16th-6th century BC*: 187–204. Athens: University of Crete and A.G. Leventis Foundation.

Crispino, A. 2018. Castelluccio (Noto, SR). *Notiziario di Preistoria e Protostoria* 5: 98–102.

Crouch, D.P. 1984. The Hellenistic water system of Morgantina (Sicily): contributions to the history of urbanization. *American Journal of Archeology* 88: 353–365.

Crouch, D.P. 1993. *Water management in ancient Greek cities*. New York: Oxford University Press.

Cultrera, F., Barreca, G., Burrato, P., Ferranti, L., Monaco, C., Passaro, S., Pepe, F. and Scarfì, L. 2017. Active faulting and continental slope instability in the Gulf of Patti (Tyrrhenian side of NE Sicily, Italy): A field, marine and seismological joint analysis. *Natural Hazards* 86 (suppl. 2): 253–272.

Curry, B., Henne, P.D., Mesquita-Joanes, F., Marrone, F., Pieri, V., La Mantia, T., Calò, C. and Tinner, W. 2016. Holocene paleoclimate inferred from salinity histories of adjacent lakes in southwestern Sicily (Italy). *Quaternary Science Reviews* 150: 67–83.

D'Amore, G., Di Marco, S., Di Salvo, R., Messina, A. and Sineo, L. 2010. The early peopling of Sicily: evidence from the Mesolithic skeletal remains from Grotta d'Oriente. *Annals of Human Biology* 37: 403–426.

D'Amore, G., Di Marco, S., Tartarelli, G., Bigazzi, R. and Sineo, L. 2009. Late Pleistocene human evolution in Sicily: comparative morphometric analysis of Grotta di San Teodoro craniofacial remains. *Journal of Human Evolution* 56: 537–550.

D'Orefice, M., Bellotti, P., Bertini, A., Calderoni, G., Censi Neri, P., Di Bella, L., Fiorenza, D., Foresi, L.M., Louvari, M.A., Rainone, L., Vittori, C., Goiran, J.P., Schmitt, L., Carbonel, P., Preusser, F., Oberlin, C., Sangiorgi, F. and Davoli, L. 2020. Holocene Evolution of the Burano Paleo-Lagoon (Southern Tuscany), Italy). *Water* 12(4): 1007.

De Angelis, F. 2002. Trade and agriculture at Megara Hyblaia. *Oxford Journal of Archeology* 21: 299–310.

De Angelis, F. 2003. *Megara Hyblaia and Selinous. The development of two Greek city-states in archaic Sicily*. Oxford: Oxford University School of Archaeology Monograph 55.

De Angelis F. 2016. *Archaic and Classical Greek Sicily: A social and economic History*. New York: Oxford University Press.

De Astis, G., Lucchi, F., Dellino, P., La Volpe, L., Tranne, C.A., Frezzotti, M.L. and Peccerillo, A. 2013. Geology, volcanic history and petrology of Vulcano (central Aeolian Archipelago). *Geological Society of London, Memoirs* 37(1): 281–349.

De Castro, O. and Maugeri, G. 2006. Molecular notes on the Mediterranean species of the genus Celtis L. (Celtidaceae). *Plant Biosystems* 140(2): 171–175.

De Fiore, O. 1922. Vulcano (Isole Eolie). *Zeitschrift für Vulkanologie* 3: 3–393.

De Guidi, G., Catalano, S., Monaco, C. and Tortorici, L. 2003. Morphological evidence of Holocene coseismic deformation in the Taormina region (NE Sicily). *Journal of Geodynamics* 36(1–2): 193–211.

De Martini, P.M., Barbano, M.S., Smedile, A., Gerardi, F., Pantosti, D., Del Carlo, P. and Pirrotta, C. 2010. A unique 4000 yrs long geological record of multiple tsunami inundations in the Augusta Bay (eastern Sicily, Italy). *Marine Geology* 276: 42–57.

De Michele, R., La Bella, F., Gristina, A.S., Fontana, I., Pacifico, D., Garfi, G., Motisi, A., Crucitti, D., Abbate, L. and Carimi, F. 2019. Phylogenetic relationship among wild and cultivated grapevine in Sicily: A hotspot in the middle of the Mediterranean Basin. *Frontiers in Plant Science* 10: 1506.

De Simone, G., Thun Hohenstein, U., Petruso, D., Forgia, V., Giannitrapani, E., Iannì, F. and Martin Rodriguez, P. 2019. Gestione e sfruttamento delle risorse faunistiche nei siti di Vallone Inferno (Palermo) e Case Bastione (Enna), in De Grossi Mazzorin, I. Fiore, C. Minniti (eds) *Atti 8° Convegno Nazionale di Archeozoologia* (Lecce, 11–14 novembre 2015): 71–78. Lecce: Università del Salento.

De Vita, S., Laurenzi, M.A., Orsi, G. and Voltaggio, M. 1998. Application of 40AR/39 and 230Th dating methods to the chronostratigraphy of Quaternary basaltic volcanic areas: The Ustica Island case history. *Quaternary International* 47–48: 117–127.

Demesure, B., Comps, B. and Petit, J.R. 1996. Chloroplast DNA phylogeography of the common beech (Fagus sylvatica L.) in Europe. *Evolution* 50: 2515–2520.

Dentici Buccellato, R.M. 1994. Il bosco nella Sicilia del basso Medioevo, in S. D'Onofrio (ed.) *Atti del Convegno 'La cultura del bosco'* (Caronia, ME, 2–4 ottobre 1987): 115–122. Palermo: Istituto di scienze antropologiche e geografiche, Facoltà di Lettere e Filosofia dell'Università.

Di Donato, V., Insinga, D.D., Iorio, M., Molisso, F., Rumolo, P., Cardines, C. and Passaro, S. 2018. The palaeoclimatic and palaeoceanographic history of the Gulf of Taranto (Mediterranean Sea) in the last 15 ky. *Global and Planetary Change* 172: 278–297.

Di Maggio, C., Agate, M., Contino, A., Basilone, L. and Catalano, R. 2009. Unità a limiti inconformi utilizzate per la cartografia dei depositi quaternari nei fogli CARG della Sicilia nordoccidentale. *Alpine and Mediterranean Quaternary* 22: 345–364.

Di Maggio, C., Madonia, G., Vattano, M., Agnesi, V. and Monteleone, S. 2017. Geomorphological evolution of western Sicily, Italy. *Geologica Carpathica* 68(1): 80–93.

Di Maida G. 2020. The earliest human occupation of Sicily: A review. *Journal of Island Coastal Archaeology*; doi.org/10.1080/15564894.2020.1803460.

Di Pasquale, G., Allevato, E., Cocchiaro, A., Moser, D., Pacciarelli, M. and Saracino, A. 2014. Late Holocene persistence of *Abies alba* in low-mid altitude deciduous forests of central and southern Italy: New perspectives from charcoal data. *Journal of Vegetation Science* 25(1): 299–310.

Di Rita, R. and Magri, D. 2019. The 4.2 ka event in the vegetation record of the central Mediterranean. *Climate of the Past* 15: 237–251.

Di Stefano, A., Foresi, L.M., Incarbona, A., Sprovieri, M., Vallefuoco, M., Iorio, M., Pelosi, N., Di Stefano, E., Sangiorgi, P. and Budillon, F. 2015. Mediterranean coccolith ecobiostratigraphy since the penultimate Glacial (the last 145,000 years) and ecobioevent traceability. *Marine Micropaleontology* 115: 24–38.

Di Stefano, A. and Branca, S. 2002. Long-term uplift rate of the Etna volcano basement (southern Italy) based on biochronological data from Pleistocene sediments. *Terra Nova* 14: 61–68.

Di Stefano, E. and Incarbona, A. 2004. High-resolution palaeoenvironmental reconstruction of ODP Hole 963D (Sicily Channel) during the last deglaciation based on calcareous nannofossils. *Marine Micropaleontology* 52: 241–254.

Drake, N.A., Blench, R.M., Armitage, S.J., Bristow, C.S. and White, K.H. 2011. Ancient watercourses and biogeography of the Sahara explain the peopling of the desert. *Proceedings of the National Academy of Science* 108(2): 458–462.

Drysdale, R.N., Zanchetta, G., Hellstrom, J.C., Fallick, A.E., Zhao, J.X., Isola, I. and Bruschi, G. 2004. Palaeoclimatic implications of the growth history and stable isotope ($\delta^{18}O$ and $\delta^{13}C$) geochemistry of a Middle to Late Pleistocene stalagmite from central-western Italy. *Earth and Planetary Science Letters* 227: 215–229.

Dumoulin-Lapègue, S., Demesure, B., Fineschi, S., Le Corre, V. and Petit, R.J. 1997. Phylogeographic structure of white oaks throughout the European continent. *Genetics* 146: 1475–1487.

Duncan, A.M., Dibben, C., Chester, D.K. and Guest, J. 1996. The 1928 eruption of Mount Etna volcano, Sicily, and the destruction of the town of Mascali. *Disasters* 20: 1–20.

Ehrmann, W., Schmiedl, G., Beuscher, S. and Krüger, S. 2017. Intensity of African humid periods estimated from Saharan dust fluxes. *PLoS ONE* 12(1): e0170989.

Esposito, V., Andaloro, F., Canese, S., Bortoluzzi, G., Bo, M., Di Bella, M., Italiano, F., Sabatino G., Battaglia, P., Consoli, P., Giordano, P., Spagnoli, F., La Cono, V., Yakimov, M.M., Scotti, G. and Romeo, T. 2018. Exceptional discovery of a shallow-water hydrothermal site in the SW area of Basiluzzo islet (Aeolian archipelago, South Tyrrhenian Sea): An environment to preserve. *PLoS ONE* 13(1): e0190710.

Essallami, L., Sicre, M.A., Kallel, N., Labeyrie, L. and Siani, G. 2007. Hydrological changes in the Mediterranean Sea over the last 30,000 years. *Geochemistry, Geophysics and Geosystems* 8(7): Q07002.

Evans, R. 2016. *Ancient Syracuse: From foundation to fourth century collapse.* Oxford – New York: Routledge.

Falkenhausen (von), V. 1980. La foresta nella Sicilia normanna, in E. Marchetta (ed.) *La cultura materiale in Sicilia.* Palermo: Istituto di scienze antropologiche e geografiche, Facoltà di lettere e filosofia dell'Università.

Favara, R., Giammanco, S., Inguaggiato, S. and Pecoraino, G. 2000. Preliminary estimate of CO_2 output from Pantelleria island volcano (Sicily, Italy): Evidence of active mantle degassing. *Applied Geochemistry* 16: 883–894.

Fernandes, D.M., Mittnik, A., Olalde, I., Lazaridis, I., Cheronet, O., Rohland, N., Mallick, S., Bernardos, R., Broomandkhoshbacht, N., Carlsson, J., Culleton, B.J., Ferry, M., Gamarra, B., Lari, M., Mah, M., Michel, M., Modi, A., Novak, N., Oppenheimer, J., Sirak, K.A., Stewardson, K., Mandl, K., Schattke, C., Özdoğan, K.T., Lucci, M., Gasperetti, G., Candilio, F., Salis, G., Vai, S., Camarós, E., Calò, C., Catalano, G., Cueto, M., Forgia, V., Lozano, M., Marini, E., Micheletti, M., Miccichè, R.M., Palombo, M.R., Ramis, D., Schimmenti, V., Sureda, P., Teira, L., Teschler-Nicola, M., Kennett, D.J., Lalueza-Fox, C., Patterson, N., Sineo, L., Coppa, A., Caramelli, D., Pinhasi, R. and Reich, D. 2020.The spread of steppe and Iranian-related ancestry in the islands of the western Mediterranean. *Nature Ecology and Evolution* 4: 334–345.

Fernández-Mazuecos, M. and Vargas, P. 2011. Historical isolation versus recent long-distance connections between Europe and Africa in bifid toadflaxes (Linaria sect. Versicolores). *PLoS ONE* 6: e22234.

Ferranti, L., Antonioli, F., Mauz, B., Amorosi, A., Dai Pra, G., Mastronuzzi, G., Monaco, C., Orrù. P., Pappalardo M., Radtke U., Renda, P., Romano P., Sansò P. and Verrubbi, V. 2006. Markers of the last interglacial sea-level high stand along the coast of Italy: Tectonic implications. *Quaternary International* 145: 30–54.

Filippidi, A. and De Lange G.J. 2019. Eastern Mediterranean deep water formation during Sapropel S1: A reconstruction using geochemical records along a bathymetric transect in the Adriatic outflow region. *Paleoceanography and Paleoclimatology* 34: 409–429.

Finné, M., Woodbridge, J., Labuhn, I. and Roberts, C.N. 2019. Holocene hydro-climatic variability in the Mediterranean: A synthetic multi-proxy reconstruction. *The Holocene* 29(5): 847–863.

Finsinger, W., Schwörer, C., Heiri, O., Morales-Molino, C., Ribolini, G., Giesecke, T., Haas, J.N., Kaltenrieder, P., Magyari, E.-K., Ravazzi, C., Rubiales, M. and Tinner, W. 2019. Fire on ice and frozen trees? Inappropriate radiocarbon dating leads to unrealistic reconstructions. *New Phytologist* 222(2): 657–662.

Fiorentino, G. and D'Oronzo, C. 2007, Analisi archeobotaniche del sito di Burgio: Elementi del paleoambiente e modalità d'uso delle strutture pirotecniche in età post-medievale, in M.C. Parello (ed.) *Le fornaci di Burgio. Indagini archeologiche nell'area delle officine*: 77–81. Palermo: Regione Siciliana.

Fiorentino, G. and Recchia, G. 2015, Archipelagos adjacent to Sicily around 2200 BC: attractive environments or suitable geo-economic locations?, in H. Meller, H.W. Arz, R. Jung and R. Risch (eds) *7th Archaeological Conference of Central Germany (October 23-26, 2014, Halle, Saale) '2200 BC: Ein Klimasturz als Ursache für den Zerfall der Alten Welt? – 2200 BC: A climatic breakdown as a cause for the collapse of the old world?'. Tagungen des Landesmuseums für Vorgeschichte Halle* 12(1): 305–319.

Fokkens, H. and Harding, A. 2013. *The Oxford handbook of European Bronze Age.* Oxford: Oxford University Press.

Foresta Martin, F., Di Piazza, A., D'Oriano, C., Carapezza, M.L., Paonita, A., Rotolo, S.G. and Sagnotti, L. 2017. New insights into the provenance of the obsidian fragments of the island of Ustica (Palermo, Sicily). *Archaeometry* 59(3): 435–454.

Forgia, V., Martín, P., López-García, J. M., Ollé, A., Vergès, M., Allué, E., Angelucci D.E., Arnone, M., Blain, H.-A., Burjachs, F., Exposito, I., Messina, A., Picornell, L., Rodriguez, A., Scopelliti, G., Sineo, L., Virruso, G., Alessi, E., Di Simone, G., Morales, J.I., Pagano, E. and Belvedere, O. 2013. New data on Sicilian prehistoric and historic evolution in a mountain context, Vallone Inferno (Scillato, Italy). *Comptes Rendus Palévolution* 12(2): 115–126.

Forgia, V. 2019. *Archaeology of uplands on a Mediterranean island. The Madonie mountain range in Sicily.* Cham, Switzerland: Springer Nature.

Forni, G. 1988. Defence policy of forest ecosystem in Magna Graecia (IVth century BC), in F. Salbitano (ed.) *Proceedings of the FERN workshop 'Human influence on forest ecosystems development in Europe' (Trento, Italy, 26-29 September 1988)*: 343-348. Bologna: Pitagora.

Francalanci, L., Lucchi, F., Keller, J., De Astis, G. and Tranne, C.A. 2013. Eruptive, volcanotectonic and magmatic history of the Stromboli volcano (northeastern Aeolian archipelago, in F. Lucchi, A. Peccerillo, J. Keller, C.A. Tranne and P.L. Rossi (eds) *The Aeolian Islands Volcanoes. Geological Society of London, Memoirs* 37(1): 397–471.

Frascaroli, F., Bhagwat, S., Guarino, R., Chiarucci, A. and Schmid, B. 2015. Shrines in Central Italy conserve plant diversity and large trees. *Ambio* 45(4): 468–479.

Freund, K.P., Tykot, R.H. and Vianello, A. 2015. Blade production and the consumption of obsidian in Stentinello period Neolithic Sicily. *Comptes Rendus Palévolution* 14(3): 207–217.

Frigola, J., Moreno, A., Cacho, I., Canals, M., Sierro, F.J., Flores, J.A., Grimalt, J.O., Hodell, D.A. and Curtis, J.H. 2007. Holocene climate variability in the western Mediterranean region from a deepwater sediment record. *Paleoceanography* 22: 1–16.

Frisia, S., Borsato, A., Mangini, A., Spötl, C., Madonia, G. and Sauro, U. 2006. Holocene climate variability in Sicily from a discontinuous stalagmite record and the Mesolithic to Neolithic transition. *Quaternary Research* 66: 388–400.

Fugazzola Delpino, M.A., Pessina, A. and Tinè, V. 2004. *Il Neolitico in Sicilia*. Firenze e Roma: Istituto italiano di Preistoria e Protostoria.

Fyfe, R.M., Woodbridge, J. and Roberts, N. 2015. From forest to farmland: Pollen-inferred land cover change across Europe using the pseudobiomization approach. *Global Change Biology* 21(3): 1197–1212.

Galadini, F., Hinzen, K.-G. and Stiros, S.C. 2006. Archaeoseismology: methodological issues and procedure. *Journal of Seismology* 10: 395–414.

Galland, M., D'Amore, G., Friess, M., Miccichè, R., Pinhasi, R., Sparacello, V.S. and Sineo, L. 2019. Morphological variability of Upper Paleolithic and Mesolithic skulls from Sicily. *Journal of Anthropological Science* 97: 151–172.

Galloway, J.H. 1989. *The sugar cane industry: An historical geography from its origins to 1914*. Cambridge: Cambridge University Press

Garilli, V., Vita, G., Mulone, A., Bonfiglio, L. and Sineo, L. 2020. From sepulchre to butchery-cooking: Facies analysis, taphonomy and stratigraphy of the Upper Palaeolithic post burial layer from the San Teodoro Cave (NE Sicily) reveal change in the use of the site. *Journal of Archaeoloical Science Report*; doi.org/10.1016/j.jasrep.2020.102191.

Gasparo Morticelli, M., Avellone, G., Sulli, A., Agate, M., Basilone, L., Catalano, R. and Pierini, S. 2017. Mountain building in NW Sicily from the superimposition of subsequent thrusting and folding events during Neogene: Structural setting and tectonic evolution of the Kumeta and Pizzuta ridges. *Journal of Maps* 13(2): 276–290.

Gassner, S.K., Gobet, E., Schwörer, C., van Leeuwen, J.F.N., Vogel, H., Giagkoulis, T., Makri, S., Grosjean, M., Panajiotidis, S., Hafner, A. and Tinner, W. 2020. 20,000 years of interactions between climate, vegetation and land use in Northern Greece. *Vegetation History and Archaeobotany* 29: 75–90.

Gerardi, F., Smedile, A., Pirrotta, C., Barbano, M.S., De Martini, P.M., Pinzi, S., Gueli, A.M., Ristuccia, G.M., Stella, G. and Troja, S.O. 2012. Geological record of tsunami inundations in Pantano Morghella (southeastern Sicily) both from near and far-field sources. *Natural Hazards and Earth System Sciences* 12: 1185–1200.

Giannitrapani, E. 2017. Paesaggi e dinamiche del popolamento di età preistorica nella Sicilia centrale, in F. Anichini and M.L. Gualandi (eds) *MAPPA Data Book 2*: 43–64. Roma: Edizioni Nuova Cultura.

Giannitrapani, E. and Iannì, F. (in press). Demographic dynamics, paleoenvironmental changes and social complexity in the late prehistory of central Sicily, in T. Lachenal, R. Roure and O. Lemercier (eds) *Proceedings of the XVIIIth UISPP World Congress 'Demography and Migration. Population Trajectories from the Neolithic to the Iron Age'*, Sessions XXXII-2 and XXXIV-8 (4-9 June 2018, Paris, France): 43–64. Oxford: Archaeopress.

Giannitrapani, E., Iannì, F., Chilardi, S. and Anguilano, L. 2014. Case Bastione: a prehistoric settlement in the Erei uplands (central Sicily). *Origini* 36: 181–212.

Giardina, G., La Mantia, T., Sala, G., Di Leo, C. and Pasta, S. 2014. Possibile origine e consistenza di un popolamento di Quercus trojana Webb subsp. trojana (Fagaceae) al Bosco della Ficuzza (Provincia di Palermo, Sicilia nord-occidentale). *Naturalista sicil.* (4) 38(2): 265–289.

Giordano, G., Porreca, M., Musacchio, P. and Mattei, M. 2008. The Holocene Secche di Lazzaro phreatomagmatic succession (Stromboli, Italy): evidence of pyroclastic density current origin deduced by facies analysis and AMS flow directions. *Bulletin of Volcanology* 70: 1221–1236.

Goedecke, F., Marcenò, C., Guarino, R., Jahn, R. and Bergmeier, E. 2020. Reciprocal extrapolation of species distribution models between two islands: Specialists perform better than generalists and geological data reduce prediction accuracy. *Ecological Indicators* 108: 105652.

Gogou, A., Triantaphyllou, M., Xoplaki, E., Izdebski, A., Parinos, C., Dimiza, M., Bouloubassi, I., Luterbacher, J., Kouli, K., Martrat, B., Toreti, A., Fleitmann, D., Rousakis, G., Kaberi, H., Athanasiou, M. and Lykousis, V. 2016. Climate variability and socio-environmental changes in the northern Aegean (NE Mediterranean) during the last 1500 years. *Quaternary Science Reviews* 136: 209–228.

Graziosi, P. 1947. Gli uomini paleolitici della Grotta di S. Teodoro (Messina). *Rivista di Scienze Preistoriche* 2: 123–233.

Guarino, R. and Pasta, S. (eds) 2017. *Botanical Excursions in Central and Western Sicily. Field Guide for the 60th IAVS Symposium* (Palermo, 20-24 June 2017). Palermo: Palermo University Press.

Guarino, R. and Pasta, S. 2018. Sicily: the island that didn't know to be an archipelago. *Berichte der Reinhold-Tüxen-Gesellschaft* 30: 133–148.

Guidoboni, E., Ferrari, G., Mariotti, D., Comastri, A., Tarabusi, G. and Valensise, G. 2007-2013. CFTI4 Med.

Catalogue of Strong Earthquakes in Italy, 461 BC-1997 and Mediterranean Area 760 BC-1500. An Advanced Laboratory of Historical Seismology. http://storing.ingv.it/cfti4med/ (last accessed 30 March 2020).

Habel, J.C., Meyer, M., El Mousadik, A. and Schmitt, T. 2008. Africa goes Europe: the complete phylogeography of the marbled white butterfly species complex Melanargia galathea/M. lachesis (Lepidoptera: Satyridae). *Organisms Diversity & Evolution* 8: 121–129.

Heinzel, C. and Kolb, M. 2011. Holocene land use in western Sicily: a geoarchaeological perspective. *Journal of the Geological Society* (London) (Special Publications) 352: 97–107.

Henne, P.D., Elkin, C., Franke, C., Colombaroli, D., Calò, C., La Mantia, T., Pasta, S., Conedera, M., Dermody, O. and Tinner, W. 2015. Reviving extinct Mediterranean forests communities may improve ecosystem potential in a warmer future. *Frontiers in Ecology and the Environment* 13(7): 356–362.

Henriquet, M., Dominguez, S., Barreca, G., Malavieille, J., Cadio, C. and Monaco, C. 2019. Deep origin of the dome-shaped Hyblean Plateau, Southeastern Sicily: A new tectono-magmatic model. *Tectonics* 38: 4488–4515.

Higounet, C. 1966. Les forêts de l'Europe occidentale du Ve au XIe siècle, in *Settimane di studio del Centro Italiano sull'Alto Medioevo 13: Agricoltura e mondo rurale in Occidente nell'Alto Medioevo* (Spoleto, April 22-25 1965): 343-398. Spoleto: Centro italiano di Studi sull'alto Medioevo.

Hofmann, A., Cibella, R., Bertani, R., Miozzo, M., Fantoni, I. and Luppi, S. 2011. *Strumenti Conoscitivi per la Gestione delle Risorse Forestali della Sicilia. Sistema Informativo Forestale.* : 208. Assessorato Territorio e Ambiente, Regione Siciliana. Città di Castello: Litograf Editor s.r.l.

Hrisrova, I., Atanassova, J. and Marinova, E. 2016. Plant economy and vegetation of the Iron Age in Bulgaria: archaeobotanical evidence from pit deposits. *Archaeological and Anthropological Sciences* 9(7): 1481–1494.

Hughes, J.D. and Thirgood, J.V. 1982. Deforestation in ancient Greece and Rome: a cause of collapse. *Ecologist* 12: 196–208.

Hughes, P.D. and Woodward, J.C. (eds) 2017. Quaternary Glaciation in the Mediterranean Mountains. *Journal of the Geological Society* (London) (Special Publications) 433: 1–23.

Incarbona, A., Di Stefano, E., Parti, B., Pelosi, N., Bonomo, S., Mazzola, S., Sprovieri, R., Tranchida, G., Zgozi, S. and Bonanno, A. 2008. Holocene millennial-scale productivity variations in the Sicily Channel (Mediterranean Sea). *Paleoceanography* 23: 1–18.

Incarbona, A., Zarcone, G., Agate, M., Bonomo, S., Di Stefano, E., Masini, F., Russo, F. and Sineo, L. 2010b. A multidisciplinary approach to reveal the Sicily climate and environment over the last 20000 years. *Central European Journal of Geosciences* 2: 71–82.

Incarbona, A., Ziveri, P., Di Stefano, E., Lirer, F., Mortyn, G., Patti, B., Pelosi, N., Sprovieri, M., Tranchida, G., Vallefuoco, M., Albertazzi, S., Bellucci, L.G., Bonanno, A., Bonomo, S., Censi, P., Ferraro, L., Giuliani, S., Mazzola, S. and Sprovieri, R. 2010a. The impact of the Little Ice Age on Coccolithophores in the Central Mediterranean Sea. *Climate of the Past* 6: 795–805.

Italiano, F., Correale, A., Di Bella, M., Foresta Martin, F., Martinelli, M.C., Sabatino, G. and Spatafora, F. 2018. The neolithic obsidian artifacts from Roccapalumba (Palermo, Italy): First characterization and provenance determination. *Mediterranean Archaeology and Archaeometry* 18(3): 151–167.

Izdebski, A., Holmgren, K., Weiberg, E., Stocker, S.R., Büntgen, U., Florenzano, A., Gogou, A., Leroy, S.G.A., Luterbacher, J., Martrat B., Masi, A., Mercuri, A.M., Montagna, P., Sadori, L., Schneider, L., Sicre, M.-L., Triantaphyllou, M. and Xoplaki, E. 2016. Realising consilience: How better communication between archaeologists, historians and natural scientists can transform the study of past climate change in the Mediterranean. *Quaternary Science Reviews* 136: 5–22.

Jordan, N.J., Rotolo, S.G., Williams, R., Speranza, F., McIntosh, W.C., Branney. M.J. and Scaillet, S. 2018. Explosive eruptive history of Pantelleria, Italy: Repeated caldera collapse and ignimbrite emplacement at a peralkaline volcano. *Journal of Volcanology and Geothermal Research* 349: 47–73.

Kaniewski, D., Van Campo, E., Guiot, J., Le Burel, S., Otto, T. and Baeteman, C. 2013. Environmental roots of the Late Bronze Age crisis. *PLoS ONE* 8(8): e71004.

Kaplan, J.O., Krumhardt, K.M. and Zimmerman, N. 2009. The prehistoric and preindustrial deforestation of Europe. *Quaternary Science Reviews* 28: 3016–3034.

Keller, J. 1970. Datierung der Obsidiane und Bimstoffe von Lipari. *Neues Jahrbuch für Geologie und Paläontologie* 1970(2): 90–101.

Keller, J. 2002. *Lipari's fiery past: Dating the Medieval pumice eruption of Monte Pelato. Abstracts of the International Conference 'The fire between air and water: Volcanic islands in science and myth. Preservation and improvement'* (Lipari, 29 September – 2 October 2002). Lipari: UNESCO and Regione Siciliana.

Kelly, T.J., Carey, S., Pistolesi, M., Rosi, M., Croff-Bell, K.L.C., Roman, C. and Marani, M. 2014. Exploration of the 1891 Foerstner submarine vent site (Pantelleria, Italy): Insights into the formation of basaltic balloons. *Bulletin of Volcanology* 76: 1–18.

Kleinen, T., Tarasov, P., Brovkin, V., Andreev, A. and Stebich, M. 2011. Comparison of modeled and reconstructed changes in forest cover through the past 8,000 years: Eurasian perspective. *The Holocene* 21(5): 723–734.

Kontakiotis, G. 2016. Late Quaternary paleoenvironmental reconstruction and paleoclimatic implications of the Aegean Sea (eastern Mediterranean) based on paleoceanographic indexes and stable isotopes. *Quaternary International* 401: 28–42.

Kouli, K., Masi, A., Mercuri, A.M., Florenzano, A. and L. Sadori, L. 2015. Regional vegetation histories: An overview of the pollen evidence from the Central Mediterranean. *Late Antique Archaeology* 11(1): 69–82.

La Mantia, T., Carimi, F., Di Lorenzo, R. and Pasta, S. 2011. The agricultural heritage of Lampedusa (Pelagie Archipelago, South Italy) and its key role for cultivar and wildlife conservation. *Italian Journal of Agronomy* 6:e17: 106–110.

La Mela Veca, D.S., Cullotta, S., Sferlazza, S. and Maetzke, F.G. 2016. Anthropogenic Influences in land use/land cover changes in Mediterranean forest landscapes in Sicily. *Land* 5(3): 1–13.

La Rosa, A., Gianguzzi, L., Salluzzo, G., Scuderi, L. and Pasta, S. 2021. Last tesserae of a fading mosaic: Floristic census and forest vegetation survey at Parche di Bilello (south-western Sicily, Italy), a site needing urgent protection measures. *Plant Sociology* 58(1): 55–74.

Laghetti, G., Hammer, K. and Perrino, P. 1996. Plant genetic resources in Pantelleria and Pelagie archipelago, Italy: collecting and conservation of local crop germplasm. *Plant Genetic Resources Newsletter* 108: 17–25.

Laghetti, G., Hammer, K., Olita, G. and Perrino, P. 1998. Crop genetic resources from Ustica Island (Italy): collecting and safeguarding. *Plant Genetic Resources Newsletter* 116: 12–17.

Lambeck, K., Rouby, H., Purcell, A., Sun, Y. and Sambridge, M. 2014. Sea level and global ice volumes from the Last Glacial Maximum to the Holocene. *Proceedings of the National Academy of Sciences* 111(43): 15296–15303.

Lambert, F., Delmonte, B., Petit, J.R., Bigler, M, Kaufmann, P.R., Hutterli, M.A., Stocker, T.F., Ruth, U., Steffensen, J.P. and Maggi, V. 2008. Dust-climate couplings over the past 800,000 years from the EPICA Dome C ice core. *Nature* 452: 616–619.

Lang, G. 1994: *Quartäre Vegetationsgeschichte Europas. Methoden und Ergebnisse*. Jena: G. Fischer.

Larrasoaña, J.C., Roberts, A.P. and Rohling, E.J. 2013. Dynamics of Green Sahara periods and their role in hominin evolution. *PLoS One* 8(10): e76514.

Lasanta-Martínez, T., Vicente-Serrano, S.M. and Cuadrat-Prats, J.M. 2005. Mountain Mediterranean landscape evolution caused by the abandonment of traditional primary activities: A study of the Spanish Central Pyrenees. *Applied Geography* 25: 47–65.

Leighton, R. 1999. *Sicily before History: An Archaeological Survey from the Palaeolithic to the Iron Age*. Ithaca (NY): Cornell University Press.

Lillios, K.T., Blanco-González, A., Drake, B.L. and López-Sáez, J.A. 2019. Mid-late Holocene climate, demography, and cultural dynamics in Iberia: A multi-proxy approach. *Quaternary Science Reviews* 135: 138–153.

Lo Presti, V., Antonioli, F., Palombo, M.R., Agnesi, V., Biolchi, S., Calcagnile, L., Di Patti, C., Donati, S., Furlani, S., Merizzi, J., Pepe, F., Quarta, G., Renda, P., Sulli, A. and Tusa, S. 2019. Palaeogeographical evolution of the Egadi Islands (western Sicily, Italy). Implications for late Pleistocene and early Holocene sea crossings by humans and other mammals in the western Mediterranean. *Earth-Science Reviews* 194: 160–181.

Lo Vetro, D. and Martini, F. 2012. Il Paleolitico e il Mesolitico in Sicilia, in R. Giglio and S. Tusa (eds) *Atti della XLI Riunione Scientifica 'Dai Ciclopi agli Ecisti: Società e territorio nella Sicilia preistorica e protostorica'* (San Cipirello, PA, 16–19 novembre 2006): 19–47. Firenze: Istituto Italiano di Preistoria e Protostoria.

Lo Vetro, D. and Martini, F. 2016. Mesolithic in central-southern Italy: Overview of lithic productions. *Quaternary International* 423: 279–302.

Lo Vetro, D., Colonese, A., Mannino, M., Thomas, K., Di Giuseppe, Z. and Martini, F. 2016. The mesolithic occupation at Isolidda (San Vito Lo Capo), Sicily. *Preistoria Alpina* 48: 237–243.

Locati, M., Rovida, A., Albini, P. and Stucchi, M. 2014. The AHEAD Portal: A gateway to European historical earthquake data. *Seismological Research Letters* 85: 727–734.

Lombard, M. 1959. Un problème cartographié: le bois dans la Méditerranée musulmane (VIIe–XIe siècles). *Annales Economies, Sociétés, Civilisations*, 14: 234–254.

Longhitano, S. and Colella, A. 2003. Stratigraphy and basin-fill architecture of a Plio-Pleistocene foredeep basin (Catania Plain, eastern Sicily): A preliminary synthesis. *GeoActa* 1: 111–130.

Lucchi, F., Gertisser R., Keller J., Forni F., De Astis G. and Tranne, C.A. 2013. Eruptive history and magmatic evolution of the island of Salina (central Aeolian archipelago), in F. Lucchi, A. Peccerillo, J. Keller, C.A. Tranne and P.L. Rossi (eds) *The Aeolian Islands Volcanoes*: 155–211. Memoirs 37. London: Geological Society.

Lumaret, R., Mir, C., Michaud, H. and Raynal, V. 2002. Phylogeographical variation of chloroplast DNA in holm oak (Quercus ilex L.). *Molecular Ecology* 11: 2327–2336.

Magny, M., Vannière, B., Calò, C., Millet, L., Leroux, A., Peyron, O., Zanchetta, G., La Mantia, T. and Tinner, W. 2011. Holocene hydrological changes in south-western Mediterranean as recorded by lake-level fluctuations at Lago Preola, a coastal lake in southern Sicily, Italy. *Quaternary Science Reviews* 30: 2459–2475.

Magri, D., Di Rita, F., Aranbarri, J., Fletcher, W. and González-Sampériz, P. 2017. Quaternary disappearance of tree taxa from Southern Europe: Timing and trends. *Quaternary Science Reviews* 163: 23–55.

Maiorano, P., Marino, M. and De Lange, G.J. 2019. Dynamic surface-water alterations during sapropel

Mallegni, F. 2005. San Teodoro, in G. Alciati, V. Pesce Delfino and E. Vacca (eds) *Catalogue of Italian human remains from the Palaeolithic to the Mesolithic. Journal of Anthropological Sciences* 83 (suppl.): 133–136.

Mangano, G. 2005. I resti faunistici del Tardiglaciale della Grotta di S. Teodoro (Acquedolci, Messina), in I. Fiore, G. Malerba and S. Chilardi (eds) *Atti del 3° Convegno Nazionale di Archeozoologia (Siracusa, 3-5 novembre 2000), Studi di Paletnologia II, Collana del Bullettino di Paletnologia*: 173–180. Roma: Istituto Poligrafico e Zecca dello Stato.

Manni, M., Coltelli, M. and Martinelli, M.C. 2019. Volcanic events that have marked the anthropic history of the Aeolian Islands. *Annals of Geophysics* 62(1): VO08.

Mannino, G. 2012. Lo 'scarico' neolitico di Castellaccio di Fiaccati Roccapalumba (PA), in R. Giglio and S. Tusa (eds) *Atti della XLI Riunione Scientifica 'Dai Ciclopi agli Ecisti: Società e territorio nella Sicilia preistorica e protostorica'* (San Cipirello, PA, 16–19 novembre 2006): 1099–1100. Firenze: Istituto Italiano di Preistoria e Protostoria.

Mannino, M.A. and Thomas, K.D. 2010. Studio preliminare del campione faunistico della Grotta Schiacciata a Levanzo (Trapani), in A. Tagliacozzo, I. Fiore, S. Marconi and U. Tecchiati (eds) *Atti del 5° Convegno Nazionale di Archeozoologia* (Rovereto, 10–12 novembre 2006): 97–99. Rovereto: Edizioni Osiride.

Mannino, M.A. and Thomas, K. 2012. Studi archeozoologici ed archeometrici sui reperti di malacofauna della Grotta dell'Uzzo (TP), in R. Giglio and S. Tusa (eds) *Atti della XLI Riunione Scientifica 'Dai Ciclopi agli Ecisti: società e territorio nella Sicilia preistorica e protostorica'* (San Cipirello, PA, 16–19 novembre 2006): 471–480. Firenze: Istituto Italiano di Preistoria e Protostoria.

Mannino, M.A., Catalano, G., Talamo, S., Mannino, G., Di Salvo, R., Schimmenti, V., Lalueza-Fox, C., Messina, A., Petruso, D., Caramelli, D., Richards, M.P. and Sineo, L. 2012. Origin and diet of the prehistoric hunter-gatherers on the Mediterranean Island of Favignana (Egadi Islands, Sicily). *PLoS ONE* 7(11): e49802.

Mannino, M.A., Di Salvo, R., Schimmenti, V., Di Patti, C., Incarbona, A., Sineo, L. and Richards, M.P. 2011a. Upper Palaeolithic hunter-gatherer subsistence in Mediterranean coastal environments: an isotopic study of the diets of the earliest directly-dated humans from Sicily. *Journal of Archaeological Science* 38: 3094–3100.

Mannino, M.A., Thomas, K.D., Leng, M.J., Di Salvo, R. and Richards, M.P. 2011b. Stuck to the shore? Investigating prehistoric hunter-gatherer subsistence, mobility and territoriality in a Mediterranean coastal landscape through isotope analyses on marine mollusc shell carbonates and human bone collagen. *Quaternary International* 244(1): 88–104.

Mannino, M.A., Talamo, S., Tagliacozzo, A., Fiore, I., Nehlich, O., Piperno, M., Tusa, S., Collina, C., Di Salvo, R., Schimmenti, V. and Richards, M.P. 2015. Climate-driven environmental changes around 8,200 years ago favoured increases in cetacean strandings and Mediterranean hunter-gatherers exploited them. *Scientific Reports* 5: 16288.

Marchal, O., Cacho, I., Stocker, T.F., Grimalt, J.O., Calvo, E., Martrat, B., Shackleton, N., Vautravers, M., Cortijo, E., Van Kreveld, S., Andersson, C., Koç, N., Chapman, M., Sbaffi, L., Duplessy, J.C., Sarnthein, M., Turon, J.L., Duprat, J. and Jansen, E. 2002. Apparent long-term cooling of the sea surface in the northeast Atlantic and Mediterranean during the Holocene. *Quaternary Science Reviews* 21: 455–483.

Marco-Barba, J., Holmes, J.A., Mesquita-Joanes, F. and Miracle, M.R. 2013. The influence of climate and sea level change on the Holocene evolution of a Mediterranean coastal lagoon: Evidence from ostracod palaeoecology and geochemistry. *Geobios* 46: 409–421.

Margaritelli, G., Cacho, I., Català A., Barra M., Bellucci L.G., Lubritto C., Rettori, R. and Lirer, F. 2020. Persistent warm Mediterranean surface waters during the Roman period. *Scientific Reports* 10: 10431; doi.org/10.1038/s41598-020-67281-2.

Mariotti Lippi, M., Florenzano, A., Rinaldi, R., Allevato, E., Arobba, D., Bacchetta, G., Bal, M., Bandini Mazzanti, M., Benatti, A., Beneš, J., Bosi, G., Buonincontri, M., Caramiello, R., Castelletti, L., Castiglioni, E., Celant, A., Clò, E., Costantin, L., Di Pasquale, G., Di Rita, F., Fiorentino, G., Furlanetto, G., Giardini, M., Grillo, O., Guido, M., Herchenbach, M., Magri, D., Marchesini, M., Maritan, M., Marvelli, S., Masi, A., Miola, A., Montanari, C., Montecchi, M.C., Motella, S., Nisbet, R., Orrù, M., Peña Chocarro, L., Pepe, C., Perego, R., Rattighieri, E., Ravazzi, C., Rottoli, M., Rowan, E., Sabato, D., Sadori, L., Sarigu, M., Torri, P., Ucchesu, M. and Mercuri, A.M. 2018. The Botanical Record of Archaeobotany Italian Network - BRAIN: A cooperative network, database and website. *Flora Mediterranea* 28: 365–376.

Marquer, L., Gaillard M.-J., Sugitac, S., Poska A., Trondman, A.-K., Mazier, F., Nielsen, A.B., Fyfeg, R.M., Jönsson A.M., Smith, B., Kaplan, J.O., Alenius, T., Birks H.J.B., Bjune A.E., Christiansen, J., Dodson, J., Edwards, K.J., Giesecken, T., Herzschuh, U., Kangur, M., Koff, T., Latałowa, M., Lechterbeck, J., Olofsson, J. and Seppä, H. 2017. Quantifying the effects of land use and climate on Holocene vegetation in Europe. *Quaternary Science Reviews* 171: 20–37.

Martinelli, M.C., Dawson, H., Lo Cascio, P., Levi, S.T. and Fiorentino, G. 2021. Blowin' in the wind: Settlement, landscape and network dynamics in the prehistory of the Aeolian Islands. *Journal of Mediterranean Archaeology* 34(1): 28–57.

Martinelli, M.C. and Lo Cascio, P. 2018. Topografia della preistoria nelle Isole Eolie, in M. Bernabò Brea, M. Cultraro, M. Gras, M.C. Martinelli, C. Pouzadoux and U. Spigo (eds) *A Madeleine Cavalier*: 65–78. Naples: Collection du Centre Jean Bérard 49.

Martinelli, M., Tykot, R.H. and Vianello, A. 2019. Lipari (Aeolian Islands) obsidian in the Late Neolithic. Artifacts, supply and function. *Open Archaeology* 5(1): 46–64.

Martini, F., Colonese, A.C., Di Giuseppe, Z., Ghinassi, M., Lo Vetro, D. and Ricciardi, S. 2009. Human-environment relationships during the Late Glacial-Early Holocene transition: Some examples from Campania, Calabria and Sicily. *Méditerranée* 112: 89–94.

Martini, F., Lo Vetro, D., Colonese, A.C., De Curtis, O., Di Giuseppe, Z., Locatelli, E. and Sala, B. 2007. L'Epigravettiano Finale in Sicilia, in F. Martini (ed.) *L'Italia tra 10.000 e 15.000 anni fa. Cosmopolitismo e regionalità nel Tardoglaciale*: 209–254. Firenze: Museo fiorentino di Preistoria 'Paolo Graziosi'.

Martini, F., Lo Vetro, D., Colonese, A.C., Di Giuseppe, Z. and Ricciardi, S. 2012. Primi risultati della campagna di scavo 2005 a Grotta delle Uccerie (Favignana, Trapani), in R. Giglio and S. Tusa (eds) *Atti della XLI Riunione Scientifica ' Dai Ciclopi agli Ecisti : società e territorio nella Sicilia preistorica e protostorica'* (San Cipirello, PA, 16–19 novembre 2006): 289–302. Firenze: Istituto Italiano di Preistoria e Protostoria.

Martrat, B., Grimalt, J.O., Lopez-Martinez, C., Cacho, I., Sierro, F.J., Flores, J.A., Zahn, R., Canals, M., Curtis, J.H. and Hodell, D.A. 2004. Abrupt temperature changes in the Western Mediterranean over the past 250,000 years. *Science* 306: 1762–1765.

Martrat, B., Jimenez-Amat, P., Zahn, R. and Grimalt, J.O. 2014. Similarities and dissimilarities between the last two deglaciations and interglaciations in the North Atlantic region. *Quaternary Science Reviews* 99: 122–134.

Masseti, M. 2009. In the gardens of Norman Palermo, Sicily (twelfth century AD). *Anthropozoologica* 44(2): 7–34.

Masseti, M. 2016. *Zoologia della Sicilia araba e normanna*. Palermo: Edizioni Danaus.

Mathieson, I., Roodenberg, S.A., Posth, C., Szécsényi-Nagy, A., Rohland, N., Mallick, S., Olalde, I., Broomandkhoshbacht, N., Cheronet, O., Fernandes, D., Ferry, M., Gamarra, B., González Fortes, G., Haak, W., Harney, E., Krause-Kyora, B., Kucukkalipci, I., Michel, M., Mittnik, A., Nägele, K., Novak, M., Oppenheimer, J., Patterson, N., Pfrengle, S., Sirak, K., Stewardson, K., Vai, S., Alexandrov, S., Alt, K.W., Andreescu, R., Antonović, D., Ash, A., Atanassova, N., Bacvarov, K., Balázs Gusztáv,, M., Bocherens, H., Bolus, M., Boroneanț, A., Boyadzhiev, Y., Budnik, A., Burmaz, J., Chohadzhiev, S., Conard, N.J., Cottiaux, R., Čuka, M,. Cupillard, C., Drucker, D.G., Elenski, N., Francken, M., Galabova, B., Ganetovski, G., Gely, B., Hajdu, T., Handzhyiska, V., Harvati, K., Higham, T., Iliev, S., Janković, I., Karavanić, I., Kennett, D.J., Komš, D., Koza, A., Labuda, D., Lari, M., Lazar, C., Leppek, M., Leshtakov, K., Lo Vetro, D., Los, D., Lozanov, I., Malina, M., Martini, F., McSweeney, K., Keller, H., Menđušić, M., Mirea, P., Moiseyev, V., Petrova, V., Price, T.D., Simalcsik, A., Sineo, L., Šlaus, M., Slavchev, V., Stanev, P., Starović, A., Szeniczey, T., Talamo, S., Teschler-Nicola, M., Thevenet, C., Valchev, I., Valentin, S., Veljanovska, F., Venelinova, S., Veselovskay, E., Viola, B., Virag, C., Zaninović, J., Zäuner, S., Stockhammer, P.W., Catalano, G., Krauß, R., Caramelli, D., Zariņa, G., Gaydarska, B., Lillie, M., Nikitin, A.J., Potekhina, I., Papathanasiou, A., Borić, D., Bonsal,l C., Krause, J., Pinhasi, R. and Reich, D. 2018. The genomic history of Southeastern Europe. *Nature* 555: 197–203.

McBride, R.A., Anderson, J.B., Buynevich, I.V., Cleary, W., Fenster, M.S., Fitzgerald, D.M., Harris, M.S., Hein, C.J., Klein, A.H.F., Liu, B., de Menezes, J.T., Pejrup, M., Riggs, S.R., Short, A.D., Stone, G.W., Wallace, D.J. and Wang, P. 2013. Morphodynamics of barrier systems: a synthesis, in J. Shroder and D.J. Sherman (eds) *Treatise on Geomorphology, vol.10, 'Coastal and Submarine Geomorphology'*: 166–244. San Diego, CA: Academic Press.

McConnell, B.E. 1997. Lo sviluppo delle prime società agro-pastorali: L'Eneolitico, in S. Tusa (ed.) *Catalogo della Mostra 'Prima Sicilia: Alle Origini della Società Siciliana'* (Palermo, 18 October – 22 December 1997): 281–297. Siracusa: Ediprint.

Médail, F., Monnet, A.-C., Pavon, D., Nikolic, T., Dimopoulos, P., Bacchetta, G., Arroyo, J., Barina, Z., Albassatneh, M.C., Domina, G., Fady, B., Matevski, V., Mifsud, S. and Leriche, A. 2019. What is a tree in the Mediterranean Basin hotspot? A critical analysis. *Forest Ecosystems* 6: 17; doi.org/10.1186/s40663-019-0170-6.

Médail, F. and Diadema, K. 2009. Glacial refugia influence plant diversity patterns in the Mediterranean Basin. *Journal of Biogeography* 36: 1333–1345.

Mercalli, G. and Silvestri, O. 1891. Le eruzioni dell'isola di Vulcano, incominciate il 3 Agosto 1888 e terminate il 22 Marzo 1890. Relazione scientifica della Commissione incaricata dal R. Governo. *Annali dell'Ufficio centrale di Meteorologia e Geodinamica* 10(4): 1–212.

Mercuri, A.M., Bandini Mazzanti, M., Florenzano, A., Montecchi, M.C., Rattighieri, E. and Torri, P. 2013. Anthropogenic Pollen Indicators (API) from archaeological sites as local evidence of human-induced environments in the Italian peninsula. *Annali di Botanica* 3: 143–153.

Mercuri, A.M. and Sadori, L. 2014. Mediterranean culture and climatic change: Past patterns and future trends, in S. Goffredo and Z. Dubinsky (eds) *The Mediterranean Sea*: 507–527. Dordrecht: Springer.

Messerli, B. 1980. Mountain glaciers in the Mediterranean area and in Africa. Proceedings of the Workshop 'World Glacier Inventory' (Riederalp, September 1978). *International Association of Hydrological Sciences Publications* 126: 197–211.

Michaud, H., Toumi, L., Lumaret, R., Li, T.X., Romane, F. and Di Giusto, F. 1995. Effect of geographical discontinuity on genetic variation in Quercus ilex L. (holm-oak). Evidence from enzyme polymorphism. *Heredity* 74: 590–606.

Morales-Molino, C., Tinner, W., Perea, R., Carrión, J.S., Colombaroli, D., Valbuena-Carabaña, M., Zafra, E. and Gil, L. 2019. Unprecedented herbivory threatens rearedge populations of Betula in southwestern Eurasia. *Ecology* 100: e02833.

Moreno, J.M., Vázquez, A. and Vélez, R. 1998. *Recent history of forest fires in Spain*, in J.M. Moreno (ed.) *Large Fires*: 159–185. Leiden: Backhuys Publishers.

Moricca, C., Nigro L., Masci L., Pasta S., Cappella F., Spagnoli F. and Sadori L. 2021. Cultural landscape and plant use at the Phoenician Motya (Western Sicily, Italy) inferred by a disposal pit. *Vegetation History and Archaeobotany*; doi.org/10.1007/s00334-021-00834-1.

Morris, I. 1996. The absolute chronology of the Greek colonies in Sicily. *Acta Archaeologica* 67: 51–59.

Mulas, M., Cioni, R., Andronico, D. and Mundula, F. 2016. The explosive activity of the 1669 Monti Rossi Eruption at Mt. Etna (Italy). *Journal of Volcanology and Geothermal Research* 328: 115–133.

Natali, E. and Forgia, V. 2018. The beginning of the Neolithic in Southern Italy and Sicily. *Quaternary International* 470(B): 253–269.

Neboit, R. 1984. Érosion des sols et colonisation grecque en Sicile et en Grande Grèce. *Bulletin de l'Association de Géographes français* 499: 5–13.

Neri, M., Coltelli, M., Orombelli, G. and Pasquarè, G. 1994. Ghiacciai pleistocenici dell'Etna: un problema aperto. *Istituto Lombardo, Accademia di Scienze e Lettere (Rendiconti Scienze)* 128: 103–125.

Nicoletti, F. 2015. *Catania antica: nuove prospettive di ricerca*. Regione siciliana, Assessorato dei Beni culturali e dell'Identità siciliana, Dipartimento dei Beni Culturali e dell'Identità siciliana.

Nicoletti, F. and Tusa, S. 2012a. L'Età del Bronzo nella Sicilia occidentale, in R. Giglio and S. Tusa (eds) *Atti della XLI Riunione Scientifica 'Dai Ciclopi agli Ecisti: società e territorio nella Sicilia preistorica e protostorica'* (San Cipirello, PA, 16–19 novembre 2006): 105–129. Firenze: Istituto italiano di Preistoria e Protostoria.

Nicoletti, F. and Tusa, S. 2012b. Nuove acquisizioni scientifiche sul Riparo del Castello di Termini Imerese (PA) nel quadro della preistoria siciliana tra la fine del Pleistocene e gli inizi dell'Olocene, in R. Giglio and S. Tusa (eds) *Atti della XLI Riunione Scientifica 'Dai Ciclopi agli Ecisti: società e territorio nella Sicilia preistorica e protostorica'* (San Cipirello, PA, 16–19 novembre 2006): 303–318. Firenze: Istituto italiano di Preistoria e Protostoria.

Niemeyer, H.G. 2006. The Phoenicians in the Mediterranean. Between expansion and colonization: a non-Greek model of overseas settlement and presence, in G.R. Tsetskhladze (ed.) *Greek colonization: An account of Greek Colonies and other settlements overseas*: 143–168. Leiden: Brill.

Nieto Feliner, G. 2011. Southern European glacial refugia: A tale of tales. *Taxon* 60(2): 365–372.

Nijboer, A.J. 2006. The Iron Age in the Mediterranean: A chronological mess or 'Trade Before the Flag', part II. *Ancient West and East* 4(2): 255–277.

Noti, R., van Leeuwen, J.F.N., Colombaroli, D., Vescovi, E., Pasta, S., La Mantia, T. and Tinner, W. 2009. Mid- and Late-Holocene Vegetation and fire history of Biviere di Gela, a coastal lake in southern Sicily. *Vegetation History and Archaeobotany* 18(5): 371–387.

Olalde, I., Brace, S., Allentoft, M.E., Armit, I., Kristiansen, K., Booth, T., Rohland, N., Mallick, S., Szécsényi-Nagy, A., Mittnik, A., Altena, E., Lipson, M., Lazaridis, I., Harper, T.K., Patterson, N., Broomandkhoshbacht, N., Diekmann, Y., Faltyskova, Z., Fernandes, D., Ferry, M., Harney, E., de Knijff, P., Michel, M., Oppenheimer, J., Stewardson, K., Barclay, A., Alt, K.W., Liesau, C., Ríos, P., Blasco, C., Menduiña García, R., Avilés Fernández, A., Bánffy, E., Bernabò-Brea, M., Billoin, D., Bonsall, C., Bonsall, L., Allen, T., Büster, L., Carver, S., Castells Navarro, L., Craig, O.E., Cook, G.T., Cunliffe, B., Denaire, A., Egging Dinwiddy, K., Dodwell, N., Ernée, M., Evans, C., Kuchařík, M., Farré, J.F., Fowler, C., Gazenbeek, M., Garrido Pena, R., Haber-Uriarte, M., Haduch, E., Hey, G., Jowett, N., Knowles, T., Massy, K., Pfrengle, S., Lefranc, P., Lemercier, O., Lefebvre, A., Heras Martínez, C., Galera Olmo, V., Bastida Ramírez, A., Lomba Maurandi, J., Majó, T., McKinley, J.I., McSweeney, K., Mende, B.G., Modi, A., Kulcsár, G., Kiss, V., Czene, A., Patay, R., Endrődi, A., Köhler, K., Hajdu,T., Szeniczey, T., Dani, J., Bernert, Z., Hoole, M., Cheronet, O., Keating, D., Velemínský, P., Dobeš, M., Candilio, F., Brown, F., Raúl Flores Fernández, R., Herrero-Corral A.-M., Tusa, S., Carnieri, E., Lentini, L., Valenti, A., Zanini, A., Waddington, C., Delibes, G., Guerra-Doce, E., Neil, B., Brittain, M., Luke, M., Mortimer, R., Desideri, J., Besse, M., Brücken, G., Furmanek, M., Hałuszko, A., Mackiewicz, M., Rapiński, A., Leach, S., Soriano, I., Lillios, K.T., Cardoso, J.L., Parker Pearson, M., Włodarczak, P., Price, T.D., Prieto, P., Rey, P.-J., Risch, R., Rojo Guerra, M.A., Schmitt, A., Serralongue, J., Silva, A.M., Smrčka, V., Vergnaud, L., Zilhão, J., Caramelli, D., Higham, T., Thomas, M.G., Kennett, D.J., Fokkens, H., Heyd, V., Sheridan, A., Sjögren, K.-G., Stockhammer, P.W., Krause, J., Pinhasi, R., Haak, W., Barnes, I., Lalueza-Fox, C. and Reich, D. 2018. The Beaker phenomenon and the genomic transformation of Northwest Europe. *Nature* 555(7695): 190–196.

Pacciarelli, M., Scarano, T. and Crispino, A. 2015. The transition between the Copper and Bronze Ages in Southern Italy and Sicily, in H. Meller, H.W. Arz, R. Jung and R. Risch (eds) *2200 BC: Ein Klimasturz als Ursache für den Zerfall der Alten Welt? - 2200 BC: A climatic breakdown as a cause for the collapse of the old world?' Tagungen des Landesmuseums für Vorgeschichte Halle* 12(1): 253–281. 7th Archaeological Conference of Central Germany (October 23–26, 2014, Halle, Saale). Halle: Landesamt für Denkmalpflege und Archäologie Sachsen-Anhalt, Landesmuseum für Vorgeschichte.

Palermo, D. and Tanasi, D. 2005. Diodoro a Polizzello, in C. Miccichè, S. Modeo and L. Santagati (eds) *Atti del Convegno di studi 'Diodoro Siculo e la Sicilia indigena'* (Caltanissetta, 21–22 maggio 2005): 89–102. Palermo: Palermo: Regione Sicilia, Assessorato dei beni culturali ed ambientali e della pubblica istruzione.

Pareschi, M.T., Boschi, E. and Favalli, M. 2006. Lost tsunami. *Geophysical Research Letters* 33: L22608.

Pasta, S. and La Mantia, T. 2004. Note sul paesaggio vegetale delle isole minori circumsiciliane. II. La vegetazione pre-forestale e forestale nelle isole del Canale di Sicilia. *Annali dell'Accademia italiana di Scienze Forestali* 52 (2003): 77–124.

Pasta, S., La Rosa, A., Garfì, G., Marcenò, C., Gristina, A.S., Carimi, F. and Guarino, R. 2020. An updated checklist of the Sicilian native edible plants: Preserving the traditional ecological knowledge of century-old agro-pastoral landscapes. *Frontiers in Plant Science* 11: 388.

Pasta, S., Sala, G., La Mantia, T., Bondì, C. and Tinner, W. 2019. Past distribution of *Abies nebrodensis* (Lojac.) Mattei: results of a multidisciplinary study. *Vegetation History and Archaeobotany* 29: 357–371.

Pausata, F.S.R., Messori, G. and Zhang, Q. 2016. Impacts of dust reduction on the northward expansion of the African monsoon during the Green Sahara period. *Earth and Planetary Science Letters* 434: 298–307.

Pessina, A. and Tinè, V. 2008. *Archeologia del Neolitico. L'italia tra sesto e quarto millennio*. Roma: Carocci Editore.

Petit, R.J., Brewer, S., Bordács, S., Burg, K., Cheddadi, R., Coart, E., Cottrell, J., Csaikl, U.M., Deans, J.D., Espinel, S., Fineschi, S., Finkeldey, R., Glaz, I., Goicoechea, P., Jensen, J.S., König, A., Lowe, A.J., Madsen, S.F., Mátyás, G., Munro, R.C., Popescu, F., Slade, D., Tabbener, H., Vries, G.M. (de), van Dam, B., Ziegenhagen, B., de Beaulieu, J.L. and Kremer, A. 2002. Identification of refugia and post-glaciation colonisation routes of European white oaks based on chloroplast DNA and fossil pollen evidence. *Forest Ecology and Management* 156: 49–74.

Petroselli, A., Vessella, F., Cavagnuolo, L., Piovesan, G. and Schirone, B. 2013. Ecological behavior of *Quercus suber* and *Quercus ilex* inferred by topographic wetness index (TWI). *Trees* 27(5): 1201–1215.

Petruso, D. and Sineo, L. 2012. *Human influence on faunal turnover during Early Holocene in Sicily. Abstract book of the AIQUA Congress, 'The transition from natural to anthropogenic-dominated. Environmental change in Italy and the surrounding regions since the Neolithic. Session Environment, Climate and Human impact: The archaeological evidence'* (Pisa, 15–17 February 2012): 71. Pisa.

Piervitali, E. and Colacino, M. 2001. Evidence of drought in western Sicily during the period 1565–1915 from liturgical offices. *Climate Change* 49: 225–238.

Piperno, M. 1985. Some 14C dates for the palaeoeconomic evidence from the Holocene levels of the Uzzo cave (Sicily), in C. Malone and S. Stoddart (eds) *Papers in Italian Archaeology IV, part III, Patterns in Protohistory*: 83–86. British Archaeological Reports International Series 245. Oxford: British Archaeological Reports.

Poggiali, F., Martini, F., Buonincontri, M. and Di Pasquale, G. 2012. *Charcoal data from Oriente cave (Favignana island, Sicily). Abstract book of the AIQUA Congress, 'The transition from natural to anthropogenic-dominated. Environmental change in Italy and the surrounding regions since the Neolithic. Session Environment, Climate and Human impact: The archaeological evidence'* (Pisa, 15–17 February 2012): 72. Pisa.

Ponte, G. 1919. *La catastrofica eruzione dello Stromboli*. Atti della reale Accademia nazionale dei Lincei 28: 89–94.

Posth, C., Renaud, G., Mittnik, A., Drucker, D.G., Rougier, H., Cupillard, C., Valentin, F., Thevenet, C., Furtwängler, A., Wißing, C., Francken, M., Malina, M., Bolus, M., Lari, M., Gigli, E., Capecchi, G., Crevecoeur, I., Beauval, C., Flas, D., Germonpré, M., van der Plicht, J., Cottiaux, R., Gély, B., Ronchitelli, A., Wehrberger, K., Grigorescu, D., Svoboda, J., Semal, P., Caramelli, D., Bocherens, H., Harvati, K., Conard, N.J., Haak, W., Powell, A. and Krause, J. 2016. Pleistocene mitochondrial genomes suggest a single major dispersal of non-Africans and a Late Glacial population turnover in Europe. *Current Biology* 26: 827–833.

Procelli, E. 2016. Ethne e facies culturali nella protostoria siciliana. *Kokalos* 53: 277–286.

Quattrocchi, A. 2017. Le foreste del Piano di Milazzo nella storia tra difesa e distruzioni, in L. Catalioto, G. Pantano and E. Santagati (eds) *Atti del Convegno 'Sicilia millenaria. Dalla microstoria alla dimensione mediterranea'* (Montalbano Elicona, ME, 9–11 ottobre 2015): 443–468. Reggio Calabria: Edizioni Leonida.

Quero, T., Martinelli, M.C. and Giordano, L. 2019. The Neolithic site of San Martino, Sicily: Working and circulation of obsidian from Lipari. *Open Archaeology* 5: 65–82.

Rabbel, W., Hoffmann Wieck, G., Jacobsen, O., Özkap, K., Stümpel, H., Suhr, W., Szalaiova, E. and Wölz, S. 2014. Seismische Vermessung der verlandeten Buchter des Modione und Gorgo Cotone. Hinweise zur Lage des Hafens der antiken Stadt Selinunt. *Römischen Mitteilungen* 120: 135–150.

Racimo, F., Woodbridge, J., Fyfeb, R.M., Sikora, M., Sjögren, K.-G., Kristiansen, K. and Vander Linden, M. 2020. The spatiotemporal spread of human migrations during the European Holocene. *Proceedings of the National Academy of Science*; www.pnas.org/cgi/doi/10.1073/pnas.1920051117.

Raneri, S., Barone, G., Mazzoleni, P., Tanasi, D. and Costa, E. 2015. Mobility of men versus mobility of goods: Archaeometric characterization of Middle Bronze Age pottery in Malta and Sicily (15th – 13th century BC). *Periodico di Mineralogia, Special Issue* 84(1): 23–44.

Rasmussen, S.O., Bigler, M., Blockley, S.P., Blunier, T., Buchardt, S.L., Clausen, H.B., Cvijanovic, I., Dahl-Jensen, D., Johnsen, S.J., Fischer, H., Gkinis, V., Guillevic, M., Hoek, W.Z., Lowe, J.J., Pedro, J.B., Popp, T., Seierstad, I.K., Steffensen, J.P., Svensson, A.M., Vallelonga, P., Vinther, B.M., Walker, M.J.C., Wheatley, J.J. and Winstrup, M. 2014. A stratigraphic framework for abrupt climatic changes during the Last Glacial period based on three synchronized Greenland ice-core records: Refining and extending the INTIMATE event stratigraphy. *Quaternary Science Reviews* 106: 14–28.

Reimer, P.J., Bard, E., Bayliss, A., Beck, J.W., Blackwell, C., Bronk Ramsey, C., Buck C.E., Cheng, H., Lawrence, R.L., Friedrich, M., Grootes, P.M., Guilderson, T.P., Haflidason, H., Hajdas, I., Hatté, C., Heaton, T.J., Hoffmann, D.I., Hogg, A.G., Hughen, K.A., Kaiser, K.F., Kromer, B., Manning, S.W., Niu, M., Reimer, R.W., Richards, D.A., Scott, E.M., Southon, J.R., Staff, R.A., Turney, C.S:M. and van der Plicht, J. 2013. Intcal13 and Marine13 radiocarbon age calibration curves 0-50,000 years Cal Bp. *Radiocarbon* 55(4): 1869–1887.

Revelles, J., Cho, S., Iriarte, E., Burjachs, F., van Geel, B., Palomo, A., Piqué, R., Peña-Chocarro, L. and Terradas, X. 2015. Mid-Holocene vegetation history and Neolithic land-use in the Lake Banyoles area (Girona, Spain). *Palaeogeography, Palaeoclimatology, Palaeoecology* 435: 70–85.

Rius, D., Vannière, B. and Galop, D. 2009. Fire frequency and landscape management in the northwestern Pyrenean piedmont, France, since the early Neolithic (8000 cal. BP). *The Holocene*, 19(6): 847–859.

Roberts, N., Fyfe, R.M., Woodbridge, J., Gaillard, M.-J., Davis, B.A.S., Kaplan, J. O., Marquer, L., Mazier, F., Nielsen, A.B., Sugita, S., Trondman, A.-K. and Leydet, M. 2018. Europe's lost forests: A pollen-based synthesis for the last 11,000 years. *Scientific Reports* 8: 716.

Roberts, N., Jones, M.D., Benkaddour, A., Eastwood, W.J., Filippi, M.L., Frogley, M.R., Lamb, H.F., Leng, M.J., Reed, J.M., Stein, M., Stevens, L., Valero-Garcés, B. and Zanchetta, G. 2008. Stable isotope records of Late Quaternary climate and hydrology from Mediterranean lakes: The ISOMED synthesis. *Quaternary Science Reviews* 27: 2426–2441.

Roberts, N., Woodbridge, J., Bevan, A., Palmisano, A., Shennan, S. and Asouti, E. 2017. Human responses and non-responses to climatic variations during the last Glacial-Interglacial transition in the eastern Mediterranean. *Quaternary Science Reviews* 184: 47–67.

Roberts, N., Woodbridge, J., Palmisano, A., Bevan, A., Fyfe, R. and Shennan, S. 2019. Mediterranean landscape change during the Holocene: Synthesis, comparison and regional trends in population, land cover and climate. *Holocene* 29: 923–937.

Rodrigo-Gámiz, M., Martínez-Ruiz, F., Rodríguez-Tovar, F.J., Jiménez-Espejo, F.J. and Pardo-Igúzquiza, E. 2014. Millennial- to centennial-scale climate periodicities and forcing mechanisms in the westernmost Mediterranean for the past 20,000 yr. *Quaternary Research* 81(1): 78–93.

Rohling, E.J. and Pälike, H. 2005. Centennial-scale climate cooling with a sudden cold event around 8,200 years ago. *Nature* 434: 975–979.

Rohling, E.J., Cane, T.R., Cooke, S., Sprovieri, M., Bouloubassi, I., Emeis, K.C., Schiebel, R., Kroon, D., Jorissen, F.J., Lorre, A. and Kemp, A.E.S. 2002. African monsoon variability during the previous interglacial maximum. *Earth and Planetary Science Letters* 202: 61–75.

Rohling, E.J., Marino, G. and Grant, K.M. 2015. Mediterranean climate and oceanography, and the periodic development of anoxic events (sapropels). *Earth-Science Reviews* 143: 62–97.

Rohling, E.J., Marino, G., Grant, K.M., Mayewski, P.A. and Weninger, B. 2019. A model for archaeologically relevant Holocene climate impacts in the Aegean-Levantine region (easternmost Mediterranean). *Quaternary Science Reviews* 208: 38–53.

Romagnoli, C., Belvisi, V., Innangi, S., Di Martino, G. and Tonielli, R. 2020. New insights on the evolution of the Linosa volcano (Sicily Channel) from the study of its submarine portions. *Marine Geology* 419: 106060.

Romano, V., Catalano, G., Bazan, G., Calì, F. and Sineo, L. 2021. Archaeogenetics and landscape dynamics in Sicily during the Holocene: A review. *Sustainability* 13: 9469.

Rosi, M., Levi, S.T., Pistolesi, M., Bertagnini, A., Brunelli, D., Cannavò, V., Di Renzoni, A., Ferranti, F., Renzulli, A. and Yoon, D. 2019. Geoarchaeological evidence of Middle Age tsunamis at Stromboli and consequences for the tsunami hazard in the southern Tyrrhenian Sea. *Scientific Reports* 9: 677.

Rotolo, S.G., Castorina, F., Cellura, D. and Pompilio, M. 2006. Petrology and geochemistry of submarine volcanism in the Sicily Channel Rift. *Journal of Geology* 114(3): 355–365.

Rotolo, S.G., La Felice, S., Mangalaviti, A. and Landi, P. 2007. Geology and petrochemistry of recent (25 ka) silicic volcanism at Pantelleria Island. *Bollettino della Società Geologica Italiana* 126(2): 191–208.

Rovida, A., Locati, M., Camassi, R., Lolli, B. and Gasperini, P. (eds) 2016. *CPTI15, the 2015 version of the Parametric Catalogue of Italian Earthquakes*. Rome: Istituto Nazionale di Geofisica e Vulcanologia.

Rühl, J., Chiavetta, U., La Mantia, T., La Mela Veca, D.S. and Pasta, S. 2005. Land cover change in the Nature Reserve 'Sughereta di Niscemi' (SE Sicily) in the 20th century, in S. Erasmi, B. Cyffka and M. Kappas (eds) *Proceedings of the 1st GGRS (Göttingen GIS & Remote Sensing Days), Environmental Studies 'Remote Sensing & GIS for Environmental Studies: Applications in Geography'* (Göttingen, Germany, 7–8 October 2004). *Göttinger Geographische Abhandlungen* 113: 54-62.

Rühl, J., Pasta, S. and Schnittler, M. 2006. A chronosequence study of vegetation dynamics on vine and caper terraces of Pantelleria Island (Sicily). *Archive of Nature Conservation and Landscape Research* 45(1): 71–90.

Russell, A. 2017. Sicily without Mycenae: A cross-cultural consumption analysis of connectivity in the Bronze Age Central Mediterranean. *Journal of Mediterranean Archaeology* 30(1): 59–83.

Sadori L., Giraudi, C., Masi, A., Magny, M., Ortu, E., Zanchetta, G. and Izdebski, A. 2016. Climate, environment and society in southern Italy during the last 2000 years. A review of the environmental, historical and archaeological evidence. *Quaternary Science Reviews* 136: 173–188.

Sadori, L., Masi, A. and Ricotta, C. 2015. Climate-driven past fires in central Sicily. *Plant Biosystems* 149: 166–173.

Sadori, L. and Narcisi, B. 2001. The Postglacial record of environmental history from Lago di Pergusa, Sicily. *The Holocene* 11: 655–671.

Sadori, L., Ortu, E., Peyron, O., Zanchetta, G., Vannière, B., Desmet, M. and Magny, M. 2013. The last 7 millennia of vegetation and climate changes at Lago di Pergusa (central Sicily, Italy). *Climate of the Past* 9(2): 1969–1984.

Salari, L. and Masseti, M. 2016. Attardamenti olocenici di *Equus hydruntinus* Regalia, 1907 in Italia, in U. Thun Hohenstein, M. Cangemi, I. Fiore and J. De Grossi Mazzorin (eds) *Atti del 7° Convegno Nazionale di Archeozoologia. Annali Univ. Ferrara (Museologia Scientifica e Naturalistica)* 12(1): 313-320.

Santiso, X., Lopez, L., Retuerto, R. and Barreiro, R. 2016. Phylogeography of a widespread species: pre-glacial vicariance, refugia, occasional blocking straits and long-distance migrations. *AoB PLANTS* 8: plw003.

Sbaffi, L., Wezel, F.C., Kallel, N., Paterne, M., Cacho, I., Ziveri, P. and Shackleton, N. 2001. Response of the pelagic environment to palaeoclimatic changes in the central Mediterranean Sea during the Late Quaternary. *Marine Geology* 178: 39–62.

Scaillet, S., Rotolo, S.G., La Felice, S. and Vita-Scaillet, G. 2011. High-resolution 40Ar/39Ar chronostratigraphy of the post-caldera (<20ka) volcanic activity at Pantelleria. *Earth and Planetary Science Letters* 309(3–4): 280–290.

Schicchi, R. and Raimondo, F.M. 2007. *I grandi alberi di Sicilia*. Palermo: Azienda Foreste Demaniali della Sicilia, Collana Sicilia Foreste.

Schicchi, R., Speciale, C., Amato, F., Bazan, G., Di Noto, G., Marino, P., Ricciardo, P. and Geraci, A. 2021. The monumental olive trees as biocultural heritage of Mediterranean landscapes: The case study of Sicily. *Sustainability* 13: 6767.

Scicchitano, G., Antonioli, F., Berlinghieri, E.F.C., Dutton, A. and Monaco, C. 2008. Submerged archaeological sites along the Ionian coast of southeastern Sicily (Italy) and implications for the Holocene relative sea-level change. *Quaternary Research* 70(1): 26–39.

Scicchitano, G., Monaco, C. and Tortorici, L. 2007. Large boulder deposits by tsunami waves along the Ionian coast of south-eastern Sicily (Italy). *Marine Geology* 238(1–4): 75–91.

Scicchitano, G., Spampinato, C.R., Ferranti, L., Antonioli, F., Monaco, C., Capano, M. and Lubritto, C. 2011. Uplifted Holocene shorelines at Capo Milazzo (NE Sicily, Italy): evidence of co-seismic and steady-state deformation. *Quaternary International* 232(1–2): 201–213.

Servizio Geologico d'Italia 2010. *Foglio 607 Corleone. Carta Geologica d'Italia alla scala 1:50.000*. Istituto Superiore per la Protezione e la Ricerca Ambientale (ISPRA). Assessorato Territorio e Ambiente, Regione Siciliana.

Servizio Geologico d'Italia 2011. *Foglio 609-596 Termini Imerese-Capo Plaia. Carta Geologica d'Italia alla scala 1:50.000*. Istituto Superiore per la Protezione e la Ricerca Ambientale (ISPRA). Assessorato Territorio e Ambiente, Regione Siciliana.

Sha, L., Ait Brahim, Y., Wassenburg, J.A., Yin, J., Peros, M., Cruz, F.W., Cai, Y., Li, H., Du, W., Zhang, H., Edwards, R.L. and Cheng, H. 2019. How far north did the African monsoon fringe expand during the African humid period? Insights from Southwest Moroccan speleothems. *Geophysical Research Letters* 46: 14093–14102.

Silenzi, S., Antonioli, F. and Chemello, R. 2004. A new marker for sea surface temperature trend during the last centuries in temperate areas: Vermetid reef. *Global and Planetary Change* 40: 105–114.

Sineo, L., Petruso, D., Forgia, V., Messina, A.D. and D'Amore, G. 2015. Human peopling of Sicily during Quaternary, in L.D. Fernández (ed.) *Geological Epochs*: 25–68. Cheyenne, WY: AcademyPublish.org.

Smedile, A., De Martini, P.M., Pantosti, D., Bellucci, L., Del Carlo, P., Gasperini, L., Pirrotta, C., Polonia, A. and Boschi, E. 2011. Possible tsunami signatures from an integrated study in the Augusta Bay offshore (Eastern Sicily, Italy). *Marine Geology* 281: 1–13.

Spampinato, C.R., Costa, B., Di Stefano, A., Monaco, C., and Scicchitano, G. 2011. The contribution of tectonics to relative sea-level change during the Holocene in coastal South-Eastern Sicily: new data from boreholes. *Quaternary International* 232(1–2): 214–227.

Spampinato, C.R., Scicchitano, G., Ferranti, L., and Monaco, C. 2012. Raised Holocene paleo-shorelines along the Capo Schisò coast, Taormina: New evidence of recent co-seismic deformation in northeastern Sicily (Italy). *Journal of Geodynamics* 55: 18–31.

Sparacello, V.S., Samsel, M., Villotte, S., Varalli, A., Schimmenti, V. and Sineo, L. 2020. Inferences on Sicilian Mesolithic subsistence patterns from cross-sectional geometry and entheseal changes. *Archaeological and Anthropological Science* 12: 101.

Spatafora, F. 2002. Sicani, Elimi e Greci. Storie di contatti e terre di frontiera, in F. Spatafora and S. Vassallo (eds) *Siculi, Elimi e Greci*: 3–11. Palermo: Assessorato regionale ai Beni Culturali e Ambientali e della Pubblica Istruzione, S. F. Flaccovio s.a.s Editore.

Speciale, C. 2021. *Human-Environment dynamics in the Aeolian Islands during the Bronze Age. A paleo-demographic model*, BAR Int. Ser. 3052, Oxford.

Speciale, C., Bentaleb, I., Combourieu-Nebout, N., Di Sansebastiano, G.P., Iannì, F., Fourel, F. and Giannitrapani, E. 2020. The case study of Case Bastione: First analyses of 3rd millennium Cal BC paleoenvironment and subsistence systems in central Sicily. *Journal of Archaeological Science Reports* 31: 102232.

Speciale, C., d'Oronzo, C., Stellati, A. and Fiorentino, G. 2016. Ubi minor… deinde summa? Archaeobotanical data from the prehistoric village of Filo Braccio (Filicudi, Aeolian Archipelago): Spatial analysis, crop production and paleoclimate reconstruction. *Scienze dell'Antichità* 22(2): 281–296.

Speciale, C., Larosa, N., Spatafora, F., Calascibetta, A.M.G., Di Sansebastiano, G.P., Battaglia, G. and Pasta, S. 2021. Archaeobotanical and historical insights on some steps of forest cover disruption at Ustica Island (Sicily, Italy) from prehistory until present day. *Environmental Archaeology*; doi.org/10.1080/14614103.2021.1962578.

Spies, T.A., Hemstrom, M.A., Youngblood, A. and Hummel, S. 2006: Conserving old-growth forest diversity in disturbance-prone landscapes. *Conservation Biology* 20: 351–362.

Spötl, C., Fairchild, I.J. and Tooth, A.F. 2005. Cave air control on dripwater geochemistry, Obir Caves (Austria): Implications for speleothem deposition in dynamically ventilated caves. *Geochimica Cosmochimica Acta* 69: 2451–2468.

Sprovieri, R., Di Stefano, E., Incarbona, A. and Gargano, M.E. 2003. A high-resolution record of the last deglaciation in the Sicily Channel based on foraminifera and calcareous nannofossil quantitative distribution. *Palaeogeography Palaeoclimatology Palaeoecology* 202: 119–142.

Stika, H.P., Heiss, A.G. and Zach, B. 2008. Plant remains from the early Iron Age in western Sicily: Differences in subsistence strategies of Greek and Elymian sites. *Vegetation History and Archaeobotany* 17: 139–148.

Stiros, S.C. 2019. On the historical role of earthquakes in Antiquity, in M. Ghilardi (ed.) *Geoarchaeology of the Mediterranean Islands*: 191–198. Paris: CNRS Éditions via OpenEdition.

Stiros, S.C. 2001. The AD 365 Crete earthquake and possible seismic clustering during the fourth to sixth centuries AD in the Eastern Mediterranean: A review of historical and archaeological data. *Journal of Structural Geology* 23(2-3): 545–562.

Stöck, M., Sicilia, A., Belfiore, N.M., Buckley, D., Lo Brutto, S., Lo Valvo, M. and Arculeo, M. 2008. Post-Messinian evolutionary relationships across the Sicilian Channel: Mitochondrial and nuclear markers link a new green toad from Sicily to African relatives. *BMC Evolutionary Biology* 8: 56.

Tagliacozzo, A. 1993. Archeozoologia della Grotta dell'Uzzo, Sicilia. Da un'economia di caccia ad un'economia di pesca ed allevamento. *Bullettino di Paletnologia Italiana*, nuova serie 84 (suppl.).

Tagliacozzo, A. 1997. Dalla caccia alla pastorizia. La domesticazione animale. Le modificazioni economiche tra il Mesolitico ed il Neolitico e l'introduzione degli animali domestici in Sicilia, in S. Tusa (ed.) *Catalogo della Mostra 'Prima Sicilia: Alle Origini della Società Siciliana'* (Palermo, 18 October–22 December 1997): 227–247. Siracusa: Ediprint.

Tainter, J. 2006. Archaeology of overshoot and collapse. *Annual Review of Anthropology* 35: 59–74.

Tanasi, D. 2010. Bridging the Gap. New data on the relationship between Sicily, the Maltese Archipelago and the Aegean in the Middle Bronze Age. *Mare Internum - Archeologia e Culture del Mediterraneo* 2: 103–112.

Tanasi, D. and Veca, C. 2019. Incontri e mobilità nel Mediterraneo preistorico: Le necropoli siciliane di Cozzo del Pantano e Matrensa. British Archaeological Reports International Series 2950. Oxford: British Archaeological Reports.

Thierry, S., Dick, S., George, S., Benoit, L. and Cyrille, P. 2019. EMODnet Bathymetry: a compilation of bathymetric data in the European waters. *OCEANS 2019-Marseille*: 1–7.

Tierney, J.E., Pausata, F.S.R. and deMenocal, P.B. 2017. Rainfall regimes of the Green Sahara. *Scientific Advancements* 3(1): e1601503.

Tiné, V. and Tusa, S. 2012. Il Neolitico in Sicilia, in R. Giglio and S. Tusa (eds) *Atti della XLI Riunione Scientifica 'Dai Ciclopi agli Ecisti : Società e territorio nella Sicilia preistorica e protostorica'* (San Cipirello, PA, 16–19 novembre 2006): 49–80. Firenze: Istituto Italiano di Preistoria e Protostoria.

Tinner, W., van Leeuwen, J.F.N., Colombaroli, D., Vescovi, E., van der Knaap, W.O., Henne, P.D., Pasta, S., D'Angelo, S. and La Mantia, T. 2009. Holocene environmental and climatic changes at Gorgo Basso, a coastal lake in southern Sicily, Italy. *Quaternary Science Reviews* 28(15–16): 1498–1510.

Tinner, W., Vescovi, E., van Leeuwen J.F.N., Colombaroli, D., Henne, P.D., Kaltenrieder, P., Morales-Molino, C., Beffa, G., Gnaegi, B., van der Knaap, W.O., La Mantia,

T. and Pasta, S. 2016. Holocene vegetation and fire history of the mountains of Northern Sicily (Italy). *Vegetation History and Archaeobotany* 25(5): 499–519.

Toucanne, S., Angue Minto'o, C.M., Fontanier, C., Bassetti, M.A., Jorry, S.J. and Jouet, G. 2015. Tracking rainfall in the northern Mediterranean borderlands during sapropel deposition. *Quaternary Science Reviews* 129: 178–195.

Trasselli, C. 1955. Produzione e commercio dello zucchero in Sicilia dal XII al XIX secolo. *Economia e Storia* 2: 325–342.

Tremetsberger, K., Ortiz, M.Á., Terrab, A., Balao, F., Casimiro-Soriguer, R., Talavera, M. and Talavera, S. 2016. Phylogeography above the species level for perennial species in a composite genus. *AoB PLANTS* 8: plv142.

Troia, A., Raimondo, F.M. and Geraci, A. 2012. Does genetic population structure of Ambrosina bassii L. (Araceae, Ambrosineae) attest a post-Messinian land-bridge between Sicily and Africa? *Flora* 207: 646–653.

Tusa, S. 1999. *La Sicilia nella preistoria* (2nd edn). Palermo: Sellerio editore.

Tykot, R.H. 2017. Obsidian studies in the prehistoric Central Mediterranean: After 50 years, what have we learned and what still needs to be done? *Open Archaeology* 3(1): 264–278.

Tykot, R.H. 2019. Geological sources of obsidian on Lipari and artifact production and distribution in the Neolithic and Bronze Age Central Mediterranean. *Open Archaeology* 5(1): 83–105.

Vacante S. 2016. Wetlands and environment in Hellenistic Sicily: Historical and ecological remarks. *Ancient History Bulletin* 30(1-2): 27–42.

Vacchi, M., Ghilardi, M., Melis, R.T., Spada, G., Giaime, M., Marriner, N., Lorscheid, T., Mohrange, C., Burjachs F. and Rovere, A. 2018. New relative sea-level insights into the isostatic history of the Western Mediterranean. *Quaternary Science Reviews* 201: 396–408.

Vacchi, M., Marriner, N., Morhange, C., Spada, G., Fontana, A. and Rovere, A. 2016. Multiproxy assessment of Holocene relative sea-level changes in the western Mediterranean: sea-level variability and improvements in the definition of the isostatic signal. *Earth-Science Reviews* 155: 172–197.

Van de Loosdrecht, M.S., Mannino, M.A., Talamo, S., Villalba-Mouco, V., Posth, C., Aron, F., Brandt, G., Burri, M., Freund, C., Radzeviciute, R., Stahl, R., Wissgott, A., Klausnitzer, L., Nagel, S., Meyer, M., Tagliacozzo, A., Piperno, M., Tusa, T., Collina, C., Schimmenti, V., Di Salvo, R., Prüfer, K., Hublin, J.-J., Schiffels, S., Jeong, C., Haak, W. and Krause, J. 2020. Genomic and dietary transitions during the Mesolithic and Early Neolithic in Sicily. *bioRxiv preprint*; doi.org/10.1101/2020.03.11.986158.

Van Dommelen, P. and Knapp, B. 2014. *The Cambridge Prehistory of the Bronze and Iron Age Mediterranean*. Cambridge: Cambridge University Press.

Vannière, B., Power, M.J., Roberts, C.N., Tinner, W., Carrión, J., Magny, M., Bartlein, P., Colombaroli, D., Daniau, A.L., Finsinger, W., Gil-Romera, G., Kaltenrieder, P., Pini, R., Sadori, L., Turner, R., Valsecchi, V. and Vescovi, E. 2011. Circum-Mediterranean fire activity and climate changes during the mid-Holocene environmental transition (8500–2500 cal. BP). *The Holocene* 21(1): 53–73.

Vanzetti, A., Castangia, G., de Palmas, A., Ialongo, N., Leonelli, V., Perra, M. and Usai, A. 2013. Complessi fortificati della Sardegna e delle isole del Mediterraneo occidentale nella protostoria, in G. Bartoloni and L.M. Michetti (eds) *Mura di legno, mura di terra, mura di pietra: fortificazioni nel Mediterraneo antico. Scienze dell'Antichità* 19(2–3): 83–124.

Vendramin, G.G., Anzidei, M., Madaghiele, A. and Bucci, G. 1998. Distribution of genetic diversity in Pinus pinaster Ait. as revealed by chloroplast microsatellites. *Theoretical and Applied Genetics* 97(3): 456–463.

Verschuuren, B., Wild, R., McNeeley, J. and Oviedo, G. 2010. *Sacred natural sites: Conserving nature and culture*. London: Earthscan.

Vettori, C., Vendramin, G.G., Anzidei, M., Pastorelli, R., Paffetti, D. and Giannini, R. 2004. Geographic distribution of chloroplast variation in Italian populations of beech (Fagus sylvatica L.). *Theoretical and Applied Genetics* 109: 1–9.

Villari, P. 1995. *Le faune della tarda preistoria nella Sicilia orientale*. Siracusa: Ente Fauna Siciliana.

Wagner, B., Vogel, H., Francke, A., Friedrich, T., Donders, T., Lacey, J.H., Leng, M.J., Regattieri, E., Sadori, L., Wilke, T., Zanchetta, G., Albrecht, C., Bertini, A., Combourieu-Nebout, N., Cvetkoska, A., Giaccio, B., Grazhdani, A., Hauffe, T., Holtvoeth, J., Joannin, S., Jovanovska, E., Just, J., Kouli, K., Kousis, I., Koutsodendris, A., Krastel, S., Lagos, M., Leicher, N., Levkov, Z., Lindhorst, K., Masi, A., Melles, M., Mercuri, A.M., Nomade, S., Nowaczyk, N., Panagiotopoulos, K., Peyron, O., Reed, J.M., Sagnotti, L., Sinopoli, G., Stelbrink, B., Sulpizio, R., Timmermann, A., Tofilovska, S., Torri, P., Wagner-Cremer, F., Wonik, T. and Zhang, X. 2019. Mediterranean winter rainfall in phase with African monsoons during the past 1.36 million years. *Nature* 573: 256–260.

Walker, M., Head, M.J., Berkelhammer, M., Björck, S., Cheng, H., Fisher, D., Gkinis, V., Long, A., Newnham, R., Olander Rasmussen, S. and Weiss, H. 2019. Subdividing the Holocene Series/Epoch: Formalization of stages/ages and subseries/subepochs, and designation of GSSPs and auxiliary stratotypes. *Journal of Quaternary Sciences* 34(3): 173–186.

Wang, P.X., Wang, B., Cheng, H., Fasullo, J., Guo, Z.T., Kiefer, T. and Liu, Z.Y. 2014. The global monsoon across timescales: Coherent variability of regional monsoons. *Climate of the Past* 10: 2007–2052.

Weiss, H. 2016. Global megadrought, societal collapse and resilience at 4.2–3.9 ka BP across the

Mediterranean and west Asia. *PAGES Magazine* 24(2): 62–63.

Weiss, S. and Ferrand, N. (eds) 2007. *Phylogeography of Southern European Refugia: Evolutionary perspectives on the origins and conservation of European biodiversity.* Dordrecht: Springer.

Zanchetta, G., Bini, M., Cremaschi, M., Magny, M. and Sadori, L. 2013. The transition from natural to anthropogenic-dominated environmental change in Italy and the surrounding regions since the Neolithic. *Quaternary International* 303: 1–9.

Zanchetta, G., Borghini, A., Fallick, A.E., Bonadonna, F.P. and Leone, G. 2007. Late Quaternary palaeohydrology of Lake Pergusa (Sicily, southern Italy) as inferred by stable isotopes of lacustrine carbonates. *Journal of Palaeolimnology* 38: 227–239.

Zanon, M., Davis, B.A.S., Marquer, L., Brewer, S. and Kaplan, J.O. 2018. European forest cover during the past 12,000 years: A palynological reconstruction based on modern analogs and remote sensing. *Frontiers in Plant Science* 9: 253.

Zohary, D., Hopf, M. and Weiss, E. 2012. *Domestication of Plants in the Old World: The origin and spread of domesticated plants in Southwest Asia, Europe, and the Mediterranean Basin* (4th edn). Oxford: Oxford University Press.

Analyse historique des variations du débit provoqué par les séismes pendant les siècles XVe–XXe: le cas de Termini Imerese (Sicile centro-septentrionale)

Patrizia Bova[†,2] Antonio Contino,[1] and Giuseppe Esposito[2]

[1] 1 Università degli studi di Palermo - Dipartimento di Scienze della Terra e del Mare (DISTEM).
[2] Accademia Mediterranea Euracea di Scienze, Lettere e Arti (AMESLA) di Termini Imerese (Palermo).

Abstract: This study was carried out in northern-central Sicily, especially in the eastern sector of Trabia-Termini Imerese Mounts (1326 m a.s.l. at Mount S. Calogero). These reliefs, an important segment of the Sicilian Maghrebid chain, consist of built tectonic units, which originate from the deformation of two distinct Meso-Cenozoic paleogeographic domains (so-called Imerese and Sicilide), emplaced during the Neogene. The Imerese Units derive from the deformation of a carbonate and carbonate-siliciclastic marine deep-water succession (basin-slope, from the Middle Triassic to the Oligocene, thickness c. 1200–1500 m). The Sicilidi Units, tectonically overlying the terrigenous covers of the Imerese (Numidian Flysch, Upper Oligocene-Lower Miocene), derive from the deformation of a marly clays and marly limestones basinal marine succession (Upper Cretaceous-Lower Oligocene). The main aquifers of Trabia-Termini Imerese Mounts are hosted in the calcareous-dolomitic (Scillato and Fanusi fms.) and calcareous (Crisanti fm.) rocks of the Imerese succession. The structural edifice is unconformably overlain by a syntectonic terrigenous, evaporitic and clastic-carbonate deposits (Serravallian–Pliocene age).

The most recent deposits regard the marine and alluvial terraces (Pleistocene) and debris cover, colluvia, alluvia, sand beach (Holocene). The tectonic thrusts (Middle-Upper Miocene), currently SW-dipping, consisting in a NE–SW-trending maximum compression. The strike-slip and/or transpressional fault system, consisting in a NNW–SSE and NE–SW N–S-trending (Miocene Pliocene interval). The extensive tectonic (Plio-Quaternaire) generated two faults system, respectively oriented E–W and N–S. Termini Imerese, the ancient and flourishing Augustan colony of Thermae Himerenses, is a picturesque town, situated on the Tyrrhenian coast, c. 30 km from Palermo. Since the Roman domination, the town was supplied through an aqueduct, an impressive water engineering project (two lines, c. 7.1 and 3 km long), that conveyed large amounts of potable water from the two Brocato-Fridda and Favara-Scamaccio spring groups (the latter bearing warm waters with low enthalpy). In the late Medieval period, the ancient aqueduct had to be restored in some parts, following structural decline, wars, and natural disasters. During the droughts of the late 15th/early 16th century, the ancient aqueduct gave indications of reduced water flow, and this forced the local civil administration to strengthen the water supply from the Favara-Scamaccio springs (which then exhibited flow rates of up to 60 l/s). Finally, in 1525, water poured again from the monumental fountain in the main square, which terminated the new aqueduct. Owing to the very serious earthquake of 1693 (which struck the eastern sector of Sicily), the Favara-Scamaccio springs suffered drastic decreases in flow and marked oscillations, with disappearances/reappearances of water that worsened in dry periods. This seismic-triggered vulnerability was attested also in the centuries that followed (1726, 1823, 1906, 1907, 1908, 1968), until the Favara-Scamaccio springs totally dried up.

The numerous data presented provide a detailed reconstruction and complete picture, since the 15th century, of the history of the public water supply of Termini Imerese, through the aqueduct of Favara-Scamaccio. The research discussed in this paper, through an integrated methodological approach, combines the findings of geological research and of an historical analysis of collected documentary records, paving the way for further investigations (e.g. seismic-triggered hydrogeological vulnerability modelling, future management, protection and planning of underground water resources, new geothermal investigations).

Keywords: seismic hydrogeology, seismic vulnerability, multi-layered aquifers, documentary data, Termini Imerese, Sicily

Introduction

Dans l'Antiquité, le phénomène de diminution/augmentation du débit ou même la disparition/réapparition des sources, lié aux séismes, était phenomène connu. Un des pionniers de ces études était le géographe grec Démétrios de Callati (2e moitié du 3e siècle avant J.C.), auteur d'un ouvrage perdu de 20 livres sur l'Europe et l'Asie, utilisée aussi par Strabon[1].

Après deux millénaires, au XIXe siècle, le célèbre géologue écossais Charles Lyell (Kinnordy Kirrimuir, Forfarshire 1797, Londres 1875) a également souligné ces phénomènes, à titre d'exemple, citant le séisme sicilien de 1693 qui a détruit plusieurs villes et bourgs en Sicile oriental[2].

Dans la littérature géologique du XIXe et du début du XXe siècle, plusieurs études traitent des relations entre les séismes et les variations des débits des sources, des puits et des cours d'eau[3]. Ces études ont mis en

---
[1] Strabon éd. 1969, I, 3, 29.
[2] Lyell 1832–1833 II: 159–160.
[3] Adinolfi Falcone *et al.* 2012; Amoruso *et al.* 2011; De Luca *et al.* 2016; Ingebritsen et Manga 2019; La Vigna *et al.* 2012; Manga et Rowland 2009; Manga et Wang 2015; Muir-Wood et King 1993; Xue *et al.* 2013; Zhang *et al.* 2019.

évidence les problèmes liés à la vulnérabilité des aquifères pendant les séismes. Ces derniers peuvent en effet influencer les équilibres hydrodynamiques des aquifères et, conséquemment, l'alimentation et le débit des sources. Dans la littérature internationale, l'hydrologie/hydrogéologie sismique (*seismic hydrology/hydrogeology*) est une nouvelle discipline hydrogéologique qui traite, respectivement, l'étude des interactions entre séismes et systèmes hydriques de surface et souterrains.

Les caractéristiques reliefs Mésozoïques calcaréo-dolomitiques et calcaréo-siliciclastiques des Monts de Termini Imerese-Trabia constituent un laboratoire naturel pour l'étude des ressources d'eau souterraine dans ce secteur de la Sicile.

Les caractéristiques hydrogéologiques de cette zone sont strictement contrôlées par les lithologies et par les facteurs structurels et géomorphologiques (surtout le processus karstiques).

Le cas d'étude des sources du groupe Favara-Scamaccio (dorénavant GFS), au moins à notre connaissance, constitue une unicité parce que nous permet d'avoir des mesures de débit dès la fin du XVIIe siècle, selon les systèmes employées en Sicile pendant cette époque. De plus, le GFS est particulièrement significatif parce que aux oscillations ordinaires du débit des sources, liées au régime pluviométrique, se superposent des augmentations ou diminutions soudaines (jusqu'à zéro) liées aux événements météo-climatiques extrêmes et aussi aux séismes. Une complication ultérieure dérive de la complexe interaction entre les aquifères plus superficiels, froids, et ceux plus profonds caractérisés par un thermalisme à basse enthalpie.

Cette étude, basée sur une approche interdisciplinaire/multidisciplinaire, menée dans un secteur clé des monts susmentionnées, s'insère dans le cadre d'une recherche qui vise à analyser les interactions entre séismes et systèmes aquifères calcaréo-dolomitiques et calcaréo-siliciclastiques fissurés et karstifiées, superficiels et profonds. Les résultats préliminaires de ces recherches sont présentés ci-dessous.

Zone d'étude

La zone étudiée constitue le secteur oriental des Monts de Termini Imerese-Trabia (Sicile centro-septentrionale). Ces montagnes, un segment de la chaîne du Maghreb sicilien, sont constituées des unités tectoniques, généralement transportées vers le Sud, formées pendant le Néogène (Figure 1). La zone étudiée a une superficie approximativement de 115 km² et une altitude moyenne de 600 m s.l.m., atteignant son élévation maximale à Monte S. Calogero ou Euraco (1326 m d'altitude). La ville plus importante est Termini Imerese (environ 30.000 habitants), partagée en deux secteurs[4]: une inférieure, sur la plaine côtière et le pente toute proche, et l'autre supérieure, sur les hautes terrasses (environ 70 m d'altitude).

Géologie et tectonique

La structure principale des monts de Termini Imerese-Trabia se compose de deux séquences méso-cénozoïques tectoniquement superposées, respectivement constituées par des dépôts appartenant à deux domaines paléogéographiques: le domaine Imerese (Trias moyen-Oligocène, épaisseur environ 1200–1500 m) et le domaine Sicilide (Crétacé supérieur-Oligocène inférieur, épaisseur environ 50–70 m), localement couvert en discordances par des dépôts synthéctoniques déformés du Mio-Pliocène[5].

Dans la zone d'étude, les unités Imeresi dérivent de la déformation d'une succession de bassin-escarpement calcaréo-dolomitiques et/ou calcaréo-siliciclastiques d'âge méso-cénozoïque (domaine Imerese).

La succession Imerese *sensu strictu*, de bas en haut, se compose de: marnes et calcaires (Formation Mufara, Trias moyen), calcaires marneux à silex (Formation Scillato, Trias supérieur), dolomies et brèches dolomitique (Formation Fanusi, Lias inférieur), radiolarites, argilites siliceuses avec intercalations calcarénitiques et/or calciruditiques resédimentés, constitués par des éléments clastiques de mer peu profonde (Formation Crisanti, Lias moyen-Crétacé supérieur), marnes et calcaires marneux (Formation Caltavuturo, Eocène-Oligocène inférieur). La couverture argilo-gréseuse et gréso-argileuse de cette succession (Flysch numidien, Oligocène supérieur-Miocène inférieur) est souvent décollée de son substrat.

Les unités Sicilidi, structurellement au-dessus du Flysch numidien, sont constituées de lambeaux d'une succession marine de bassin, argilo-marneuse et marno-calcaire, du Crétacé-Éocène.

Dans le secteur Sud de la zone étudiée, les affleurements du Mio-Pliocène, de bas en haut, présentent: dépôts marins marneux ou marno-sableux (Formation Castellana, Serravallien supérieur-Tortonien inférieur), conglomératiques de faciès continental et gréso-argileux marins (Formation Terravecchia, Tortonien-Messinien supérieur), évaporitiques (Groupe gypseux-solfifère, Messinian) et au sommet franchement marins, marno-calcaires (Trubi, Zancleano). Dans les alentours proche de la côte, des dépôts marins terrassés et continentaux du Quaternaire (éboulis, éboulis ordonnées, dépôts colluviales, alluviales terrassé etc.)

---
[4] Contino 2019: 19–20.
[5] Catalano *et al.* 2004.

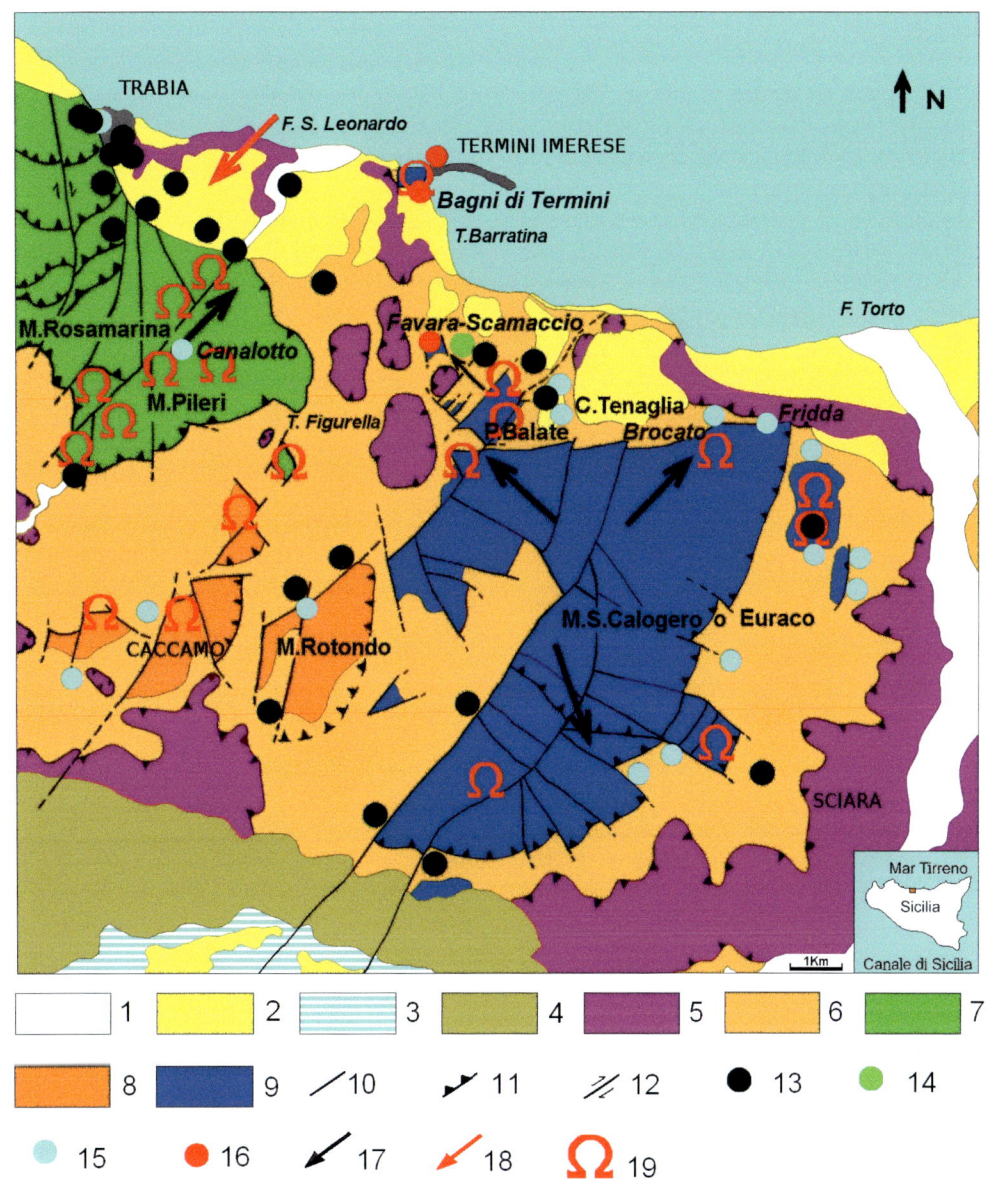

Figure 1: Modèle hydrostructural de la zone d'étude (modifié et révisé d'après Contino, 2004 et 2007).
1. Alluvions et dépôts littoraux 2. Dépôts carbonatés et clastiques (Plio-Pléistocène); 3. Evaporites (Miocène, Messinien); 4. Successions clastiques-terrigènes (Miocène, Serravallien-Tortonien); 5. Unités Sicilidi (Crétacé supérieur-Oligocène); 6. Unité du Flysch numidien (Oligocène supérieur-Miocène inférieur); Unité Imeresi: 7. Unité Monte Rosamarina-Monte Pileri; 8. Unité Monte S. Onofrio-Monte Rotondo; 9. Unité Pizzo di Cane-Monte S. Calogero; 10. failles normales; 11. failles inverses et/ou chevauchements (les triangles sont du côté chevauchant); 12. Failles à composantes horizontales (les flèches indiquent la direction du mouvement); 13. Sondages hydrologiques et géologiques; 14. Sondage VIGOR 'Termini Imerese'; 15. Sources froides; 16. Sources thermales; 17. Direction du flux des eaux souterraines; 18. Zones affectées par intrusion marine; 19. Cavités karstiques.

recouvrent en discordance directement la succession méso-cénozoïque.

La tectonique de compression (Miocène moyen-supérieur) a généré des redoublements internes dans les unités Imeresi[6] avec une direction de compression maximale orientée NE–SO. La subséquente tectonique de transpression (du Mio-Pliocène, NNO–SSE pour les décrochements droits et NE–SO pour les systèmes antithétiques gauches), a produit un raccourcissement supplémentaire dans les unités Imeresi et Sicilidi. La tectonique cassante (Plio-Quaternaire) a généré des failles à direction E–O coupées de celles N–S, plus récentes.[7]

---

[6] Contino et al. 2004; Catalano et al. 2011.

[7] Gennaro et al. 2012.

Géomorphologie

Le schéma géomorphologique des Monts de Termini Imerese-Trabia présente deux paysages différents, respectivement calcaréo-dolomitiques et/ou calcaréo-siliciclastiques et argilo-marneux. Le premier présente des pentes abruptes et de nombreuses morphologies karstiques et de morphosélection, le second est caractérisé par des pentes irrégulières et faiblement inclinées, principalement affectées par des glissements de terrain et, localement, par l'érosion accélérée.

L'alternance des lithologies à différents degrés d'érodibilité et les structures tectoniques sont responsables du développement des morphologies structurales, particulièrement évidentes dans les pentes NE et SO des Monts de Termini Imerese-Trabia. Les escarpements de failles, orientés NO–SE, dont la hauteur a été amplifiée par l'érosion sélective, dominent le paysage, en particulier du côté tyrrhénien.

Les couvertures numidiennes ont été progressivement érodés dans les reliefs montueux, permettant l'exhumation des roches calcaréo-dolomitiques. Sur la base de l'existence des notables lacunes stratigraphiques, qui ont déterminé le dépôt des marnes du Pliocène inférieur (Trubi) sur les terrains mésozoïques, on a émis l'hypothèse que ce processus ait commencé pendant la phase tectonique du Messinien[8]. Pendant le Pliocène supérieur (?)-Pléistocène inférieur, les phases d'émersion ont été vraisemblablement accentuées par le *uplift* de ce secteur de la chaîne. Les oscillations eustatiques du niveau de la mer au cours du Quaternaire, les processus intenses de dégradation météorique (en particulier pendant le Würmienne) et la remontée post-glaciaire ont certainement influcié l'évolution du réseau karstique et l'érosion des terrains calcaréo-dolomitiques et calcaréo-siliciclastique.[9]

Les campagnes de terrain et l'interprétation aéro-photogrammétrique montrent que le karst s'est développé à petite et grande échelle, produisant des formes superficielles et souterraines. Les facteurs structurels contrôlent significativement l'évolution des formes de dissolution. Les formes épigées reconnues sont principalement des karren, tandis que celles hypogées sont des grottes, disposées le long des couches ou des fractures et généralement peu développé en horizontal. La plupart des formes karstiques superficielles et profondes se trouvent principalement dans les roches calcaréo-dolomitiques (formations Scillato et Fanusi), fissurées et/ou karstifiées. Certaines grottes sont situées chez un horizon marneux imperméable, juste au passage entre les deux formations, niveau de base local du karst[10]. Dans la supérieure formation Crisanti, les corps carbonatés ré-sédimentés dans la séquence du bassin, montrent souvent à la base, en contact avec les niveaux siliclastiques relativement imperméables sous-jacents, des formes karstiques épigées et hypogées. Les cavités karstiques sont donc liées au réseau des discontinuités tectoniques et aux limites imperméables (horizons calcaires-marneux ou siliceux-marneux). Le réseau des fractures ouvertes et des conduits karstiques joue un rôle hydrogéologique important: favoriser le développement du karst souterrain et conditionner l'écoulement des eaux souterraines notamment dans les roches calcaires et calcaréo-dolomitiques.

La grande élévation morphostructural du Mont S. Calogero[11], correspond au sommet d'un grand anticlinorium, dont les structures anticlinales sont encore largement reconnaissables (S. Calogero, Balate, Favara et Termini Imerese), tranchées par des failles, à la fois inverses et normales, notamment dans le flanc N–O, qui ont donné lieu à une succession de hauts et de bas morphostructuraux.

Hydrogéologie

Les roches calcaréo-dolomitiques (complexe calcaréo-dolomitique, formations Scillato et Fanusi) et calcaires (complexe calcaire, formation Crisanti) de la succession Imerese, milieu perméable fissuré et/ou karstifié, abritent les principaux aquifères des monts de Trabia-Termini Imerese. La grande éminence morphostructurale calcaire-dolomitique du Mont S. Calogero est la principale zone de recharge des aquifères du territoire étudié.

Quatre unités hydrostructurales Imeresi ont été reconnues dans cette région:[12] *Monte Cane-Monte S. Calogero; Monte S. Onofrio-Monte Rotondo, Monte Rosamarina-Monte Pileri* et *Capo Grosso-Torre Colonna*. Dans la zone d'étude, sont visibles seulement les parties orientales des trois premières unités hydrostructurales (hydrostructures de S. Calogero, Rotondo et Rosamarina-Pileri).

La circulation de l'eau souterraine est conditionnée par les caractéristiques hydrostratigraphiques de la succession Imerese, par les géométries structurales (notamment les *ramp-flat* et les discontinuités tectoniques) et par la présence de formes karstiques.[13]

Les terrains des unités Sicilidi (recouvrant tectoniquement les Imeresi) jouent généralement un rôle hydrogéologique modeste pour plusieurs raisons :

---

[8] Contino *et al.* 2015.
[9] Contino *et al.* 2007.
[10] Contino *et al.* 2007.
[11] Di Maggio *et al.* 2017; Cappadonia *et al.* 2019.
[12] Contino *et al.* 2004.
[13] Contino *et al.* 2007.

ils ont une extension réduite ; ils sont disjoints en klippes (ou îlots) ; et ils sont constitués d'horizons argilo-siliceux et calcaires-marneux.

Certaines *horsts* (Favara et Termini Imerese) abritent des manifestations hydrothermales superficielles à basse enthalpie, généralement liées à des systèmes de failles qui jouent le rôle des voies préférentielles pour la montée de l'eau chaude, tandis que dans les *grabens*, les fluides thermales présents dans les réservoirs carbonatés profonds sont scellés par les couvertures argileux imperméables du Flysch numidique.[14] De plus, sur la côte de Termini Imerese, sont connus des flux d'eaux chaudes sous-marines, particulièrement riches en chlorures.[15]

Caractéristiques climatiques et hydrologiques

La zone étudié présente un climat méditerranéen typique avec des étés longues, chaudes et secs (en raison du Sirocco) et des hivers courts, parfois très froids. La température moyenne du mois le plus chaud est supérieure à 22 °C, tandis que les moyennes hivernales sont supérieures à 5 °C mais le minimum descend au-dessous de 0 °C sur les montagnes. Le régime des précipitations atmosphériques se caractérise par une concentration des pluies en particulier au semestre automne-hiver et par une raréfaction au semestre printemps-été.[16]

Le bilan hydrique moyen pour la période 1982–1996 a été établi pour évaluer l'infiltration annuelle moyenne[17]. À cette fin, les auteurs ont utilisé les données de précipitations (P) et de température (T) pour les suivantes stations météorologiques représentatives:

- stations pluviométriques de Casteldaccia, Misilmeri, Monumentale et Caccamo;

- stations thermométriques de Ciminna et Cefalù.

Les précipitations annuelles moyennes sont de 744 mm, la température moyenne de 15,6 °C. Pour le calcul de l'évaporation potentielle (ETP), ils ont utilisé la formule de Thornthwaite[18]. Le bilan hydrique mensuel au sol, pour une capacité de champ de 20 mm, a permis d'estimer la valeur mensuelle de l'évapotranspiration réelle (ETR) égale à 366 mm. Étant admis que la zone est principalement drainée par le fleuve S. Leonardo, le ruissellement de surface a été évalué à l'aide des données (période 1982–1994) de la station hydrométrique *Monumentale* (pont monumental sur le S. Leonardo). Les valeurs moyennes de ruissellement de surface (Ds) représentent le 10% des précipitations annuelles (P). Sur la base des calculs de bilan, il en résulte 303 mm (pour Ds = 10% P), ce qui correspond à environ 1450 l/s.

Le groupe Favara-Scamaccio et son aqueduc

À l'époque romaine, *Thermae Himerenses* était alimentée par un imposant aqueduc, une œuvre vraiment de haute ingénierie (nommée *Aqua Cornelia*, longueur: 7,1 km), qui captait les sources de Brocato et Fridda, situées sur les pentes N, calcaire-dolomitiques, du S. Calogero et, vraisemblablement, aussi celles du GFS (au SE de l'actuel Termini Imerese). Ces ouvrages, à travers un complexe système de décanteurs, cuves, canaux, puits de ventilation, ponts, tuyaux (dans certaines sections sous pression) et tunnels, ont traversé plusieurs torrents, conduisant les eaux potables vers la ville antique.

Une deuxième ligne d'alimentation (longueur: 3 km), remontait la vallée de la Barratina, traversant les contrées Figurella et Mazzarino vers la ville antique. Selon Belvedere[19], il n'est pas clair si cet autre aqueduc était une œuvre autonome utilisant le GFS, ou réalisé en remplacement partiel du précédent, pour remédier au dysfonctionnement d'un secteur de l'*Aqua Cornelia*.

À défaut de preuve, nous pouvons seulement avancer des hypothèses sur la continuité d'utilisation de l'ancien aqueduc entre la décadence de l'Empire romain et le bas moyen-âge. Pendant les siècles, les antiques œuvres hydrauliques ont souffert peut-être la détérioration structural, le ruine produite par les calamités naturels et les invasions.

Dans la première moitié du XIVe siècle, les Angevins, pendant lors siège destructeur de Termini (qui a duré deux mois du 19 juin au 27 août 1338[20]), coupèrent les conduites des eaux courantes (*vivis aquis*) pour contraindre les habitants à la soif. À l'état actuel des connaissances, il n'est pas possible d'attribuer à cet événement de guerre, ni à des épisodes sismiques ou de glissements de terrain, la reconstruction des arcs (en ogive gothiques) de l'antique aqueduc dans la contrée sous Tenaglia.[21]

Un acte perdu du notaire Tommaso La Viola de Termini, daté Octobre 1496, rappelle l'existence de la contrée *canalium* (aujoud'hui *Piazza S. Antonio* et rue *Falcone-Borsellino*), c'est-à-dire des canalisations du vieux aqueduc.[22] L'*aquaeductus* est aussi mentionnée, en passant, par Arezzo.[23]

---

[14] Abate *et al.* 2015; Contino 2019: 139.
[15] Ciofalo 1924: 11.
[16] Contino 2019: 127.
[17] Contino *et al.* 2004; Contino *et al.* 2007.
[18] Contino *et al.* 2007.
[19] Belvedere 1986: 79.
[20] *Michaelis Platensis, Historia Sicula* dans Gregorio 1791–92 I: 545–546.
[21] Romano 1838: 157, note no. 35.
[22] De Michele (éd.) 1950: 29.
[23] Arezzo 1537: 27.

Figure 2: Contrada Favara. Calcaires à Rudistes du Crétacé supérieur (photo: Carmelo Gennaro).

Avant de détailler l'histoire de l'aqueduc d'époque moderne du GFS, nous voulons fournir un encadrement physiographique, géologique-structural, hydrogéologique et hydronymique des émergences.

Le GFS, constitué d'une source principale (Favara) et d'une secondaire (Scamaccio), depuis le XVIe siècle il s'appelait *Sette Zappe d'Acqua* (hydronyme dérivant d'une mesure de débit, dont sept *zappe* correspond à peu près 60 l/s). Le GFS, actuellement inactif, était situé à 100 m d'altitude environ, entre les pentes nord-ouest du S. Calogero et la ville de Termini Imerese, et ses eaux étaient hypothermales (T = 20°–24° C) et potables, probablement suite au mélange entre eaux froides (alimentées par l'aire d'infiltration du S. Calogero) et thermales (alimentées par les aquifères profonds). Les eaux jaillissaient du flanc nord de la structure anticlinale faillée de Favara, entourée des argiles numidiques. Cette structure, avec plongement périclinalique, est constituée spécialement par des calcaires du Crétacé supérieur (membre des brèches à rudistes, Formation Crisanti, Figure 2) et bordée sur trois côtés par deux systèmes de failles normales (les plus anciennes WSW–ENE et le plus récents N–S). En septentrion la structure est taillé par des petites failles inverses de direction E–W, en accord avec les données de la prospection sismique.[24]

Favara est un hydronyme islamique: *fawwāra* '(source qui) fait des bulles ou gicle', 'jet d'eau' (cf. *fawwār*, 'gargouillement', effervescent'); il s'agissait donc d'une source liée aux aquifères au moins partiellement confinés et cela s'accorde également avec les événements ultérieurs, tels que le régime d'écoulement, avec une notable réduction de débit en conséquence des séismes, soulignée par des disparitions/réapparitions répétées.[25] Le hydronyme Scamaccio, cependant, est sicilien: *Scamàcciu* 'trop-plein', 'déversoir'.

Vers le XVe siècle, la Favara s'appelait aussi *Fontana d'Incapo* (du sicilien *'ncapu*, 'supérieur'). Les actes des XVIe-XVIIe siècles, la désignent *Fontana Superiore* ou *Fontana Vecchia*.[26]

Pendant la grande sécheresse qui a affecté l'île entre la seconde moitié du XVe siècle et le milieu du XVIe siècle de notre ère,[27] l'ancien aqueduc a montré des problèmes de débit, probablement liés à une diminution des eaux fournies par Brucato-Fridda. Au début du XVIe siècle, les *giurati* (administrateurs municipaux) de Termini Imerese, pour intégrer la diminution d'alimentation, ont décidé d'améliorer les œuvres hydraulique du GFS, créant une nouvelle prise d'eau avec relative

---

[24] Abate *et al.* 2015: 94.

[25] Contino *et al.* 2013.
[26] Contino 2019: 23.
[27] Trasselli 1982.

canalisation (qui, au moins en partie, reprendrait le chemin du pont romain de Figurella ou *archi vecchi*, c'est à dire l'antique aqueduc). Le nouveau aqueduc, en particulier dans la section de la contrée Mazzarino (encore partiellement existante), a maintenu toujours sa valeur monumentale, devenant une caractéristique typique du paysage (Figures 3, 4).

Au parlement général du 10 août Xe indiction 1508, tenue à Palerme dans la grande salle du *Steri* (du latin *osterium* 'palais', aujourd'hui siège du recteur de l'Université locale), l'ambassadeur de la ville de Termini a présenté une supplique au vice-roi Raimondo Cardona, concernant diverses arguments, parmi lesquelles de donner la priorité au nouvel aqueduc (selon une

Figure 3 : Ruines de l'aqueduc de Favara (XVIe siècle) et de la tour romaine de décompression, Photo Michele Salvo, XXe siècle (courtoisie de Michele et Francesco Ciofalo).

Figure 4: Ruines de l'aqueduc de Favara (XVIe siècle) dans la contrée Mazzarino (aujourd'hui rue Falcone Borsellino), Photo Michele Salvo, XXe siècle (courtoisie de Michele et Francesco Ciofalo).

précédente inconnue autorisation). L'aqueduc était financé avec les recettes de trois *denari* (000,3 lires) pour chaque *salma* (2,74 hectolitres) et *cantàro* (80 kg) des provisions extraites du *Caricatore* (complexe des entrepôts pour le stockage temporaire des fournitures avant l'embarquement) et qui a été enregistré quatre jours plus tard[28]. La pétition (qui expressément déclare une réduction du débit affluent dans l'aqueduc vieux) a été réitérée le 4 février XIVa Indiction 1511, peut-être en prévision du parlement successif du 13 août.

En 1511–12, étant l'*universitas* (c'est-à-dire la municipalité) sans fonds, les possédant étaient obligé par les *giurati* à avancer l'argent pour les nouveaux ouvrages hydrauliques.[29]

Une fois réalisé, le nouvel aqueduc du GFS, était adorné dans la place principale (actuelle *Piazza Duomo*) par une fontaine monumentale avec quatre canaux en bronze, mentionnée par Maurolico,[30] Alberti,[31] Solìto[32] se souviennent. Cette fontaine publique est idéalement représentée dans le grand fresque représentant 'Sthenius qui s'oppose à la rapacité du propréteur *Caius Cornelius Verres*', parti e d'un cycle imposant (1609–10) qui orne la chambre du magistrat dans l'édifice municipal local, œuvre de Vincenzo La Barbera (Termini Imerese, 1576/1577-Palerme, 1642), peintre maniériste et architecte.[33]

L'inscription en distiques, placée sur ce monument, racontait de 305 lustres (pendent la naissance du Rédempteur), c'est-à-dire de l'année 1525, date d'accomplissement (Battaglia 1887–90), tandis que Solìto fait remonter la construction au 1500.[34]

Les actes des *giurati* de Termini Imerese (*Acta Magnificorum Juratorum Splendidissimae et Fidelis Civitatis Thermarum*, dorénavant AMJ), œuvres manuscrites de la local Bibliothèque municipal sont la source documentaire principale que nous avons utilisé pour ces recherches.

Ce conduit a été construit sous la direction du *gouverneur de l'eau* (responsable des travailles hydrauliques), le *Magister* Angelo Amodeo, qui il a aussi aménagé et nettoyé la tour piézométrique de compression de l'*Aqua Cornelia*.[35]

Un acte du 5 octobre Xa Indiction 1551 (Anonyme XVIII siècle: 183r–188v), stipulé en faveur d'Alessandro

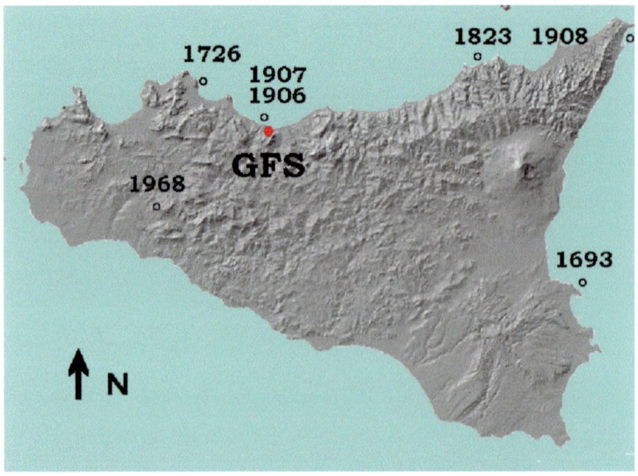

Figure 5: Emplacement des épicentres macrosismiques des tremblements de terre ayant entraîné des dommages dans le GFS.

Panzica et Francesco Russo, avec le *discretus* Giovanni Antonio Garofalo, tous de Termini, documente l'existence des ouvrages hydrauliques de la fontaine supérieure, appelée *cuba aquarum*, pour la forme carrée, avec une voûte en dôme,[36] ainsi que les sous-jacents jardins maraîchers ou d'agrément, qui utilisaient un système d'irrigation composé de citerne[37] et d'un canal de distribution pour les tours d'eau, avec les jours d'irrigation de chacun propriétaire (*vicissitudine aquarum*).

La plus ancienne attestation documentaire du thermalisme du GFS est présente dans un acte des *giurati* de Termini du 28 avril XV Indiction 1587 (AMJ 1586–1587). À cette date, les *giurati* ont envoyé aux autorités vice-royales la demande de concession des œuvres hydrauliques pour le nouvel projet de canalisation de la source de *Canalotto* (dans la vallée du S. Leonardo). L'idée était de intégrer, avec cette source, le débit de l'aqueduc municipal alimenté par le GFS, d'où sortait, comme spécifie le document, de l'eau thermale (*et est acqua calda*). Cette œuvre hydraulique était maintenue grâce à l'annone de deux *grani* (0,04 lires) sur chaque *salma* et *cantàro* des provisions extraites du *Caricatore*. Malheureusement des nouveau travaux hydraulique de restauration et de soutien étaient nécessaires pour les dommages provoqués par les récents glissements de terrains (*lavanchi*).

Entre la fin du XVIe siècle et le début du XVIIe siècle, il y a eu plusieurs années sans pluie qui ont provoqué des sécheresses.[38] En 1609, les documents rappellent une

---

[28] Musso 1760.
[29] Trasselli 1982.
[30] Maurolico 1562: 51.
[31] Alberti 1596, *Descrittione della Isola di Sicilia. Valle di Mazzara, Therme de Himera*: 44r.
[32] Solìto 1671: 99.
[33] Contino 2019: 29.
[34] Solìto 1671: 99.
[35] AMJ 1532–1533; Contino 2019: 29–30.

[36] En Sicilien *cubba* de l'arabe *qubbah*. Caracausi 1983: 195–196. *Cantàru* (arabe *qinṭār*) est une ancienne mesure de poids. Caracausi 1983: 155–156. *Dinàru* (arabe *dīnār*) est une ancienne mesure d'eau (Traina, 1868: 309).
[37] En Sicilien *gebbia* de l'arabe *ǧābiyah*, voir Caracausi 1983: 240–241.
[38] Contino et Mantia 2006: 205–206.

Figure 6: Contrada Favara. Prise d'eau du Scamaccio (photo: Antonio Contino).

diminution générale du débit des aqueducs, provoqué par la prolongée pénurie des pluies dans les temps passés, qui a provoqué le tarissement de l'aqueduc. Pendant ce temps, des sources ont été découvertes dans le périmètre urbain, sous les églises de l'hôpital (aujourd'hui musée municipal) et de *S. Vito* (aujourd'hui maison privé), c'est-à-dire dans la zone à présent situé entre les rues Cicerone, Roma et Ugdulena. Avec ces sources on pouvait fournir aisément la ville inférieure. À cet égard, les *giurati*, ont envoyé à l'autorité vice-royale une demande d'autorisation (ratifiée le 20 octobre VIII Indiction 1609) pour rechercher et collecter ces eaux (jamais effectuée), en utilisant les revenus déjà destinés aux aqueducs, c'est-à-dire deux *grani* et demi (0,05 lires) sur chaque *salma* et *cantàro* extraites du *Caricatore*.[39]

En conséquence du grand séisme du 9 janvier 1693 (Figure 5), qui a ravagé la Sicile orientale (intensité épicentrale, MCS, c'est-à-dire en degrés de l'échelle Mercalli, Cancani, Sieberg: VIII–IX; magnitude équivalente, Me: 6,1),[40] le GFS a subi une réduction d'un tiers du débit,[41] à tel point que les habitants de Termini et de la garnison du château royal souffraient la soif.[42]

Avant du séisme, les sources du GFS se trouvaient toutes les deux à la mémé altitude, et les eaux étaient collectées à l'intérieur de la ville supérieure pour être après conduites vers la ville inférieure. Après le séisme, au contraire, la Scamaccio (Figure 6) jaillissait aux pentes du relief de Favara.[43] De plus, le séisme a causé des dégâts dans les ouvrages hydrauliques sur une longueur supérieure aux 300 m, aggravées par les pluies qui ont activé des glissements de terrain. En juillet, Scamaccio a fourni 15 *denari* d'eau (2 l/s), suffisantes pour alimenter trois fontaines de la ville inférieure.

En 1701, la Favara était totalement disparue et cela s'est répété dans la sécheresse estivale de 1724. En 1724–1725, le *Supremo Tribunale del Real Patrimonio* (magistrature concernent le patrimoine domanial) a envoyé l'ingénieur hydraulique Francesco Kochler pour concevoir les remèdes appropriés.[44] Le nouveau adducteur a été réalisée sous la direction de *Mastro Diego La Cavara* (qui pour ces mérites devenait *gouverneur de l'eau*) et achevée au cours de l'année de l'indiction suivante. Le nouveau ouvrage hydraulique a été inauguré lors du 33e anniversaire du séisme, le 11

---

[39] AMJ 1609–1610: 46v–47r no. 13.
[40] Boschi *et al.* 1997.
[41] Contino *et al.* 2003; Contino *et al.* 2012.
[42] Contino 2019: 197.

[43] Gargotta 1830: 56. Jusqu'à ce jour, la prise d'eau de Scamaccio, situé au dessous de la route, est surmonté par une voûte en berceau, en Sicilien *dammùsu* (de l'arabe *dammūs*. Caracausi 1983: 203–204).
[44] Contino 2005: 157.

janvier 1726, comme proclamait une inscription murale (détruite en 1885) dans le bâtiment municipal.[45]

Le caractère fragmentaire ou lacuneux de la série des actes des *giurati* di Termini (1726–1727; 1730–1732; 1734–1735; 1739–1753; 1756–1766; 1767–1768; 1770–1788) ne permet pas une reconstruction ponctuelle des vicissitudes du GFS pendant cette époque.

Heureusement, nous avons la précieuse source documentaire d'un historien local, Giovanni Andrea Guarino (1703–1777). Guarino, en effet, témoigne que le 1er septembre 1726, la pénurie d'eau était terminée parce-que le séisme destructive de Palerme (MCS VI–VII; Me 5.7) avait provoqué une augmentation du débit de la Favara. Cette augmentation fut telle que le comte Ferrer dut travailler davantage pour améliorer les œuvres hydrauliques, afin que les eaux soient abondantes aussi bien dans la ville supérieure, y compris dans le château royal, que dans la ville inférieure.[46] Les fontaines publiques ont été rénovées, comme celle très importante de la place principale, et autres. Des abreuvoirs pour les bétails ont été construits *extra moenia* au coin des plus importantes portes de la ville (de Caccamo, de Palerme et de Messine).

Pendant l'automne 1728, malgré l'incrément de débit de la Favara, la quatrième partie de la ville était sans eau, surtout celle inférieure (spécialement les populeuses quartiers appelées *Marina*, *Bagni* et *Pescheria*) approvisionnées par la Scamaccio. Curieusement, aussi le quartier centré sur la *Strada del Cavaliere* (actuels rues Cicerone et Ugdulena), où pendant le siècle précédent on avait découvert des sources, était sans eau (selon une supplique des *giurati* au vice-roi, daté 2 Octobre VIIe Indiction 1728, cf. AMJ, 1728–1729 sans numération). Les fonds ratifiées pour la manutention de l'aqueduc, c'est-à-dire *once* 100, ont été reversés pour remédier à la pénurie de blé.

Un deuxième séisme c'est alors produit en août 1732, mais nous ne savons pas quels effets cela a eu sur le GFS.

Selon Guarino, au cours des années suivantes, le débit de la Favara a progressivement diminué.[47] Un certain Frasciana, un agriculteur au service des seigneurs Ugdulena, a eu l'idée de réserver comme eau potable pour la ville inférieure les eaux de Scamaccio qui alimentaient les sous-jacents jardins de la contrée S. Girolamo, et, à travers un petit canal, également les tanneries, jusqu'à leurs déversement dans le torrent Barratina. La nouvelle ouvrage hydraulique a été commencé en 1751 et a dû grimper sur le torrent à travers un pont. À cette occasion, au niveau de la ville inférieure, dans le quartier des *Botteghelle* (boutiques de charcutier), fut construite une fontaine avec deux canaux et nombreuses maisons ont reçu l'eau courante.

En 1755, cependant, les pluies ont provoqué une diminution d'une partie de l'adduction des eaux.[48]

En 1761, un période de nouvelle sécheresse a commencé.[49] Trois ans plus tard, il y avait encore l'assèchement de la Favara et seule la ville inférieure (dont les fontaines publiques restaient le seules en activité) restait alimentée par le Scamaccio, tandis que la partie restante, y compris le château, était sans eau.

Guarino affirme que cela se serait produit en raison de certains travaux (avec un coût de 600 *once*) exécutés pour essayer de réunir dans Scamaccio aussi les sources de la Favara, en supposant, à tort, de pouvoir atteindre débit majeur dans la ville supérieure.[50]

Une pétition des *giurati* de Termini et des *deputati dell'acqua* (fonctionnaires pour la gestion des eaux), datée du 11 octobre 1765 et un rapport du 19 novembre de la même année,[51] documentent un débit de 5-6 *pinne* (0,16–0,18 l/s) pour la Favara, juste assez pour baigner le début des œuvres hydrauliques. Les mauvaises travaux dans la prise d'eau proposés par le maître Giacomo Di Pasquale et l'ingénieur Nicolò Palma sont mentionnés, en opposition avec le maître Pasquale Cardinale. En Novembre, une grande inondation a provoqué l'obstruction de l'aqueduc de Scamaccio par la boue et les débris.[52] D'après un rapport établi le 8 février 1766, nous savons que le débit de la Favara venait de remonter à environ 10 *pinne* (0,3 l/s), même si on espérait sa croissance ultérieure en raison de l'abondance des pluies hivernales de cette année.[53]

Le 10 février 1788, le *Supremo Tribunale del Real Patrimonio* a autorisée les nouveaux travaux de maintenance et restauration de l'ouvrage (partiellement obstruée par les concrétions calcaires) déjà existante de Favara (à accomplir pendant les cinq ans successifs), laquelle a été fissuré et endommagé (danger de croulement d'un arc) par les pluies torrentielles saisonnières qui ont provoqué des glissements de terre, avec le risque de pollution, surtout pour le ruissellement superficiel, dans les parties les plus fragiles.[54] Parallèlement, le prédit tribunal a autorisée un nouveau conduit

---

[45] Battaglia 1887–1890: 116.
[46] Guarino 1759–1770: 239–240, no. 487–489.
[47] Guarino 1759–1770: 239–240, no. 487–489.
[48] AMJ 1755–1756, Contino, 2019: 219.
[49] Contino et Mantia 2006: 208.
[50] Guarino 1759–1770: 240, no. 488. Une visite au GFS (9 Mars 2020) as montré la réelle réalisation des travailles mentionné par Guarino, avec l'excavation des plusieurs mètres cubes de calcaire, pour essayer de réunir dans Scamaccio la Favara.
[51] Contino 2019: 219.
[52] AMJ 1765–1766: 11–12, 30–31.
[53] AMJ 1765–1766: 33.
[54] AMJ 1788, voir surtout l'inédit document essentiel, c'est-à-dire la relation de maître Diego La Cavara: 58r–79v.

(*catusato*) de distribution des eaux de Scamaccio dans la ville inférieure[55] et de Favara dans la ville supérieure.[56]

Le séisme du 5 mars 1823 (épicentre dans le NE de la Sicile nord-orientale près de Barcellona Pozzo di Gotto et Naso, mais il a été averti jusqu'à Palerme; MCS VIII; Me 5.9[57]), a provoqué l'augmentation de la température (jusqu'à 47,5° C), le quadruplement du débit et, pendant deux jours, une coloration rougeâtre des eaux thermales des *Bagni di Termini*.[58]

Le tremblement de terre, selon toute probabilité, a dû causer des fissures dans le conduit des eaux de Favara, qui se sont étendues avec les ultérieures *aftershock*, jusqu'à la défaillance structurelle. Les glissements de terrain induits par le tremblement de terre sont rappelés par deux relations détaillés.[59]

Les actes des *decurioni* (nouvelle dénomination des administrateurs locales) de Termini Imerese (*Atti dei Decurioni di Termini* dorénavant ADT), œuvres manuscrites de la local bibliothèque municipal, sont la source documentaire principale que nous avons utilisé pour le période antérieur au 1860.

Le 22 octobre 1824, la chute d'environ 134 m de l'aqueduc de Favara, qui a eu lieu à côté de la terre des Coniglio, déterminait la sécheresse de toute la ville et de *Real Forte*, à tel point que les travaux de réhabilitation ont été commencé déjà le 3 novembre.[60] Le 24 avril 1830, la réhabilitation des arches de l'aqueduc de Favara dans la contrée Mazzarino a été ordonnée en raison de l'état de ruine imminente des structures avec le danger pour la population de rester sans eau.[61]

Depuis l'annexion de la Sicile au Règne d'Italie la local municipalité voulait la réalisation d'un aqueduc en tuyaux en fer. L'acte du notaire Mormino de Termini Imerese, daté 29 décembre 1866, épreuve l'adjudication de cet conduit du GFS.[62]

La double série séismique de Termini Imerese du 1906 (11 septembre; MCS VII; Me 5.0) et du 1907 (21 janvier; MCS V-VI; Me 4.2),[63] a provoqué sur les thermes des effets similaires au séisme du 1823[64] et en combinaison avec l'ultérieur événement de Messina du 28 décembre 1908 (MCS XI; Me 7.1) a provoqué une réduction de 2/3 du débit de Favara.[65]

Figure 7: Source de Favara. Ouvrages de protection et déversoir (d'après Cingolani *et al.* 1935–1937), Bibliothèque municipale de Termini Imerese (publication autorisée le 4 Février 2020).

Le grand recensement des sources italiennes, relatif à la Sicile, certifie que le 23 septembre 1930 le GFS présentait un débit de 36,0 l/s et une température de 20° C. Il était utilisé pour l'aqueduc municipal et l'irrigation.[66]

Cingolani *et al.*[67] montrent les débits (1934), respectivement pour les sources du GFS: 40,5 l/s (Favara, Figures 7, 8) et 6 l/s (Scamaccio). En novembre 1955, le débit maximal de 53,3 l/s a été mesuré,[68] tandis qu'au même mois de 1961, le minimal était de 17 l/s.[69]

Le séisme destructive du 14–16 janvier 1968 (MCS VIII–IX; magnitude instrumentale Ms 4.9–5.9) de la vallée du Belice (dans le W de la Sicile),[70] a provoqué des effets dans le GFS. Cinq an depuis, à l'occasion de l'achèvement d'une recherche hydrique commandée par l'administration municipale de Termini Imerese, la diminution considérable du débit du GFS a été mise

---

[55] AMJ 1789: 91r–96v, 105–108.
[56] AMJ 1789: 38v–39v.
[57] Guidoboni *et al.* 2003: 1654–1656.
[58] Gargotta 1830; Scinà 1823.
[59] Gargotta 1830; Scinà 1823.
[60] ADT 1823–1824: 73, art. 2; Contino 2005.
[61] ADT 1827–1834: 351, no. 8; Contino 2019: 224.
[62] Ciofalo 1868: 18.
[63] Rigano *et al.* 1998.
[64] Ciofalo 1909; 1924.
[65] Ciofalo 1924: 27.

[66] Ministero dei Lavori Pubblici - Servizio Idrografico 1934: 316–317.
[67] Cingolani *et al.* 1935–1937: 912–913.
[68] Ministero dei Lavori Pubblici 1973.
[69] Regione Sicilia Assessorato Territorio e Ambiente 1987: 553.
[70] Rigano *et al.* 1998.

Figure 8: Termini Imerese, aqueduc de Favara, montée d'eau dans la ville (d'après Cingolani *et al*. 1935–1937), Bibliothèque municipale Termini Imerese (publication autorisée le 4 Février 2020).

en évidence,[71] tandis que dans le recensement des sources souhaité par la *Cassa del Mezzogiorno* (Office de l'État pour le développement du Midi d'Italie), il y a la dernière mesure officiel de débit (août 1979), c'est-à-dire 8 l/s.[72] Dans l'état actuel des recherches, on ne sait pas quand la disparition totale du GFS s'est produite.

Récemment, les nouvelles recherches géologiques et géophysiques, dans le cadre du projet VIGOR, menées par le Conseil national de la recherche italien, avec le soutien scientifique du DISTEM (Université de Palerme), dans cette zone géothermique ont permis la première enquête géognostique pour la recherche de fluides thermaux à basse enthalpie. En effet, le sondage Termini Imerese (140 m d'altitude, 351,30 m de profondeur), situé dans les parages du GFS, a permis d'intercepter un aquifère confiné, thermale à basse enthalpie (22 °C, à -280 / -285 m du plan de campagne), à l'intérieur du complexe calcaire multicouche (Brèche à Rudiste, formation Crisanti), probablement lié aux sources disparues.[73]

Conclusions et perspectives

Les données historiques collectées jusqu'à présent et relatives au GFS, malgré la discontinuité des séries d'archives étudiées (AMJ, ADT), depuis le début du XVIIe siècle documentent une certaine vulnérabilité des sources aux épisodes de sécheresse, fréquents et sur des périodes prolongées, qui ont provoqué des diminutions notables du débit. Au moins à partir du séisme de 1693, à cette vulnérabilité s'est également superposée celle relative aux variations de débit induites par les tremblements de terre. Les séismes les plus forts (dont certains avec des épicentres relativement éloignés) ont généré fréquemment une diminution du débit et, plus rarement, des augmentations inattendues (par exemple en 1726).

Une prochaine enquête archivistique (par exemple dans le fonds des notaires défunts de la section de Termini Imerese de l'archive d'État de Palerme) devrait enrichir les données collectées jusqu'à présent.

L'étude discutée dans cet article, à travers une approche méthodologique intégrée, combine les résultats de la recherche géologique et d'une analyse historique des documents collectés, ouvrant la voie à des investigations supplémentaires (par exemple, modélisation de la vulnérabilité hydrogéologique déclenché par les séismes, future gestion, protection et planification des ressources en eau souterraine, nouvelles études géothermiques).

Contributions des auteurs et remerciements

Recherche géologique, interprétation des données et discussion des résultats: PB, AC, GE; recherche

---

[71] SIPAC 1973.
[72] Regione Sicilia Assessorato Territorio e Ambiente 1987: 553.
[73] Abate *et al.* 2013: 99–101; VIGOR 2010–2015.

historique et archivistique: PB, AC. Nous sommes très reconnaissants aux organisateurs du colloque 'Watertraces. De l'hydrogéologie à l'archéologie hydraulique en Méditerranée antique' d'avoir voulu publier, avec une grande disponibilité, les résultats de nos études. Nous voudrions remercier sincèrement les chercheurs suivants pour leur contribution indispensable: Oscar Belvedere (Université de Palerme); Francesco Ciofalo (Termini Imerese); Michele Ciofalo (Université de Palerme); Donaldo Di Cristofalo (géologue, municipalité de Termini Imerese), Carmelo Gennaro (Petralia Soprana); Enzo Giunta (Termini Imerese); Francesco La Mantia (Termini Imerese), Claudia Raimondo (Bibliothèque municipale de Termini Imerese). Cette recherche est dédiée à Gioacchino Cusimano (22 novembre 1947; 9 juin 2018), grand connaisseur de l'hydrogéologie, professeur sans égal, collègue inoubliable et ami fraternel.

Bibliographie

Abate, S., Albanese, C., Angelino, A., Balasco, M., Bambina, A., Bellani, S., Bertini, G., Botteghi, S., Bruno, P.P., Caielli, G., Caiozzi, F., Calvanese, L., Calvi, E., Caputi, A., Cardellicchio, N., Catalano, R., Catania, M., Contino, A., De Franco, R., De Rosa, D., Desiderio, G., Destro, E., Di Fiore, V., Di Sipio, E., Donato, A., Doveri, M., Fedi, M., Ferrari, E., Di Gregorio, G., Di Leo, M., Galgaro, A., Gennaro, C., Gianelli, G., Gibilaro, C., Giocoli, A., Giorgi, C., Gola, G., Gueguen, E., Iorio, M., La Manna, M., Lavarone, M., Lombardo, G., Maggi, S., Manzella, A., Maraio, S., Menghini, A., Minissale, A., Montanari, D., Montegrossi, G., Monteleone, S., Mussi, M., Norini, G., Pelosi, N., Perrone, A., Piemonte, C., Pierini, S., Piscitelli, S., Punzo, M., Rizzo, E., Romano, G., Sabatino, M., Santilano, A., Scotto di Vettimo, P., Tamburrino, S., Tarallo, D., Teza, G., Tranchida, G., Trifirò, S., Trumpy, E., Varriale, F., Viezzoli, A. e Votta, M. 2015. *VIGOR: Sviluppo geotermico nella regione Sicilia-Studi di fattibilità a Mazara del Vallo e Termini Imerese, Valutazione geotermica con geofisica elitrasportata*. Progetto VIGOR Valutazione del Potenziale Geotermico delle Regioni della Convergenza, POI Energie Rinnovabili e Risparmio Energetico 2007–2013. Pisa: CNR-IGG.

Adinolfi Falcone, R., Carucci, V., Falgiani, A., Manetta, M., Parisse, B., Petitta, M., Rusi, S., Spizzico, M. e Tallini, M. 2012. Changes on groundwater flow and hydrochemistry of the Gran Sasso carbonate aquifer after 2009 L'Aquila earthquake. *Italian Journal of Geosciences* 131(3): 459–474; doi.org/10.3301/IJG.2011.34.

Alberti, L. 1596. *Descrittione di tutta l'Italia. Et isole pertinenti ad essa*. Venetia: Ugolino.

Acta Magnificorum Juratorum Splendidissimae et Fidelis Civitatis Thermarum. Mss. sec. XVI–XIX Biblioteca comunale di Termini Imerese *III 10 a 1 e segg*.

Atti del Decurionato di Termini Imerese. Mss. sec. XIX Biblioteca comunale di Termini Imerese *III 10 i 6 e segg*.

Amoruso, A., Crescentini, L., Petitta, M., Rusi, S., Tallini, M. 2011. Impact of the 6 April 2009 L'Aquila earthquake on groundwater flow in the Gran Sasso carbonate aquifer, Central Italy. *Hydrological Processes* 25(11): 1754–1764; doi.org/10.1002/hyp.7933.

Anonimo sec. XVIII. *Scritture di Sceusa di N[umer]o 4*. Ms. sec. XVIII. Collezione privata.

Arezzo, V.M. 1537. *De situ insulae Siciliae libellus*. Messana: Spira.

Battaglia, A. 1887–1890. *Sui bagni termo-minerali in Termini Imerese*. Termini Imerese: Amore.

Belvedere, O. 1986. *L'acquedotto Cornelio di Termini Imerese*. Roma: L'Erma di Bretschneider.

Boschi, E., Ferrari, G., Gasperini, P., Guidoboni, E., Smriglio, G. e Valensise, G. 1997. *Catalogo dei Forti Terremoti in Italia dal 461 a.C. al 1990*. Istituto Nazionale di Geofisica, Storia Geofisica Ambiente, Ozzano Emilia: Istituto Nazionale di Geofisica.

Cappadonia, C., Confuorto, P., Sepe, C., Di Martire, D. 2019. Preliminary results of a geomorphological and DInSAR characterization of a recently identified Deep-Seated Gravitational Slope Deformation in Sicily (Southern Italy). *Rendiconti Online Società Geologica Italiana* 49: 149–156.

Caracausi, G. 1983. *Arabismi medievali di Sicilia*. Centro di Studi filologici e linguistici siciliani (suppl. 5). Palermo: Luxograf.

Catalano, R., Sulli, A., Abate, B., Basilone, L. 2004. The crust in western and central eastern Sicily, in R. Catalano, A. Sulli, B. Abate, M. Agate, G. Avellone and L. Basilone (eds) *Field Trip Guide Book P45*: 3–40. 32nd International Geological Congress. Florence: APAT.

Catalano, R., Avellone, G., Basilone, L., Contino, A. and M. Agate 2011. *Note illustrative della Carta Geologica d'Italia alla scala 1.50000 dei Fogli 609 e 596 'Termini Imerese-Capo Plaia'*. Progetto CARG, Istituto Superiore per la Protezione e la Ricerca Ambientale. Rome: Servizio Geologico d'Italia Roma: System Cart.

Cingolani, E., Di Castelnuovo, G., Lucci, V. 1935–1937: *Acquedotti fascisti* (2 vol.) Roma: Luzzati.

Ciofalo, S. 1868. *Topografia di Termini Imerese e suoi dintorni*. Palermo: Perino.

Ciofalo, M. 1909. Orografia, geologia e tectonica della zona scossa dai terremoti di Termini Imerese del Settembre 1906. *Bollettino della Società Sismologica Italiana* 13: 153–183.

Ciofalo, M. 1924. *Studio idrogeologico delle sorgenti termominerali di Termini Imerese*. Palermo: Tip. Di Carlo.

Contino, A., Monteleone, S. and Sabatino, M. 2015. Water resource assessment in karst and fractured aquifers of Termini Imerese-Trabia Mts. (Northern central Sicily, Italy), in G. Lollino, M. Aratano, M. Rinaldi, O. Giustolisi, J.-C. Marechal and G.E. Grant (eds) *Engineering Geology for Society and Territory* (vol. 3): 573–577. Cham, Switzerland: Springer.

Contino, A. 2005. *Geologia Urbana dell'abitato e della Zona Industriale di Termini Imerese (Sicilia settentrionale)*:

214. Unpublished PhD. dissertation, University of Palermo, Italy.

Contino, A., Cusimano, G., Frias Forcada, A. 2004. Assetto Idrostrutturale dei Monti di Trabia e Termini Imerese (Sicilia Settentrionale) e valutazione delle risorse idriche immagazzinate. Attività esplorativa e ricerca nelle aree carsiche siciliane, in R. Ruggieri (ed) *Atti del 4° Convegno di Speleologia della Sicilia, 1° – 5 Maggio 2002, Custonaci (TP)*: 99–106. Ragusa: CIRS.

Contino, A., Cusimano, G., Gatto, L., Hauser, S. et Pisciotta, A. 2007. Gli acquiferi costieri dei Monti di Trabia-Termini Imerese: modalità di circolazione idrica, ruolo del carsismo e vulnerazione per fenomeni di ingressione marina, in R. Ruggieri (ed) *Atti del 2° Seminario Internazionale di studi sul Carsismo negli Iblei e nell'area sud-mediterranea, 28-30 maggio 2004, Castello di Donnafugata (RG)*: 85–93. Ragusa: CIRS.

Contino, A. e Mantia, S. 2006. Sugli eventi siccitosi nell'area di Tèrmini Imerese dal XVII al XVIII secolo. *Gli Apoti* 2 (1): 204–210.

De Luca, G., Di Carlo, G. and Tallini, M. 2016. Hydraulic pressure variations of groundwater in the Gran Sasso under-ground laboratory during the Amatrice earthquake of August 24, 2016. *Annals of Geophysics* 59 FAST TRACK 5; doi.org/10.4401/AG-7200.

De Michele, M. (éd.) 1950. *L'Ebreismo di Termini Imerese*. Candioto I (ed). Termini Imerese: ASCI.

Di Maggio, C., Madonia, G., Vattano, M., Agnesi, V., Monteleone, S. 2017. *Geomorphological evolution of western Sicily, Italy*. Geologica Carpathica 68(1): 80–93.

Gargotta, A. 1830. *Su i Bagni termo-minerali di Termini-Imerese*. Palermo: Dato.

Gennaro, C., Contino, A., Avellone, G. and Di Maggio, C. 2012. A Multidisciplinary approach to define the Plio-Quaternary tectonics in the Termini Imerese Mts. (Northern central Sicily), in S. Critelli, F. Muto, F. Perri, F.M. Petti, M. Sonnino e A. Zuccari (eds) *Note brevi e riassunti dell'86° Congresso Nazionale della Società Geologica Italiana Arcavacata di Rende (CS) 18-20 settembre 2012*: 170–171. Roma: Società Geologica Italiana.

Gregorio, R. 1791–1792. *Bibliotheca scriptorum qui res in Sicilia gestas sub Aragonum imperio retulere* (2 vol). Panormi: Regio Typographeo.

Guarino, G.A. 1759–1770. *Libro in cui si descrive l'antichità del Ven[erabil]e Reclusorio delle Donzelle vergini, sotto Tit[ol]o del Principe d[ell]a Chiesa catolica S[an]: Pietro etc*. Ms. Biblioteca comunale di Termini Imerese AR e α 10: 346.

Guidoboni, E., Mariotti, D., Giammarinaro, M.S. and Rovelli, A. 2003. Identification of Amplified Damage Zones in Palermo, Sicily (Italy), during the Earthquakes of the Last Three Centuries. *Bulletin of the Seismological Society of America* 93(4): 1649–1669.

Ingebritsen, S.E. and Manga, M. 2019. Earthquake hydrogeology. *Water Resources Research* 55: 5212–5216; doi.org/10.1029/2019WR025341.

La Vigna, F. 2013. Earthquake hydrology. *Italian Journal of Groundwater* 2/2013: 45–46; doi.org/10.7343/AS-030-13-0055.

La Vigna, F., Carucci, V., Mariani, I., Minelli, L., Pascale, F., Mattei, M., Mazza, R. and Tallini, M. 2012. Intermediate-field hydrogeological response induced by L'Aquila earthquake: the Acque Albule hydrothermal system (Central Italy). *Italian Journal of Geosciences* 131(3): 475–485; doi.org/10.3301/IJG.2012.05.

Lyell, C. 1832–1833. *Principles of Geology, Being an Attempt to Explain the Former Changes of the Earth's Surface, by reference to causes now in operation* (2nd edn, 3 vols). London: John Murray.

Manga, M. and Rowland, J.C. 2009. Response of Alum Rock springs to the October 30, 2007 Alum Rock earthquake and implications for the origin of increased discharge after earthquakes. *Geofluids* 9: 237–250; doi.org/10.1111/j.1468-8123.2009.00250.x.

Manga, M. and Wang, C.Y. 2015. Earthquake Hydrology, in G. Schubert (ed.) *Treatise on Geophysics* (2nd edn, vol. 4): 305–328. Oxford: Elsevier; doi.org/10.1016/B978-0-444-53802-4.00082-8.

Maurolico, F. 1562. *Sicaniarum Rerum Compendium*. Messanae: Spira.

Ministero dei Lavori Pubblici 1973. *Piano Regolatore Generale degli Acquedotti (Legge 4 Febbraio 1963 n. 129). Indice dei Comuni ed elenco delle acque da riservare*. Regione Sicilia. Roma: Edigraf.

Ministero dei Lavori Pubblici Servizio Idrografico 1934. *Le sorgenti italiane. Sicilia* (vol. II), Servizio Idrografico di Palermo. Roma: Istituto Poligrafico dello Stato.

Molin, P., Acocella, V. and Funiciello, R. 2003. Structural seismic and hydrothermal features at the border of an active intermittent block: Ischia island (Italy). *Journal of Volcanology and Geothermal Research* 121: 65–81.

Musso, A.M. 1760. *Codice de' Privilegi e Consuetudini della Splendidissima e Fedele Città di Termini*. Ms. Biblioteca comunale Liciniana di Termini Imerese AR e α 2: 328.

Regione Sicilia Assessorato Territorio e Ambiente 1987. *Piano Regionale di Risanamento delle Acque* (vol. 6). Censimento corpi idrici. Palermo: LIS.

Rigano, R., Antichi, B., Arena, L., Azzaro, R. e Barbano, M.S. 1998. Sismicità e zonazione sismogenetica in Sicilia occidentale. *Atti del 17° Convegno Nazionale GNGTS*, 12.04: 1–11.

Romano, B. 1838. *Antichità termitane*. Palermo: Lao.

Scinà, D. 1823. Rapporto sulle frane avvenute a Termini. *Giornale Letterario di Sicilia* 1: 120–132, 136–139.

SIPAC 1973. *Indagini geoidrologiche sulle sorgenti Favara e Brocato*. Comune di Termini Imerese, dattiloscritto inedito dell'archivio ufficio tecnico 74.

Solìto, V. 1669–1671. *Termini Himerese posta in Teatro etc.* I, Palermo: Dell'Isola: 128; II, Messina, Bisagni: 158.

Strabon, éd. 1969. *Géographie*. G. Aujac, F. Lasserre, R. Baladié et B. Laudenbach (eds), vol. I. Paris: Belles Lettres.

Traina, A. 1868. *Vocabolario Siciliano-Italiano* (reprinted 1991). Milano: Reprint s.a.s.

Trasselli, C. 1982. *Da Ferdinando il Cattolico a Carlo V. L'esperienza siciliana 1475-1525* (2 vol.) Soveria Mannelli: Rubettino.

VIGOR 2010–2015. CNR-IGG, Pisa, dernier accès 20 Février 2020, http://www.vigor-geotermia.it/geo-portal/.

Xue, L., Li, H.B., Brodsky, E.E., Xu, Z.Q., Kano, Y., Wang, H., Mori, J.J., Si, J.L., Pei, J.L., Zhang, W., Yang, G., Sun, Z.M. and Huang, Y. 2013. Continuous Permeability Measurements Record Healing Inside the Wenchuan Earthquake Fault Zone. *Science* 340: 1555–1559 ; doi.org/ 10.1126/science.1237237.

Le risorse idriche nel territorio di Alesa

Aurelio Burgio[1]

[1] Università degli Studi di Palermo, Dipartimento Culture e Società, Laboratorio di Topografia Antica

Abstract: This contribution examines the complex archaeological data produced by the prospections carried out in the region of Alesa, in an attempt to recognise the essential elements of the system of collection and distribution of water in rural settlements. The archaeological data are integrated with what can be obtained from the well-known 'Table of Alesa', and with other toponymical, environmental, morphological, and geological information. The results are compared with known data in areas with similar characteristics along the northern coast of Sicily.

Keywords: survey, Sicily, Halaesa, aqueduct, pipeline

Che l'acqua sia fondamentale per l'esistenza e la sopravvivenza di una città è un fatto ben noto, e in antico città e insediamenti minori facevano fronte alle loro esigenze attraverso pozzi e cisterne che sfruttavano soprattutto le acque piovane, probabilmente in un contesto ambientale contraddistinto da maggiore piovosità rispetto ad oggi.[1] È altrettanto noto però che l'adduzione di acque sorgive, lo smaltimento ed il drenaggio di quelle superficiali, e più in generale la gestione della risorsa idrica incidono talvolta nella selezione dei luoghi dove fondare città, ne condizionano l'assetto urbano anche in funzione della necessità di irreggimentazione e smaltimento. Si tratta di un tema dalla forte connotazione interdisciplinare,[2] da tempo anche in Sicilia oggetto di studi e ricerche in città e comprensori caratterizzati da articolati sistemi idrici urbani, cui sono talvolta connesse cisterne ed edifici monumentali, e da acquedotti e condotte che solcano per chilometri il territorio.

A titolo esemplificativo, ma non esaustivo, basti ricordare Akragas,[3] Camarina,[4] Selinunte,[5] Siracusa,[6] Solunto,[7] Taormina,[8] Termini Imerese,[9] Tindari.[10] A Termini, oltre alle monumentali terme[11] e all'Acquedotto Cornelio, è documentata parte di un sistema di smaltimento idrico in un settore della città alta – presso l'Anfiteatro – dove le trasformazioni di età imperiale hanno purtroppo azzerato strutture e livelli preesistenti: si tratta di un imponente condotto databile alla metà del II sec. a.C., intercettato all'interno della Chiesa di S. Caterina d'Alessandria, costituito da due canali larghi circa 60 cm e conservati in altezza fino a 120 cm, convergenti in un unico condotto che doveva scaricare le acque all'esterno delle mura della città.[12]

Frequenti sono anche gli edifici termali nelle campagne, nelle residenze di lusso di età imperiale e nelle *stationes* dislocate direttamente lungo i principali assi della viabilità,[13] nonché su *diverticula*.[14] Non sempre tuttavia è nota l'ubicazione di questo genere di edifici, in assenza di dati archeologici, come nel caso della *statio* di *Aquae Labodes*, a dispetto dell'enfasi che ha nella *Tabula Peutingeriana*.

Nell'agro delle città greche e romane esisteva inoltre una fitta trama di sorgenti e prese di captazione, cui si collegavano condutture che alimentavano fontane e fattorie, e in qualche caso anche i centri urbani. Si tratta in genere di opere non monumentali, che scorrono in pendio incassate nel terreno poche decine di cm sotto il piano di campagna, costituite da canalette scavate nel banco roccioso o realizzate con tratti in muratura, talora alternati a canali in terracotta, che in età romana potevano contenere e proteggere tubi in piombo. Molto limitata è ovviamente la documentazione archeologica di strutture di questo tipo, e non solo perché facilmente deteriorabili, una volta venuta meno la loro funzione: il costante riutilizzo nel tempo di sorgenti e prese di captazione, la collocazione delle condutture lungo i pendii, naturalmente soggetti e fenomeni erosivi, ne hanno infatti condizionato lo stato di conservazione, insieme alla sistematica distruzione dovuta alle colture agricole e alle attività antropiche post-antiche.

Una presenza, quella delle acque, che permea il paesaggio in ogni tempo, documentata anche dalla toponomastica e da edifici non meno importanti, come vasche, fontane

---

[1] Wilson 2000.
[2] In ultimo, Bouffier *et al.* 2019a; Caminneci *et al.* 2020.
[3] Furcas 2020.
[4] Di Stefano 2000.
[5] Furcas 2019.
[6] Bouffier 1987; Bouffier *et al.* 2019b.
[7] Polizzi 2019; Polizzi *et al.* 2020.
[8] Campagna e La Torre 2008; Castrianni *et al.* 2018; Muscolino 2020; Rapisarda 2020.
[9] Belvedere 1986.
[10] Fasolo 2014: 98–99, fig. 109: il tracciato proposto è largamente ipotetico.
[11] Belvedere e Forgia 2017.
[12] Burgio 1997.
[13] Sofiana e Vito Soldano sulla via Catania-Agrigento: Sfacteria 2018; La Torre 2020: 212–216.
[14] Cignana: Fiorentini 1993; Rizzo e Zambito 2007: 271; Burgio 2021.

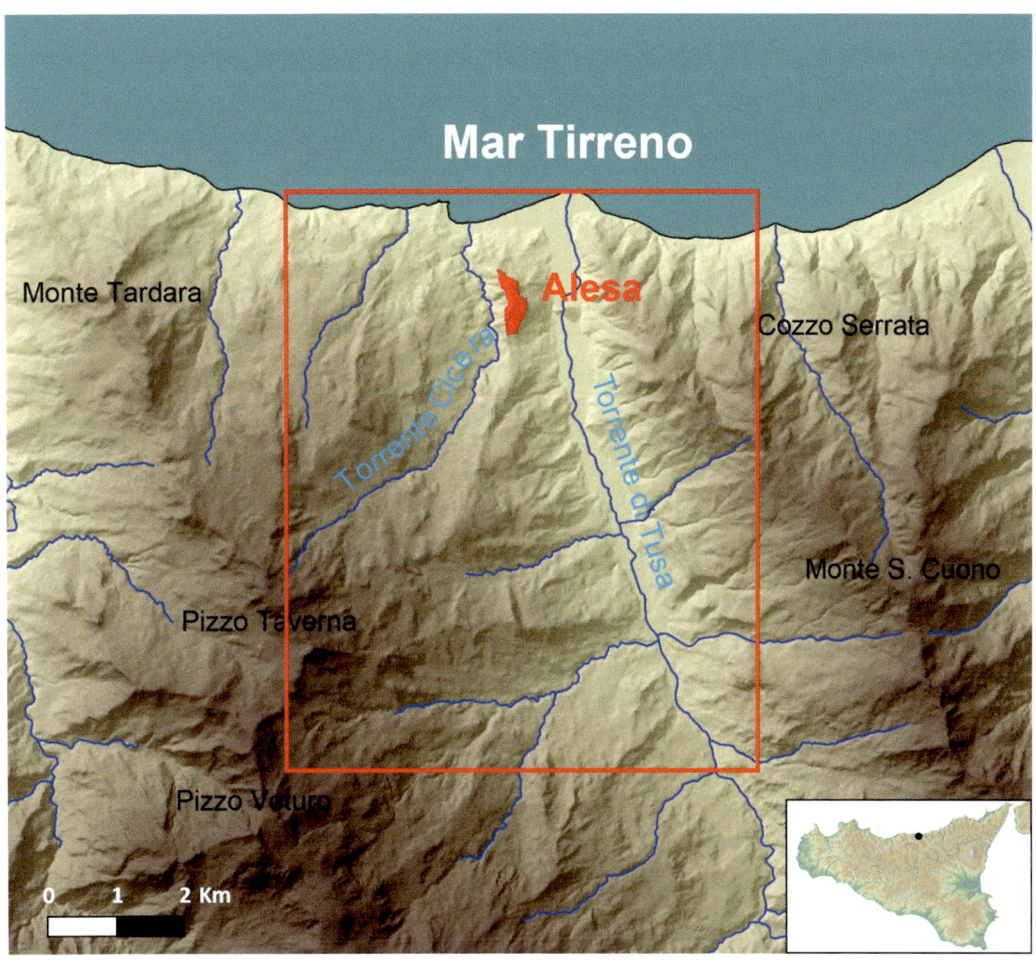

Figure 1 – Il comprensorio di Halaesa

e abbeveratoi,[15] essenziali per chiunque si muova da un luogo all'altro. Peraltro, proprio la dislocazione di queste strutture lungo direttrici viarie contraddistinte dalla presenza di insediamenti da età antica a età moderna indizia l'esistenza di un substrato geologico che consente l'agevole captazione delle acque. A titolo di esempio si segnalano, in una zona interna della Sicilia ad elevata piovosità e ricca di sorgenti, lungo un importante percorso viario che segue lo spartiacque tra i fiumi Imera settentrionale, meridionale e Platani, i siti pluristratificati di Chiesazza e San Giacinto,[16] sui due versanti opposti di Monte Catuso. Toponimo di matrice araba, correlato alla (elevata) disponibilità di acqua, affiorante sia sul versante meridionale (località Fontanelle) che su quello settentrionale del Monte, a contatto tra calcari della Fm. Gessoso-solfifera e depositi detritici,[17] dove affiora una delle più ricche sorgenti di età moderna del comprensorio, che – non a caso – alimentava fino a pochi decenni fa un importante acquedotto ('acquedotto di Marianopoli', su IGM F. 260 III SO).

In questo contributo sarà presa in esame solo questa tipologia di 'opere minori', limitatamente ad un'area della Sicilia settentrionale, l'entroterra di *Halaesa Arconidea* (Figure 1) indagato con ricerche di superficie a carattere intensivo e sistematico.[18] Si tratta come vedremo di pochissimi dati, correlati forse anche alla abilità degli artigiani e alle esigenze di una comunità cui era imposta dalle condizioni ambientali – il gran numero di sorgenti e l'elevata piovosità che contraddistingue la fascia tirrenica della Sicilia, alle pendici della catena montuosa dei Nebrodi e delle Madonie – una gestione sapiente della risorsa idrica. Non è un caso, probabilmente, che strutture simili siano note nei vicini comprensori di Calacte,[19] S. Marco d'Alunzio (in esposizione al locale *Museo della cultura e delle arti figurative bizantine e normanne*, ma senza indicazione dell'area di provenienza) e Tindari,[20] dove la piovosità, associata alle accentuate pendenze e a

---

[15] In alcuni casi sembra che mantengano il rivestimento in cocciopesto, nonostante siano in uso fino ad età moderna: Zambito 2020.
[16] Burgio 2002, UT 8, 40: 53–57, 94–97; Burgio 2017: 113.
[17] Burgio 2002: fig. 2.
[18] Burgio 2008.
[19] Contrada Serralisa-Samperi: Fiore 1971; Collura 2016: 293.
[20] Fasolo 2014: 98–99, 101–102, fig. 117.

Figure 2 – Panorama da S. Marco d'Alunzio: al centro il Monte Vecchio di San Fratello.

suoli con elevata componente argillosa, potrebbe avere favorito più che altrove la realizzazione di condutture che servissero anche ad incanalare e gestire acque sorgive e di ruscellamento (Figure 2). Al contrario, strutture analoghe non sono state individuate in altre aree della Sicilia settentrionale – in parte assimilabili a quella in esame – dove la ricerca sul campo è stata effettuata sempre in modo intensivo e sistematico, il territorio Himera/Termini Imerese[21] e i dintorni del moderno centro di Baucina.[22]

Halaesa Arconidea, fondata alla fine del V sec. a.C., fu tra le più vitali città della Sicilia da età ellenistica alla media età imperiale, finché con il IV secolo sembra decadere rapidamente; sede vescovile, è verosimile che abbia avuto un ruolo importante anche in età bizantina, finché fu abbandonata e il centro abitato si spostò, probabilmente in età islamica, sul vicino colle dove sorge la moderna Tusa.[23] Insediamenti di età tardo-antica sono all'esterno dell'area urbana, lungo la fascia costiera[24] e sul versante orientale del bacino del fiume Tusa, lungo uno degli assi viari di collegamento tra la costa e l'entroterra.[25]

Ad *Halaesa* l'esigenza di raccogliere, regimentare e smaltire le acque è comprovata dalla presenza di un articolato sistema di canalizzazione e gestione delle acque: fontane pubbliche, cisterne e grandi serbatoi ad un uso collettivo erano dislocati in punti diversi della città, presso l'agorà, le mura di fortificazione e in aree di compluvio.[26] Rilevante il serbatoio descritto da Gianfilippo Carettoni poco a Ovest della Torre C, 'costruito con blocchetti di pietra di tipo simile alle mura', che 'presenta verso monte una parete curvilinea rivestita di cocciopesto';[27] non indagato nella sua interezza, meriterebbe un'indagine mirata ed estensiva che riguardi tutta l'area circostante, tanto più che nei pressi un serbatoio analogo è stato individuato nel 2004.[28] Inoltre, sotto il tessuto viario correva una fitta rete di canali[29] realizzati con particolari accorgimenti funzionali e in stretta relazione con le linee di pendenza delle colline, che trovava il suo compimento nello smaltimento all'esterno del centro urbano attraverso le mura di fortificazione, dotate di sbocchi di forma generalmente rettangolare. Tutto il sistema, sembra sottoposto a manutenzione fino a fine II/inizi III sec. d.C., è da riferire essenzialmente ad età ellenistica, per quanto non si possa escludere – mancano tuttavia al

---

[21] Alliata *et al.* 1988; Belvedere *et al.* 2002; Burgio 2002; Lauro 2009.
[22] Bordonaro 2012.
[23] Per le vicende storiche Facella 2006; per i dati ascrivibili al IV sec. d.C., Portale 2009.
[24] Tigano 2009.
[25] Burgio 2009; Di Maggio 2012.

[26] Tigano e Burgio 2020: 222–225, fig. 4; recentissimo il rinvenimento di una cisterna sulla c.d. acropoli meridionale: Costanzi *et al.* 2019: 28.
[27] Carettoni 1959: 327, fig. 1.
[28] Tigano e Burgio 2020: 223.
[29] Tigano e Burgio 2020: 225–227, fig. 2b, 5, 6.

momento dati cronologici – che nella sua progettazione possa risalire al primo impianto urbano. Altre strutture sono di età imperiale, come il complesso termale a monte della porta di SO.[30]

Un importantissimo documento epigrafico a carattere catastale (e non solo), la c.d. *Tabula Halaesina*,[31] di età repubblicana, ci restituisce informazioni non solo sul paesaggio agrario (con implicazioni di carattere economico e giuridico) della città antica, ma anche su alcuni aspetti del tessuto idrografico del comprensorio, contraddistinto da corsi d'acqua di portata diversa.[32] Alcuni hanno origine da canali di scolo della fortificazione, e vengono menzionati poiché fanno da limite tra i lotti compresi tra le mura e il corso dell'*Alaisos*, ricorrente anche come *rhous Alaisos* (attuale fiume di Tusa); è inoltre menzionato un *potamos*, alimentato in destra idrografica da un altro corso d'acqua, l'*Opikanos*. Corsi d'acqua che rappresentavano elementi del paesaggio ben noti e visibili agli abitanti della città, riconoscibili e sicuramente denotati da nomi specifici, microtoponimi che (come avviene ancora oggi) contribuivano – insieme agli edifici extraurbani (sacri e non), alle strade e ai boschi – a fissare luoghi identitari per la comunità.[33]

Il paesaggio era anche arricchito da fonti (la fonte *Ipyrra* e quella presso il santuario del *Meilichieion*) e da un acquedotto (*ocheton*), la cui salvaguardia era garantita dal divieto di coltivare il terreno su una fascia di 70 piedi intorno alla condotta. Prescrizione di estremo interesse, se si considera che le coltivazioni di età moderna hanno portato alla luce numerosi tratti di un acquedotto – di seguito descritto (Figure 3) – che alimentava la città,[34] favorendone tuttavia la progressiva distruzione. Naturalmente non si tratta della stessa struttura menzionata nella *Tabula*, peraltro da collocare sul versante destro del fiume di Tusa,[35] dove non si è rinvenuta alcuna traccia di condutture; è verosimile però che l'acquedotto coincida con quello ricordato alla metà del 1500 da Tommaso Fazello,[36] che segnalava 'mirabili acquedotti' ed una sorgente copiosa; quest'ultima potrebbe corrispondere all'attuale località Acquacitita (Figure 3, UT 106), a tre miglia dalla costa e a quasi 700 metri di quota, nella zona di captazione del ramo principale dell'acquedotto. Si noti però che due secoli dopo Jean Houel, che soggiornò a Tusa nel 1776,[37] non sembra conosca il manufatto, o non ritiene opportuno parlarne, attratto da ben altri monumenti, statue, epigrafi e sepolture.

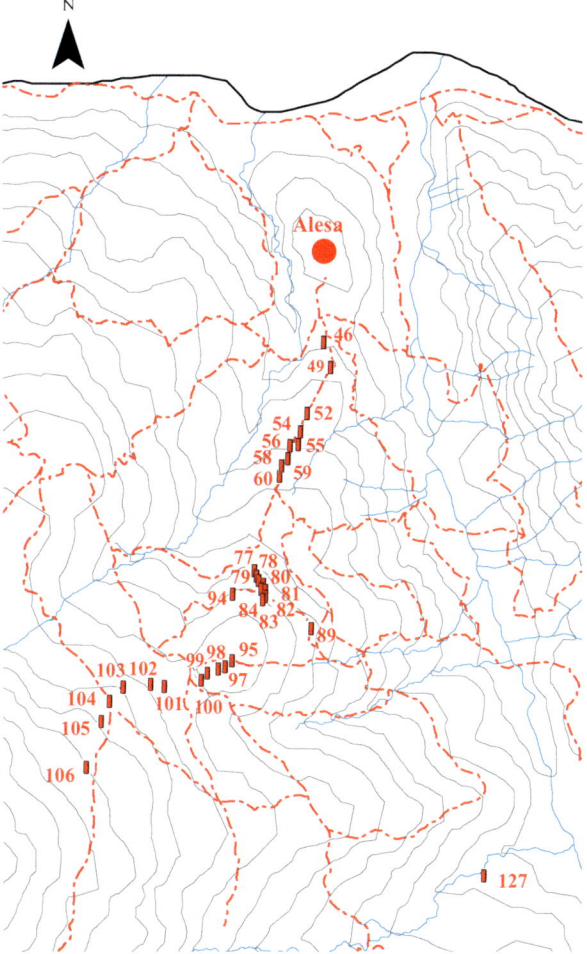

Figure 3 – Il tracciato dell'acquedotto (da Burgio 2018, fig. 181).

Anche ad *Halaesa* gestione delle acque e viabilità erano in stretta connessione: nell'area dell'*ocheton* transitava un percorso di ambito non solo locale, la *odos xenis*, la cui sopravvivenza è probabilmente da riconoscere nella Regia Trazzera Motta d'Affermo-Mistretta (Figure 4), sul versante destro del fiume di Tusa.[38] Ne deriva la stretta relazione tra questa direttrice e un paesaggio connotato da insediamenti di varia entità, ascrivibili a periodi differenti; tra questi spicca la vasta necropoli di età tardo-antica individuata – sulla dorsale che risale in quota verso il centro moderno di Motta d'Affermo – a Cozzo Sorbo, da collegare al vicino sito di contrada Belvedere.[39]

La *odos xenis* si raccordava dunque all'approdo e alla *via Valeria* che si snodava lungo la stretta piana costiera ai piedi della città, dove sorgeva la *mansio* di *Halaesa*, nota sia dall'*Itinerarium Antonini* che dalla *Tabula Peutingeriana*. Mansio, dotata certamente di

---

[30] Tigano e Burgio 2020: 223, 228.
[31] Arena 2020, cui si rinvia per l'ampia bibliografia.
[32] Capasso 1989; Prestianni Giallombardo 2004-2005; Prestianni Giallombardo 2012: 382–385.
[33] Belvedere 2008.
[34] Burgio 2008: 242–245.
[35] Arangio Ruiz e Olivieri 1925: 59.
[36] Fazello 1558, I: 9, 4.
[37] Facella 2006: 30–32.

[38] Di Maggio 2008; Di Maggio 2012.
[39] Burgio *et al.* 2008, UT 131–132: 151–152; Burgio 2009; Messina, Petruso e Sineo 2009.

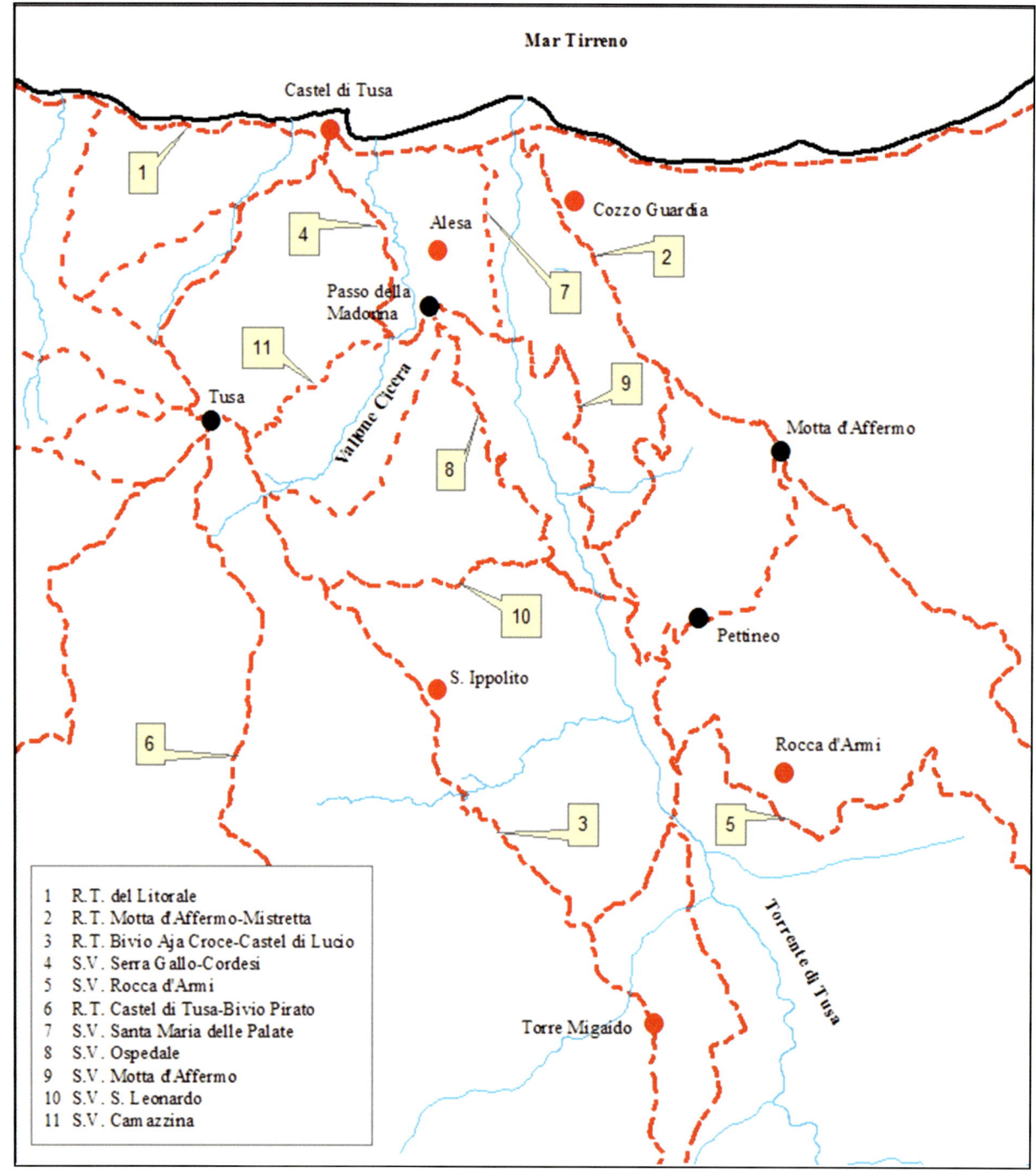

Figure 4 – La viabilità storica nel bacino del Tusa (da Di Maggio 2018, fig. 167).

ambienti termali. Impossibile tuttavia dire se la *mansio* possa corrispondere all'edificio denominato 'Bagni'[40] ricordato da Fazello ai piedi del colle su cui sorgeva la città, oppure se vada cercata (anche) tra gli edifici venuti alla luce circa 500 m più ad Ovest, parzialmente coperti da una necropoli di età tardo-antica, dove sono affiorati anche resti di un acquedotto.[41] Senza dubbio lungo la fascia costiera si sviluppò, soprattutto in età imperiale, un vasto agglomerato,[42] servito da condotti idrici. Più a occidente (in contrada Canale, toponimo più che trasparente) si segnala il rinvenimento di un frammento di tubo fittile, identico a quelli – descritti più avanti – del citato acquedotto.[43] Ancora una volta degna

---

[40] Forse lo stesso inglobato nelle case Salamone, in contrada Bagni, vicino alla foce del fiume di Tusa, al cui interno si trovano ambienti con mosaici: Burgio *et al.* 2008, UT 1: 53.

[41] Burgio *et al.* 2008, UT 3: 55-57; Tigano 2009.
[42] Burgio *et al.* 2008, UT 4–5: 57.
[43] Burgio *et al.* 2008, UT 12: 66-67.

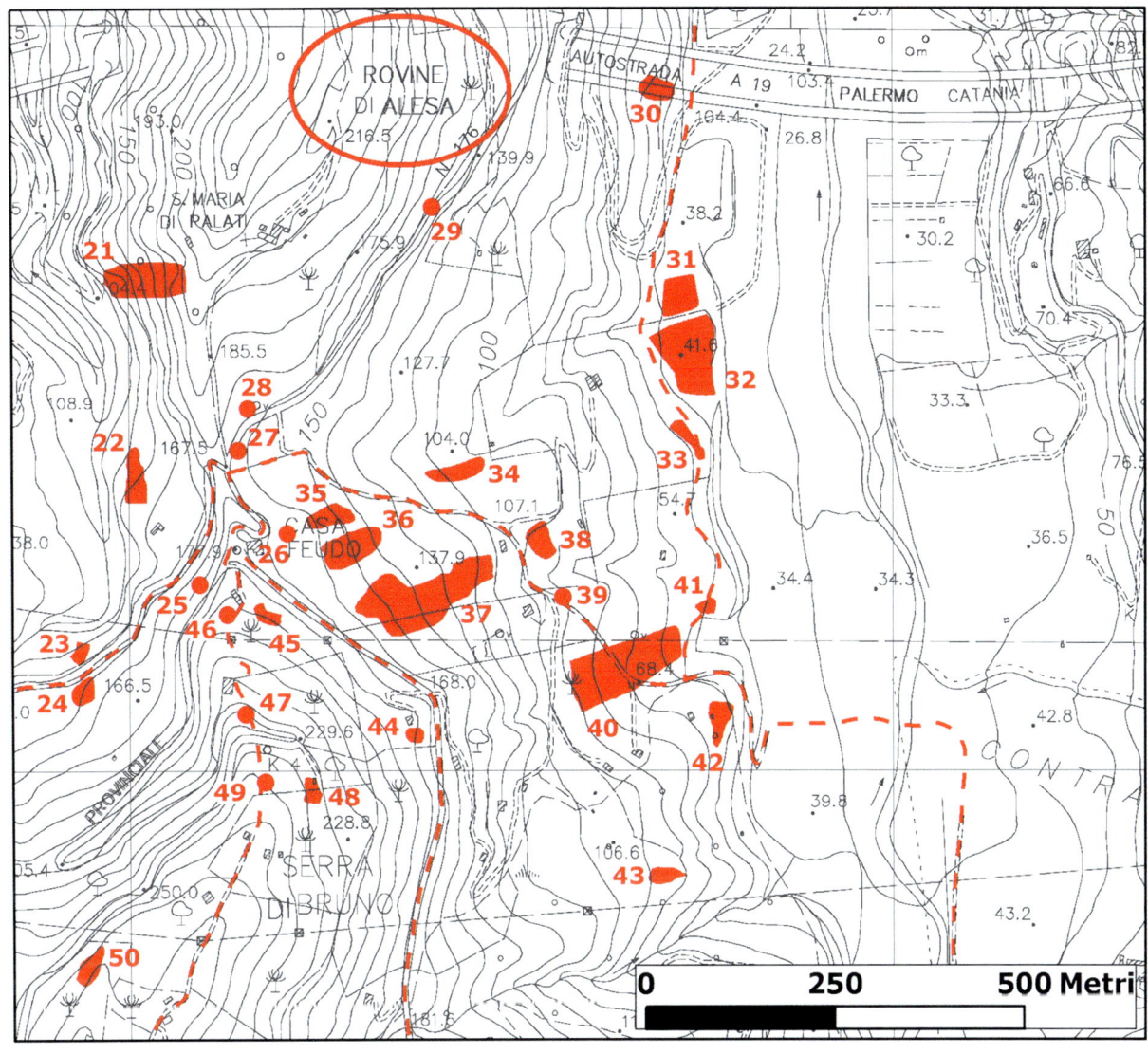

Figure 5 – Insediamenti a SE di Halaesa (da Burgio 2018, fig. 49).

di interesse la descrizione di Fazello, con la menzione di acquedotti che alimentavano sia i 'Bagni' che il vicino Castello (Castel di Tusa), segno dell'esistenza di un'ampia rete di distribuzione.

L'ambiente circostante la città antica è probabile non fosse molto diverso dall'attuale, con sorgenti che alimentano corsi d'acqua a regime torrentizio e aree boschive a roverella e querce da sughero (contrada Suvarello, lungo il vallone Cicera, a SO di *Halaesa*) anche a bassa quota. Paesaggio comune all'intero comprensorio alesino, non solo – come sopra richiamato – nelle contrade solcate da valloni che dalla zona intorno al colle di Tusa raggiungono direttamente il mare (contrade Canale, Giardino, Murro, Piano Fontane, che alimentano il vallone Nacchio; contrade Vagna e Favara, che alimentano il Langalosi-S. Angelo, ancora più a occidente), ma anche lungo il vallone Cicera (che scorre subito a Ovest di *Halaesa*) e soprattutto nel bacino del fiume di Tusa, dove vere e proprie aree boschive si sviluppano alla testata dei principali corsi d'acqua del basso e medio corso (contrada Giardinello, vallone S. Giovanni, Fosso Filesse, sulla destra; contrade Lavanche-Margibuffa, vallone del Leone-S. Pantaleo sulla sinistra).

Importanti modifiche sembra però che abbia subito l'alveo del fiume di Tusa, il cui tratto finale è molto più ampio rispetto a pochi secoli fa, come dimostra la posizione del rudere di un ponte eretto forse nel XV secolo, il Ponte Riggieri,[44] poco a SE di *Halaesa* (Figures 5, 6). Esso dista infatti oltre 300 metri dalla riva destra del fiume, e doveva pertanto essere caratterizzato da un ampio terrazzo alluvionale, smantellato dalle continue piene. D'altra parte, nei documenti della municipalità di Tusa ricorrono stanziamenti finalizzati a ripristinare il ponte, a seguito di eventi parossistici, come si dirà più avanti. Analogamente, la captazione delle acque

---

[44] Burgio 2008: 21–22.

Figure 6 – Il Ponte Riggieri.

sorgive in varie zone dell'agro di Tusa per gli usi urbani, ampiamente attestata almeno dal 1600 in poi nell'area di Piano Fontane, avrà influito sul tessuto idrografico, generando differenze tra l'età antica e quella moderna.

L'acquedotto di Halaesa

Nonostante tale ricchezza di risorse idriche, segni relativi alla gestione delle acque sono riscontrabili soltanto sul versante sinistro del Tusa, sull'asse – una dorsale in moderata pendenza da Sud a Nord (Figures 1, 3) – che collega Pizzo Taverna (Acquacitita) a Serra di Cuozzo, Serra di Bruno e la collina di S. Maria, dove sorge *Halaesa*, facendo da spartiacque con il vallone Cicera (vallone del Ponte nella zona di testata, contrada Acquatico).

La dorsale si caratterizza per ampie zone con spiccata erosione, sia areale, sia alla testata dei valloni, a causa della configurazione geomorfologica a crinale e dei litotipi prevalenti (Flysch di Troina-Tusa); qui, proprio la moderata pendenza, solo a tratti interrotta da piccole balze, e gli affioramenti rocciosi lungo il crinale potrebbero aver indotto a collocare lungo la cresta il tracciato dell'acquedotto che riforniva la città antica.

Lungo la dorsale sono venuti alla luce, in occasione di ripetuti sopralluoghi effettuati nell'arco di una decina di anni a partire dal 1993, una serie di tubi in terracotta, in qualche caso ancora *in situ*, incastrati nella roccia, in altri portati alla luce dai lavori agricoli (ulteriori sopralluoghi, tra l'estate del 2018 e la primavera del 2019, hanno rilevato tuttavia la scomparsa della maggioranza dei tubuli visti anni prima, né sono stati riscontrati nuovi affioramenti). Solo i tubi incastrati in roccia a Serra di Bruno (Figures 7, a-b), e gli spezzoni riutilizzati nei muri a secco sono ancora oggi al loro posto, e insieme ai manufatti di superficie hanno permesso di ricostruire sia il tracciato (Figure 3), sia – grazie alla collaborazione con l'allora Dipartimento di Ingegneria Idraulica dell'Università di Palermo, e alla disponibilità dei proff. Goffredo La Loggia, Giuseppe Ciraolo e Antonino Maltese – il suo profilo altimetrico (Figure 8).

Nel dettaglio, sono state individuate due distinte condotte, sui versanti occidentale ed orientale di Serra di Cuozzo (m 582), alimentate da due diverse sorgenti, rispettivamente nelle contrade Follia (quota 760, località Acquacitita) e Margivite (quota 550).[45] È verosimile che la condotta che attraversa Margivite – toponimo del sostrato arabo che indica l'abbondanza di acqua[46] – confluisse nell'altra, che rappresenta pertanto quella principale, alimentata dalla sorgente Acquacitita, la più copiosa del comprensorio. Peraltro, proprio all'interno del *bottisco* (la galleria, moderna, che permette di raggiungere la 'testa dell'acqua', cioè la sorgente vera e

---

[45] Qui si trova una fontana monumentale di età moderna, il c.d. 'Viviere', che secondo la *vulgata* si troverebbe sul sito di una sorgente o fontana di età antica, della quale tuttavia non vi è traccia; in Ragonese e Bono 1989: 35–36 è indicato – ma senza fondamento – un bacino di raccolta delle acque che alimentavano l'acquedotto di Alesa.
[46] Caracausi 1993, *s.v. Margio*.

Figure 7 – L'acquedotto: rinvenimenti *in situ*.

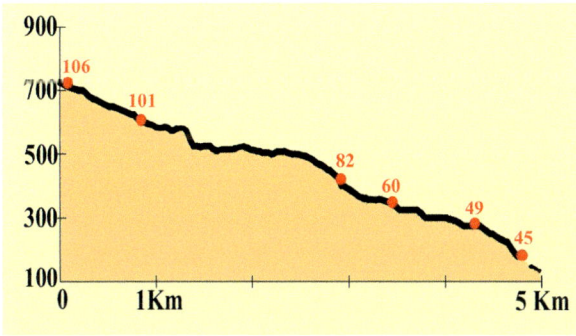

Figure 8 – L'acquedotto: profilo altimetrico del ramo principale.

propria) di Acquacitita sono state rinvenute – secondo la *vulgata* – lucerne di età ellenistica, in esposizione all'*antiquarium* Badia di Tusa (le lucerne esposte sono tuttavia di età imperiale, ma ciò non esclude che la sorgente, se effettivamente adoperata in età ellenistica, sia stata in uso per lungo tempo).

Ciò non implica continuità d'uso da età antica ad oggi, ma dal punto di vista di una prospettiva temporale di lunga durata è significativo che, nonostante le trasformazioni subite, il comprensorio continui ad essere una delle principali fonti di approvvigionamento idrico dell'agro e della città di Tusa, e che una condotta in ferro, ormai dismessa, solchi tutta la dorsale fino a Castel di Tusa, seguendo lo stesso tracciato dell'acquedotto antico, in alcuni tratti (Serra di Bruno e collina di Santa Maria, bene evidente ad Ovest del santuario di Apollo) correndo affiancata a quest'ultimo (non si può escludere che la realizzazione della condotta moderna abbia portato alla luce, distruggendola in più punti, quella antica). Non è insolita peraltro una simile continuità d'uso del territorio fino ad età moderna: non solo in riferimento alle tracce di centuriazioni e suddivisioni agrarie (quasi per nulla attestate in Sicilia), ma anche riguardo agli acquedotti, come dimostra l'Acquedotto Cornelio di Termini Imerese, affiancato per lunghi tratti, specie nel settore più vicino al centro abitato, all'acquedotto moderno,[47] o a Taormina i resti individuati presso la Chiesa dei Cappuccini.[48]

Tornando all'acquedotto alesino, la condotta principale si snoda per oltre 4.500 m da Acquacitita alla dorsale Serra di Mezzo-Serra di Cuozzo (versante occidentale)-Serra di Bruno, fino all'ingresso in città. A questa vanno aggiunti i ca. 1.500 m del tracciato Margivite-Serra di

---

[47] Belvedere 1986.
[48] Muscolino 2020: 57, fig. 5.

Cuozzo-Strina, che aggira sul versante orientale Serra di Cuozzo. Mancano dati precisi sulla quota di posa del manufatto, ad eccezione di un breve tratto sul versante SE di Serra di Bruno,[49] dove in sezione sono ancora visibili alcuni tubi a sezione circolare incassati nel banco roccioso pochi cm sotto il piano di campagna (Figure 7, a-b).

Mancano dati anche sulla pendenza,[50] che poteva essere spezzata attraverso vasche alternate a tubazioni chiuse tutte le volte che ci si imbatteva in marcate rotture di pendio. Nel suo insieme era certamente una condotta inclinata, con alternanza di tratti a forte, media e bassa pendenza, costruita in parte con canalette di terracotta dalla sezione ad 'U' (Figure 9), forse coperte da tegole piane o da mattoni, copertura mai riscontrata tuttavia nei tratti identificati, ovvero da tubi a sezione cilindrica. In questo caso il condotto era alloggiato in trincea, poche decine di cm sotto il piano di campagna, come documentato altrove:[51] una struttura analoga a Philosophiana sembra avere 'a conduit of overlapping, upturned roof tiles...'; nell'entroterra di Eraclea Minoa un ampio taglio in roccia poteva accogliere un acquedotto di età romana, forse a servizio dell'insediamento rurale di Campanaio, distante ben 2,5 km; altre segnalazioni riguardano i territori di Cassibile e di Selinunte.

I tubi cilindrici avevano le estremità conformate a maschio e femmina, e ciò consentiva un incastro perfetto, senza la necessità di ricorrere ad anelli di tenuta esterni.[52] Nell'unico caso (Figure 7c)[53] in cui l'estremità si è conservata per intero, risulta lunga cm. 10,2, pari a ca. 1/6 del tubo, mentre lo spessore varia da metà a 2/3 circa rispetto a quello delle pareti; il profilo del tubo è pertanto leggermente diverso rispetto a quello rinvenuto da Cultrera a Selinunte.[54] Al suo interno, e per tutta la lunghezza, si è conservato un sottile strato di incrostazioni calcaree (spess. cm 0,9), frutto quasi certamente dell'azione di deposito delle acque correnti; altre incrostazioni visibili sull'elemento di raccordo potrebbero essere dovute alla presenza di un collante: sappiamo infatti che i giunti erano talvolta impermeabilizzati con un cemento composto da carbonato di calcio, che poteva anche essere mescolato con acqua e olio,[55] e Vitruvio (VIII, 6, 8) ricorda l'uso di calce viva impastata con olio.

Figure 9 – L'acquedotto: canale ad 'U'.

Le misure degli elementi fittili che compongono l'acquedotto non sono ricostruibili con certezza e variano con il tipo di conduttura: la lunghezza dei tubi cilindrici doveva probabilmente essere intorno ai 60 cm, il diametro esterno tra 20 e 24 cm.[56] Le canalette con sezione ad U potevano essere più lunghe, fino ad almeno 76 cm,[57] e lo spessore delle pareti oscilla tra 3,8 e 5 cm, misure compatibili con quanto sappiamo da Vitruvio.[58] Nel centro urbano di Morgantina[59] le canalette hanno misure simili, in particolare quelle di grande diametro, ma sono quasi tutte dotate di foro ellissoidale, con coperchio, per ispezione, in nessun caso ipotizzabile tra i reperti di *Halaesa*.

L'acqua scorreva dunque all'interno di una condotta inclinata, in prevalenza sotterranea; non si può escludere che in una delle aree più in quota, a Serra di Mezzo,[60] dove è visibile per pochi metri un canale intagliato nel banco roccioso, largo 23/25 cm e dotato di spallette alte ca. 15 cm (Figure 10), procedesse in superficie, ovvero – più probabilmente – che il canale fosse fiancheggiato da più alte spallette in pietra e malta idraulica non più conservate, coperto da lastre di pietra o da mattoni. Nel territorio di Selinunte, a Torre Bigini, il canale di adduzione, realizzato con blocchi regolari, pare fosse largo 52 cm, mentre del canale di abduzione sono ancora visibili tratti intagliati nel banco roccioso.[61]

---

[49] Burgio *et al.* 2008, UT 5: 54–56; 58-60: 101–104.
[50] Tölle-Kastenbein 1993: 53–54, ricorda che nelle condotte greche e romane variava in singoli tratti tra 0,0025% e 10%, e che a Priene la condotta che portava l'acqua in città attraverso una discesa di circa 2 km aveva una pendenza del 10%.
[51] Wilson 2000: 24-25, fig. 16.
[52] Tölle-Kastenbein 1993: 99–103, figg. 56–58, rileva che in acquedotti di età romana i tubi potevano avere una sezione interna ed un profilo esterno delle stesse dimensioni.
[53] Burgio *et al.* 2008, UT 60: fig. 75.
[54] Furcas 2019: fig. 8.
[55] Belvedere 1994: 28.
[56] Burgio *et al.* 2008, UT 60: 127, figg. 75, 119.
[57] Burgio *et al.* 2008, UT 81.
[58] Vitr. VIII, 6, 8: i tubi di terracotta hanno parete spessa 'non meno di due dita', e una delle due estremità è 'sottile, a forma di lingua, per potere inserirsi l'uno nell'altro incastrandosi'; trad. E. Romano; cfr. anche Gros 1997: 1187, nota 256: due dita di spessore equivalgono a 3,6 cm.
[59] Bruno-Renna 2000: 71, 74, fig. 2b-c, tab. 1b-c.
[60] Burgio *et al.* 2008, UT 101: 129–130.
[61] Salinas 1885: 289; Fourmont 2012: 129–131, figg. 24–25, 32–33.

Figure 10 – L'acquedotto: tratto di canale intagliato nel banco roccioso.

Il sistema descritto – condotte inclinate, sotterranee o superficiali, con impiego di tubi o condutture a pelo libero, scavate nella roccia e talvolta dotate di spallette, con o senza copertura – era il più comune. In Sicilia conosciamo solo brevi condotte, sia in contesti rurali[62] che in aree urbane, come a Morgantina.[63] Nell'agro di Camarina le fattorie Castalia e Capodicasa sono dotate di canali a pelo libero, scavati nella roccia, larghi 30 m e profondi rispettivamente 40 e 20 cm; diversa la struttura dei c.d. 'muri Orsi', canale dotato di spallette in muratura realizzate con malta idraulica, largo 70 cm.[64] Come si è detto, a Calacte e *Aluntium* sono stati rinvenuti tubi in terracotta molto simili a quelli di *Halaesa*, ma dotati di foro di ispezione nella parte superiore, mentre ben diversi sono quelli degli acquedotti di Eraclea di Lucania e Amendolara, in Calabria, in entrambi i casi riferiti ad età imperiale per la tipologia costruttiva delle cisterne a questi connesse.[65]

I più simili ai manufatti di *Halaesa* sono i rinvenimenti dell'entroterra di Calacte. A monte del centro antico, nelle contrade Serralisa, Sampieri e Piano della Chiesa, è stato individuato agli inizi degli anni '70 del secolo scorso un acquedotto che sembra si sviluppasse per alcuni km, dalle sorgenti di Piano Piraino-Pizzo Castellano-Piano della Chiesa fino alla località Serra Lisi-Acqua Ramusa.[66] I canali in terracotta, lunghi cm 74 e dal profilo ad 'U',[67] erano incastrati tra loro e coperti da mattoni, alcuni dei quali recano un bollo il cui testo suggerisce l'intervento della comunità dei Calactini nella realizzazione del manufatto. L'acquedotto – che è stato riferito, ma senza dati probanti, ad età ellenistica – doveva raggiungere *Solusapre*,[68] località attestata nell'*Itinerarium Antonini*.

Tra i rinvenimenti connessi all'acquedotto alesino merita infine particolare attenzione una struttura,[69] di cui rimane solo una piccola parte (circa 2,90 x 0,80 m): si tratta di struttura isolata, visibile in elevato per non oltre 1 metro, costruita con pietre informi di medie dimensioni, legate con malta biancastra molto compatta, nei cui dintorni erano altri grossi spezzoni in malta e pietre, e di malta e cocciopesto (Figure 11). Tale struttura, ai piedi di un tratto in forte pendenza all'estremità settentrionale di Serra di Bruno, potrebbe essere interpretata come vasca di disconnessione, destinata a spezzare la pressione dell'acqua che scorreva entro la condotta forzata costituita dai tubi cilindrici di terracotta, sopra descritti, visibili in sezione nel settore più a monte di Serra di Bruno. Non si può escludere che altre vasche con medesima funzione esistessero lungo il percorso dell'acquedotto, benché al momento non individuate. Strutture simili non sono note al momento nelle zone vicine, né a Calacte né ad *Aluntium*; soltanto a Selinunte, lungo la condotta di Torre Bigini si è ipotizzata la presenza di almeno un bacino con questa funzione.[70]

Altro tema è la cronologia dell'acquedotto di *Halaesa*, per la quale non disponiamo di dati concreti. È chiaro che lunghezza e caratteristiche tecniche dimostrano che si trattò di un impegno economico importante per la città, e lo stesso vale per la manutenzione; né possiamo escludere restauri di singoli tratti o eventuali fasi cronologiche distinte tra i due tracciati identificati intorno a Serra di Cuozzo.[71] Si può ipotizzare che il manufatto risalga al periodo più florido per la città, l'età ellenistico-romana, in considerazione della vitalità del centro urbano. E per quanto si tratti di un'opera meno incisiva sul paesaggio rispetto agli imponenti manufatti di età romana, non c'è dubbio che la sua realizzazione

---

[62] Dove potevano anche avere la funzione di drenare e regolarizzare le acque del sottosuolo: Di Stefano 1997: 292-300; Di Stefano 2000; Bouffier 2006: 191-194.
[63] Bruno e Renna 2000: 71; Bouffier 2006: 190.
[64] Di Stefano 2000.
[65] Quilici 1967: 113-114, fig. 252; Settembrini 1993: 195-198. figg. 9-10.

[66] Fiore 1971: 37-38; Scibona 1987: 10-11; Collura 2016: 293, fig. 48.
[67] Fiore 1971: 38, figg. 1-3; Fiore 1976: fig. 4.
[68] Fiore 1976: 47.
[69] Burgio 2008: 98-99, UT 49, fig. 67.
[70] Fourmont 2012: 130.
[71] Ma canali e tubuli adoperati sono sempre dello stesso tipo.

Figure 11 – L'acquedotto: struttura *in situ*.

richiese adeguata conoscenza delle caratteristiche morfologiche del territorio in cui si operava, oltre che maestranze capaci.

La presenza di alcune fattorie e altre installazioni rurali – di età ellenistico-romana ed imperiale[72] – a brevissima distanza dal tracciato suggerisce la possibilità che l'acquedotto potesse alimentare singoli insediamenti nel territorio. Inoltre, rinvenimenti di tubi in aree ben distinte rispetto ai due tracciati[73] attestano l'esistenza di brevi condotte che collegavano sorgenti a singoli insediamenti rurali. Infine, delle acque sorgive ci si poteva servire per alimentare vasche, come sembrerebbe indicare la stretta connessione tra una sorgente, attuale, ed un manufatto rivestito di intonaco all'interno di una villa in contrada Feudo.[74] Proprio qui, alle pendici meridionali di *Halaesa* (Passo della Madonna), giungono i resti ancora visibili in superficie, ma non è chiaro se l'acquedotto permettesse di rifornire davvero la città, considerato il notevole dislivello tra il Passo e la porta meridionale del centro urbano, dove forse poteva trovarsi una fontana e/o cisterne.

Dunque quanto descritto suggerirebbe una forma di continuità d'uso delle risorse del territorio – per l'approvvigionamento idrico della città e per gli insediamenti rurali – che ha paralleli a Torre Bigini, dove alla vasca 'di opera classica' se ne sovrappone una moderna;[75] nella vasca sono state rinvenute lucerne di età imperiale, e nei dintorni – significativamente 'contrada Fontana' – altri manufatti e lucerne di età imperiale.[76] Peraltro, come osserva Salinas,[77] dalla descrizione di Tommaso Fazello si deduce che alla metà del XVI secolo fossero visibili acquedotti che raggiungevano Selinunte. Sembra possibile cogliere ancora un parallelo con l'agro di *Halaesa*, nella continuità d'uso delle risorse.[78]

Un'ultima considerazione, strettamente legata al sistema idrico del comprensorio, riguarda la possibilità che si possano localizzare i settori in cui ricadevano i gruppi di lotti descritti nella *Tabula Halaesina*. Sul tema

---

[72] Burgio *et al.* 2008, UT 51: 62, 85, 86.
[73] Contrade Surda, S. Ippolito e Foieri: Burgio *et al.* 2008, UT 93: 120, 127.
[74] Burgio *et al.* 2008, UT 40.

[75] Un vivaio fu realizzato nel '500 e opere di captazione idrica in età moderna: Fourmont 2012: 109-112, figg. 3, 5.
[76] Fourmont 2012: 110–113, fig. 9, riporta passi della corrispondenza tra Bruno Varvaro e Jole Bovio Marconi – anni 1949–1953 – che attestano il rinvenimento di altre due lucerne africane di età tardoantica.
[77] Salinas 1885: 288.
[78] L'edificio del 'Viviere', sopra ricordato.

divergono le opinioni, la cui corretta formulazione deve tenere conto non solo delle trasformazioni che il tessuto idrico ha subito nel corso del tempo, bene evidente come si è visto prima nel tratto finale del fiume di Tusa, ma anche dei dati archeologici e del contesto ambientale, paesaggistico e strategico in cui si inserisce la città: il medio bacino del fiume fu da subito, anche sul versante orientale,[79] parte integrante del territorio di *Halaesa*, almeno fino a S. Ippolito e Rocca d'Armi[80] e alla strettoia valicata dal Ponte di Pettineo.[81]

Se nessun dubbio sussiste sulla collocazione di un gruppo di lotti, tra le fortificazioni orientali e il fiume, la difficoltà principale riguarda i lotti ricadenti nell'area in cui scorrono il *potamos* e l'*Opikanos*, affluente di destra del primo. L'idea che il *potamos* non vada collocato sul versante destro del Tusa e che corrisponda all'attuale vallone Cicera[82] confligge tuttavia con i dati archeologici e con il contesto generale del territorio di *Halaesa*, poiché parte dalla premessa che il territorio non includesse il versante orientale del Tusa. Inoltre, l'identificazione del *potamos* con il Cicera non risolve il rapporto tra *potamos*, *krana Ypirra* e *hodos Xenis*,[83] poiché sia la contrada Ospedale, nella quale si collocherebbe la *hodos Xenis*, sia l'area di Acquacitita (ammesso che sia corretto collocare in questo settore la *krana Ypirra*) non hanno alcuna relazione con il Cicera, gravitando invece sul Tusa. Naturalmente non si può escludere che la zona di testata del Cicera (contrada Acquatico) sia stata sede di una delle sorgenti menzionate nell'epigrafe (e dunque rientri tra i settori oggetto della suddivisione catastale), ma tale considerazione investe anche varie zone del bacino del Tusa (contrada Margibuffa; valloni del Leone-S. Pantaleo e San Giovanni), nonché un'altra area ricchissima di acqua, le contrade Piano Fontane e Piano Vagna, subito ad Ovest del centro moderno, dove è stata individuata una villa di età imperiale.[84]

Proprio la zona di Piano Fontane si rivela di grande interesse, in una prospettiva che miri ad esaminare e comprendere l'antico anche attraverso l'età moderna, ineludibile prospettiva di metodo quando oggetto di ricerca è il territorio. Sorgenti e abbeveratoi, luoghi di sosta destinati a rifocillare uomini e armenti sono aspetti costanti di un paesaggio antropizzato, e numerose testimonianze menzionano realizzazione e manutenzione di abbeveratoi e fontane. Le notizie disponibili[85] si datano dalla fine del 1500, e ricordano fino al 1720 le strutture di Cubba, Ancinè, Conzaria, 'Gurga dell'orto della folia', tutte tra Tusa e Acquacitita; la proposta (del 1629) di spostare un abbeveratoio a causa di un movimento franoso che aveva danneggiato il precedente conferma il consistente grado di dissesto che caratterizzava – e caratterizza – un comprensorio a prevalente componente argillosa solcato da innumerevoli rivoli di varia portata. Di grande interesse sono gli interventi migliorativi riguardanti la distribuzione dell'acqua nella contrada Lanciné: nel 1777 (la contrada si chiamava Santa Sofia) fu realizzato un grande abbeveratoio, alimentato da un nuovo acquedotto proveniente da Cubba, e nel XIX secolo una condotta e un abbeveratoio nuovi; si potrebbe pensare ad una relazione con la presenza di concerie, attestate nei secoli XVIII e XIX,[86] che come è noto richiedono grandi quantitativi di acqua, oltre che disponibilità di legna.[87] Lavorazione delle pelli, disponibilità di acqua e legna, eccentricità rispetto all'area abitata, inducono a pensare alla prescrizione contenuta della *Tabula* circa il divieto di effettuare la concia delle pelli in zone vicine ad aree sacre; prescrizione che potrebbe estendersi anche ad aree rurali abitate, e non sfugge – ma è solo una suggestione – la marginalità della zona in esame rispetto al centro urbano di *Halaesa*, al punto che i dati archeologici testimoniano qui un solo insediamento, la già menzionata villa di Piano Fontane.

Ad ogni modo, non è questa la sede in cui affrontare il tema dell'identificazione delle aree del catasto alesino. Ciò che importa rilevare è che gestione e irreggimentazione dell'acqua coinvolgono tutto il territorio, e che i dati archeologici relativi ad un acquedotto appartengono – non sappiamo se per caso o per continuità d'uso e sopravvivenza dall'antico – soltanto alla dorsale che da Acquacitita raggiunge la costa.

Sembra di cogliere – in ultimo – elementi di continuità tra *Halaesa* e la sua erede, frazionata tra il centro abitato in altura e la borgata marinara, Castel di Tusa: il primo connotato dalla prossimità ad aree boschive e dalla presenza di sorgenti e fontane pubbliche; la seconda dotata di un piccolo approdo, direttamente legato alla viabilità paracostiera e ai percorsi di collegamento con l'interno dell'isola, innervati sul bacino del fiume di Tusa, da cui si perviene alla zona di Gangi e Nicosia e a quella Sicilia interna con cui in età romana, e certamente in ogni tempo, uomini e merci scambiavano relazioni e prodotti.

---

[79] Siti già in età tardo-arcaica/classica: Burgio *et al.* 2008, UT 137, 148–149: 221-224, figg. 174-175.
[80] Rilievi da cui si vede perfettamente la città: Burgio *et al.* 2008, UT 117, 147-149: 139-140, 163-166.
[81] Eretto nel 1571, è stato oggetto di spese ricorrenti negli anni 1715, 1717, 1725 da parte dell'amministrazione di Tusa, tese a gestire e convogliare i numerosi valloni e torrenti attraverso poderose attività di sbancamento, conseguenti ai periodici mutamenti del corso del fiume: Pettineo 2012: 40-42.
[82] Prestianni Giallombardo 2012: 381, 385-386.
[83] Prestianni Giallombardo 2012: 391 e nota 55.
[84] Burgio *et al.* 2008, UT 64.

[85] Pettineo 2012: 28-33.
[86] Pettineo 2012: 56.
[87] Molto vicino è il bosco di Tardara, oggi la più estesa area boschiva presso Tusa.

Riferimenti bibliografici

Alliata, V., Belvedere, O., Cantoni, A., Cusimano, G., Marescalchi, P., Vassallo, S. 1988. *Himera* III.1. *Prospezione archeologica nel territorio*. Roma: L'Erma di Bretschneider.

Arangio Ruiz, V. Olivieri, A. 1925 [1965]. *Inscriptiones Grecae Siciliae et Infimae Italiae ad Ius Pertinentes*. Roma: L'Erma di Bretschneider (ed. anastatica).

Arena, E. 2020. *Nuove epigrafi greche da Halaesa Archonidea. Dati inediti sulle* Tabulae Halaesinae *e su una città della Sicilia tardo-ellenistica*. British Archaeological Reports International Series 3017. Oxford: Archaeopress.

Belvedere, O., Bertini, A., Boschian, G., Burgio, A., Contino, A., Cucco, R.M., Lauro, D. 2002. *Himera* III.2. *Prospezione archeologica nel territorio*. Palermo: Roma: L'Erma di Bretschneider.

Belvedere, O. 1986. *L'acquedotto Cornelio di Termini Imerese*. Roma: L'Erma di Bretschneider.

Belvedere, O. 1984. s.v. Acquedotto, *Enciclopedia dell'Arte Antica* II Suppl., vol. I: 27–34. Roma: Istituto della Enciclopedia Italiana.

Belvedere, O. 2008. Paesaggio catastale, paesaggio letterario e archeologia del paesaggio. Tre percezioni a confronto, in A. Burgio (ed) *Il paesaggio agrario nella Sicilia ellenistico-romana: Alesa e il suo territorio*: 1–10. Roma: L'Erma di Bretschneider.

Belvedere, O., Forgia, V. 2017. Termini Imerese. Indagini nell'edificio termale. *Sicilia Antiqua* XIV: 23–32.

Bordonaro, G. 2012. *Carta archeologica e Sistema Informativo Territoriale del Comune di Baucina*. Palermo: Comune di Baucina.

Bouffier, S. 1987. L'alimentation en eau de la colonie grecque de Syracuse. Réflexion sur la cité et sur son territoire. *Mélanges de l'École française de Rome. Antiquité* 99: 661–691.

Bouffier, S. 2006. La gestion des ressources hydriques de la cité antique de Camarina, in P. Pelagatti, G. Di Stefano, L. de Lachenal (eds) *Camarina. 2600 anni dopo la fondazione. Nuovi studi sulla città e sul territorio*, Atti del Convegno Internazionale (Ragusa, 7 dicembre 2002/7-9 aprile 2003): 183–196. Roma: Centro studi F. Rossitto-Istituto poligrafico e zecca dello Stato.

Bouffier, S., Belvedere, O., Vassallo, S. (eds) 2019a. *Gérer l'eau en Méditerranée au premier millénaire avant J.-C.* Aix-en-Provence: Presses Universitaires de Provence.

Bouffier, S., Dumas, V., Lenhardt, Ph., Paillet, J-L., Turci, M. 2019b. Qui est l'auteur de l'aqueduc du Galermi? Nouvelle pistes sur un aqedc plurimillénaire en province de Syracuse (Italie), in S. Bouffier, O. Belvedere, S. Vassallo (eds) *Gérer l'eau en Méditerranée au premier millénaire avant J.-C.*: 65–85. Aix-en-Provence: Presses Universitaires de Provence.

Bruno, G., Renna, C.E. 2000. La rete idrica di Morgantina: tentativo di definizione del livello dell'acqua all'interno delle condotte in terracotta, in C.G.M. Jansen (ed.) *Cura aquarum in Sicilia* (Proceedings of the Tenth International Congress on the History of Water Management and Hydraulic Engineering in the Mediterranean Region, Syracuse May 16-22, 1998): 69–78. BABesch Supplement 6. Leiden: Peeters.

Burgio, A. 1997. Saggio archeologico nella Chiesa di S. Caterina d'Alessandria di Termini Imerese, in Aa.Vv. *Archeologia e Territorio*: 237-249. Palermo: Soprintendenza BB.CC.AA.

Burgio, A. 2002. *Resuttano, Forma Italiae 42*. Firenze: Olschki.

Burgio, A. 2008. *Il paesaggio agrario nella Sicilia ellenistico-romana: Alesa e il suo territorio*. Roma: L'Erma di Bretschneider.

Burgio, A. 2009. Indagini archeologiche nella valle del Fiume Tusa. Scavi a Cozzo Sorbo e Contrada Belvedere (Motta d'Affermo), in G. Scibona, G. Tigano (eds) *Alaisa-Halaesa. Scavi e ricerche (1970-2007)*: 221-232. Messina: Sicania.

Burgio, A. 2017. Persistenze e trasformazioni nel sistema viario tra Castronovo e le Madonie: la 'via Francigena' tra *xenodochia* e *itineraria peregrinorum*, in G. Mellusi, R. Moscheo (eds) Ktema es aiei. *Studi e ricordi in memoria di Giacomo Scibona*: 109–117. Messina: Biblioteca Archivio Storico Messinese 46, Analecta 21.

Burgio, A. 2021. La viabilità minore nella Sicilia centro-meridionale: il comprensorio di Cignana tra la via Selinuntina e la Catina-Agrigentum. *Atlante Tematico di Topografia Antica* 31: 417–436.

Burgio, A., Di Maggio A., Tigano, G. 2008. Carta archeologica, in A. Burgio *Il paesaggio agrario nella Sicilia ellenistico-romana: Alesa e il suo territorio*: 53–185. Roma: L'Erma di Bretschneider.

Caminneci, V., Parello, M.C., Rizzo, M.S. (eds) 2020. *Le forme dell'acqua. Approvvigionamento, raccolta e smaltimento nella città antica*. Atti delle Giornate Gregoriane XII Edizione (Agrigento, 1–2 dicembre 2018). Bologna: Ante Quem.

Campagna, L., La Torre, G.F. 2008. Ricerche sui monumenti e sulla topografia di *Tauromenion*: una stoà ellenistica nell'area della Naumachia. *Sicilia Antiqua* V: 115–146.

Capasso, I. 1989. Corsi d'acqua come indicazione di confine nella grande iscrizione di Alesa. *La Parola del Passato* CCXLVII: 281–285.

Caracausi, G. 1993. *Dizionario onomastico della Sicilia: repertorio storico-etimologico di nomi di famiglia e di luogo*. Palermo: Centro di Studi Filologici e Linguistici Siciliani.

Carettoni, G. 1959. Tusa (Messina) - Scavi di Halaesa (prima relazione). *Notizie degli Scavi di Antichità* S. VIII, XIII: 293–349.

Carettoni, G. 1961. Tusa (Messina) - Scavi di Halaesa (seconda relazione), *Notizie degli Scavi di Antichità* S. VIII, XV: 266–321.

Castrianni, L., Di Giacomo, G., Ditaranto, I., Miccoli, I., Rapisarda D.A., Scardozzi, G. 2018. Gli Acquedotti romani di Taormina. *Quaderni di Archeologia dell'Università di Messina* VIII: 81–128.

Collura, F. 2016. Studia Calactina I. *Ricerche su una città greco-romana di Sicilia: Kalè Akté - Calacte*. British Archaeological Reports International Series 2813. Oxford: Archaeopress.

Costanzi, M., Gerber, F., Lamare, N., Mouny, S. 2019. Halaesa: bilan des activités de la mission archéologique francçaise 2016–2019. *Kokalos* LVI: 9–97.

Di Maggio, A. 2008. *La viabilità*, in A. Burgio *Il paesaggio agrario nella Sicilia ellenistico-romana: Alesa e il suo territorio*: 199–213. Roma: L'Erma di Bretschneider.

Di Maggio, A. 2012. La prospezione archeologica e l'indagine topografica applicate alla ricostruzione della viabilità della Sicilia antica: i collegamenti tra la costa e l'entroterra lungo la valle del Torrente di Tusa (Halaisos), in *Per la conoscenza dei Beni Culturali IV, Ricerche del Dottorato in Metodologie Conoscitive per la Conservazione e la Valorizzazione dei Beni Culturali 2007-2011*: 11–21. S. Maria Capua Vetere: Spartaco.

Di Stefano, G. 1997, Il fiume e la città nella Sicilia meridionale: il caso di Camarina. Le testimonianze archeologiche, in S. Quilici Gigli (ed.) *Uomo acqua e paesaggio*, Atti dell'incontro di studio sul tema 'Irregimentazione delle acque e trasformazione del paesaggio antico' (S. Maria Capua Vetere, 22-23 novembre 1996): 292–300. Atlante Tematico di Topografia Antica, Suppl. II. Roma: L'Erma di Bretschneider.

Di Stefano, G. 2000. Sistemi idraulici nella Chora della Sicilia Greca. Il caso di Camarina, in C.G.M. Jansen (ed.) *Cura aquarum in Sicilia* (Proceedings of the Tenth International Congress on the History of Water Management and Hydraulic Engineering in the Mediterranean Region, Syracuse May 16–22, 1998): 231–239. BABesch Supplement 6. Leiden: Peteers.

Facella, A. 2006. *Alesa Arconidea. Ricerche su un'antica città della Sicilia tirrenica*. Pisa: Edizioni della Normale.

Fasolo, M. 2014. *Tyndaris e il suo territorio* (Vol. II). Carta archeologica del territorio di Tindari e materiali. Roma: mediaGEO.

Fazello, T. 1558. *De rebus Siculis decades duae*, Panormi 1558 (trad. a cura di A. de Rosalia e G. Nuzzo, 1990). Palermo: Regione Siciliana, Assessorato dei Beni Culturali e Ambientali e della Pubblica Istruzione.

Fiore, P. 1971. Acquedotto sacro a Demetra. *Sicilia Archeologica* IV, 14: 37–39.

Fiore, P. 1976. Sull'antico acquedotto calactino. *Sicilia Archeologica* IX, 31: 43–48.

Fiorentini, G. 1993-94. Attività di indagini archeologiche della Soprintendenza Beni Culturali e Ambientali di Agrigento. *Kokalos* XXXIX–XL, II, 1: 717–733.

Fourmont, M. 2012. La vasca selinuntina di Torre Bigini. *Sicilia Archeologica* 106: 109–135.

Furcas, G.L. 2019. Sistemi di approvvigionamento idrico e drenaggio urbano a Selinunte, in R. Atria, G.L. Bonanno, A. Curti Giardina, G. Titone (eds) *Selinunte. Produzioni ed economia di una colonia greca di frontiera* (Atti del Convegno internazionale, Selinunte, 15–16 aprile 2016). *Sicilia Archeologica* 111: 76–119.

Furcas, G.L. 2020. Le infrastrutture idrauliche nella Valle dei Templi: per una rilettura delle opere di Feace, in V. Caminneci, M.C. Parello, M.S. Rizzo (eds) *Le forme dell'acqua. Approvvigionamento, raccolta e smaltimento nella città antica*, Atti delle Giornate Gregoriane XII Edizione (Agrigento, 1–2 dicembre 2018): 141–155. Bologna: Ante Quem.

Gros, P. 1997. *Vitruvio, De Architectura* (trad. di E. Romano).Torino: UTET.

Jansen, C.G.M. (ed.) 2001. *Cura aquarum in Sicilia* (Proceedings of the Tenth International Congress on the History of Water Management and Hydraulic Engineering in the Mediterranean Region, Syracuse May 16–22, 1998). BABesch Supplement 6. Leiden: Peteers.

La Torre, G.F. 2018. Sofiana: storia di un sito della Sicilia interna tra età augustea e tardo-antico, in O. Belvedere, J. Bergemann (eds) *La Sicilia romana: Città e Territorio tra monumentalizzazione ed economia, crisi e sviluppo. Atti del Seminario di Gottingen (25-27 novembre 2017)*: 115–126 Palermo: University Press.

La Torre, G.F. 2020. Edifici termali e viabilità nella Sicilia romana, in V. Caminneci, M.C. Parello, M.S. Rizzo (eds) *Le forme dell'acqua. Approvvigionamento, raccolta e smaltimento nella città antica*, Atti delle Giornate Gregoriane XII Edizione (Agrigento, 1–2 dicembre 2018): 211–218. Bologna: Ante Quem.

Lauro, D. 2009. *Sambuchi (IGM 259 IV SE), Forma Italiae* 45. Firenze: Olschki.

Messina, A., Petruso, D., Sineo, L. 2009. Gli inumati di Cozzo Sorbo. Nota paleobiologica, in G. Scibona, G. Tigano (eds) *Alaisa-Halaesa. Scavi e ricerche (1970-2007)*: 221–238. Messina: Sicania.

Muscolino, F. 2020. Le cisterne di Taormina in età ellenistico-romana, in V. Caminneci, M.C. Parello, M.S. Rizzo (eds) *Le forme dell'acqua. Approvvigionamento, raccolta e smaltimento nella città antica*, Atti delle Giornate Gregoriane XII Edizione (Agrigento, 1–2 dicembre 2018): 51–61. Bologna: Ante Quem.

Pettineo, A. 2012, *Tusa. Dall'Universitas Civium alla Fiumara d'Arte*. Messina: Armando Siciliano.

Polizzi, G. 2019. Gestione e uso dell'acqua a Solunto: le cisterne, in S. Bouffier, O. Belvedere, S. Vassallo (eds) *Gérer l'eau en Méditerranée au premier millénaire avant J.-C.*: 193–209. Aix-en-Provence: Presses Universitaires de Provence.

Polizzi, G., Ollivier, V., Bellier, O., Pons-Branchu, E., Fontugne, M. 2020. Evidenze di un sistema idraulico alternativo in Sicilia e in Grecia, in V. Caminneci, M.C. Parello, M.S. Rizzo (eds) *Le forme dell'acqua. Approvvigionamento, raccolta e smaltimento nella città antica*, Atti delle Giornate Gregoriane XII Edizione (Agrigento, 1–2 dicembre 2018): 305–315. Bologna: Ante Quem.

Portale, E.C. 2009. Le sculture da Alesa, in G. Scibona, G. Tigano (eds) *Alaisa-Halaesa. Scavi e ricerche (1970-2007)*: 67–92. Messina: Sicania.

Prestianni Giallombardo, A.M. 2004–2005, Ambiente e paesaggio nella Sicilia ellenistico-romana. I percorsi dell'acqua nel territorio di Halaesa Archonidea. *Minima Epigraphica et Papirologica* VII–VIII, 9–10: 229–248.

Prestianni Giallombardo, A.M. 2012. L'acqua come elemento fondamentale nell'organizzazione e nel controllo del territorio e dello spazio urbano. Il caso di Alesa, in A. Calderone (ed.) *Cultura e religione delle acque, Atti del Convegno interdisciplinare* (Messina 29–30 marzo 2011): 375–398. Roma: Giorgio Bretschneider.

Quilici, L. 1967. *Siris-Heraclea, Forma Italiae* 10. Firenze: Olschki.

Ragonese, A., Bono, G.A. 1989. *Alesa e Tusa. Memoria di un popolo*. Palermo: Tip. A.C.P.

Rapisarda, D.A. 2020. Gli acquedotti di *Tauromenion*: approvvigionamento e gestione delle risorse idriche, in V. Caminneci, M.C. Parello, M.S. Rizzo (eds) *Le forme dell'acqua. Approvvigionamento, raccolta e smaltimento nella città antica*, Atti delle Giornate Gregoriane XII Edizione (Agrigento, 1–2 dicembre 2018): 245–253. Bologna: Ante Quem.

Rizzo, M.S., Zambito, L. 2007. Novità epigrafiche siciliane. I bolli di contrada Cignana (Naro, Ag). *Zeitschrift für papyrologie und epigraphik* 162: 271–277.

Salinas, A. 1885. Gli acquedotti di Selinunte, in V. Tusa (ed) *A. Salinas, Scritti scelti*, II: 83–96. Palermo: Edizioni Regione Siciliana.

Scibona, G. 1987. s.v. *Caronia*, in V. Gabba, G. Vallet, *Bibliografia Topografica della Colonizzazione Greca Italia* V: 8–15. Pisa – Roma: Scuola normale superiore – École française de Rome.

Scibona, G., Tigano, G. (eds) 2009. *Alaisa-Halaesa. Scavi e ricerche (1970-2007)*. Messina: Sicania.

Settembrini, A. 1993. L'acquedotto romano di Amendolara in Calabria. *Rivista di Topografia antica* III: 195–200.

Sfacteria, M. 2018. *Un approccio integrato al problema della ricostruzione della viabilità romana in Sicilia: la via Catania-Agrigento*. British Archaeological Reports International Series 2883. Oxford: Archaeopress.

Tigano, G. 2009. Brevi note sul cimitero bizantino nell'area dell'agorà, in G. Scibona, G. Tigano (eds) *Alaisa-Halaesa. Scavi e ricerche (1970-2007)*: 45–60. Messina: Sicania.

Tigano, G., Burgio, R. 2020. Prime considerazioni sul sistema di approvvigionamento idrico e di drenaggio nell'antica Alesa, in V. Caminneci, M.C. Parello, M.S. Rizzo (eds) *Le forme dell'acqua. Approvvigionamento, raccolta e smaltimento nella città antica*, Atti delle Giornate Gregoriane XII Edizione (Agrigento, 1–2 dicembre 2018): 219–230. Bologna: Ante Quem.

Tölle-Kastenbein, R. 1993. *Archeologia dell'acqua*. Milano: Longanesi.

Wilson, R.J.A. 2000. Aqueducts and water supply in Greek and Roman Sicily: the present *status quaestionis*, in C.G.M. Jansen (ed.) *Cura aquarum in Sicilia* (Proceedings of the Tenth International Congress on the History of Water Management and Hydraulic Engineering in the Mediterranean Region, Syracuse May 16–22, 1998): 5–36. BABesch Supplement 6. Leiden: Peteers.

Zambito, L. 2020. A proposito di alcune strutture idrauliche nel territorio agrigentino, per un primo inventario dei dati. Temi e prospettive per la ricostruzione della viabilità nell'agrigentino, in V. Caminneci, M.C. Parello, M.S. Rizzo (eds) *Le forme dell'acqua. Approvvigionamento, raccolta e smaltimento nella città antica*, Atti delle Giornate Gregoriane XII Edizione (Agrigento, 1–2 dicembre 2018): 417–424. Bologna: Ante Quem.

Aquae caldae.
Note sparse sul termalismo e lo sfruttamento delle acque sulfuree nel territorio a est di Agrigento tra archeologia e paletnologia

Luca Zambito[1]

[1] Independent researcher

Abstract: An excellent paper by M. Paoletti on the travelers of the Grand Tour in Sicily refers to the ways in which Jean-Pierre Houël's Voyage Pittoresque was advertised through a handbill, a notice, which explained the destination not only to a French audience, but to a wider European one. In this sort of advertising poster ante litteram, Houël put forward a summary of the contents of the work's four volumes. After palaces, city walls and bridges and before cemeteries, the reader would encounter '*acquedotti, cisterne, pozzi scavati nella roccia con le loro comunicazioni sotterranee, altri pozzi [...], impianti termali di diversi tipi*'. The particular geological features of the island were often given in summary form in comparison to Etna and its phenomena; this however did not prevent Houëll from highlighting the importance of natural resources, and water in particular, through the rest of the island.

In this paper I want to illustrate, through sample cases, some sites that, although part of the sulphur production network, are directly linked to springs of thermal or sulphurous waters, which were used for health, cult or, more simply, for domestic use and that, precisely due to these characteristics, determined the location and also conditioned the viability of the settlement.

Keywords: sulphurous waters, thermalism, water management

Introduzione

La presenza di filoni solfiferi nel territorio agrigentino ha dato vita a una fitta rete insediativa che è mio oggetto di studio da anni. Le miniere e gli impianti di raffinazione, assieme ai siti residenziali e alle connesse aree di sepoltura segnano profondamente il paesaggio, ne condizionano l'evoluzione e ne hanno nel tempo determinato i cambiamenti. Accanto alle antiche zolfare, tuttavia, e alla possibilità di sfruttamento agricolo, si osserva come sia la presenza di risorse idriche e il loro sfruttamento, ad avere un ruolo centrale nel successo o meno di un insediamento (Figure 1).[1]

Certo il tema dell'eduzione e dell'irreggimentazione idrica nelle zolfare romane presenta numerosi aspetti di interesse, ma al momento è difficile andare oltre le ipotesi di lavoro, mancando interventi di scavo stratigrafico in contesti minerari, al contrario di quello che è avvenuto per esempio in Francia, nelle *aurifodinae*, oppure in Spagna, nelle miniere di piombo e rame. In età moderna il problema dell'intercettazione di corsi d'acqua sotterranei e della loro gestione, era primario. Occupava gran parte delle risorse economiche dei capitalisti e dei loro affittuari, e ne condizionava infine le scelte e le sorti delle loro attività imprenditoriali.[2] Questo aspetto doveva essere altrettanto importante nelle zolfare romane, sappiamo infatti da Plinio (Nat. Hist, XXXV, 177) che il minerale *effoditur e cunicolis*.

In provincia di Agrigento sono numerosi i fenomeni legati al vulcanesimo secondario che hanno come loro manifestazione la presenza di acque termali o semplicemente acque che contengono, fra gli altri minerali, anche lo zolfo.[3]

Un testimone importante di questo fenomeno è senz'altro il passo di Solino (V.22–23). Il testo di Solino è noto, vale però la pena riportarlo:

'Sopra la superficie di un lago nei pressi di Agrigento, galleggia una sostanza oleosa: questa morchia aderisce alle foglie delle canne con una patina persistente, dalle cui cime si raccoglie un unguento medicamentoso contro i morbi del bestiame. Non lungi dal lago, il colle di Vulcano, sul quale coloro che compiono sacrifici ammucchiano sarmenti sopra

---
[1] Un'indagine analoga David, Rossetti and Frigato 2009 consente di mettere in rapporto le dinamiche di occupazione del territorio, e in particolare l'esistenza di vici/pagi, con le chiese cui facevano riferimento e con aree destinate al culto delle acque o alle divinità preposte alle guarigioni in periodo pre-cristiano.

[2] Se nelle miniere inglesi tanto l'eduzione delle acque quanto quella del minerale prodotto era già meccanizzata alla fine del '700, nelle zolfare siciliane non si ebbero i primi impianti se non almeno un secolo dopo, Barone 1989: 32.

[3] Manfredi 2016 sul rapporto tra siti minerari, insediamenti e aree di culto in Nord Africa. Gli studi sul rapporto tra insediamento e territorio e risorse termominerali sono molto sviluppati in Italia settentrionale. In particolare si vedano: Bassani, Bressan, Ghedini 2012; Annibaletto, Bassani, Ghedini 2014. Inoltre per altri contesti geografici: Mattias, Guerra 2008 e Chellini 2002.

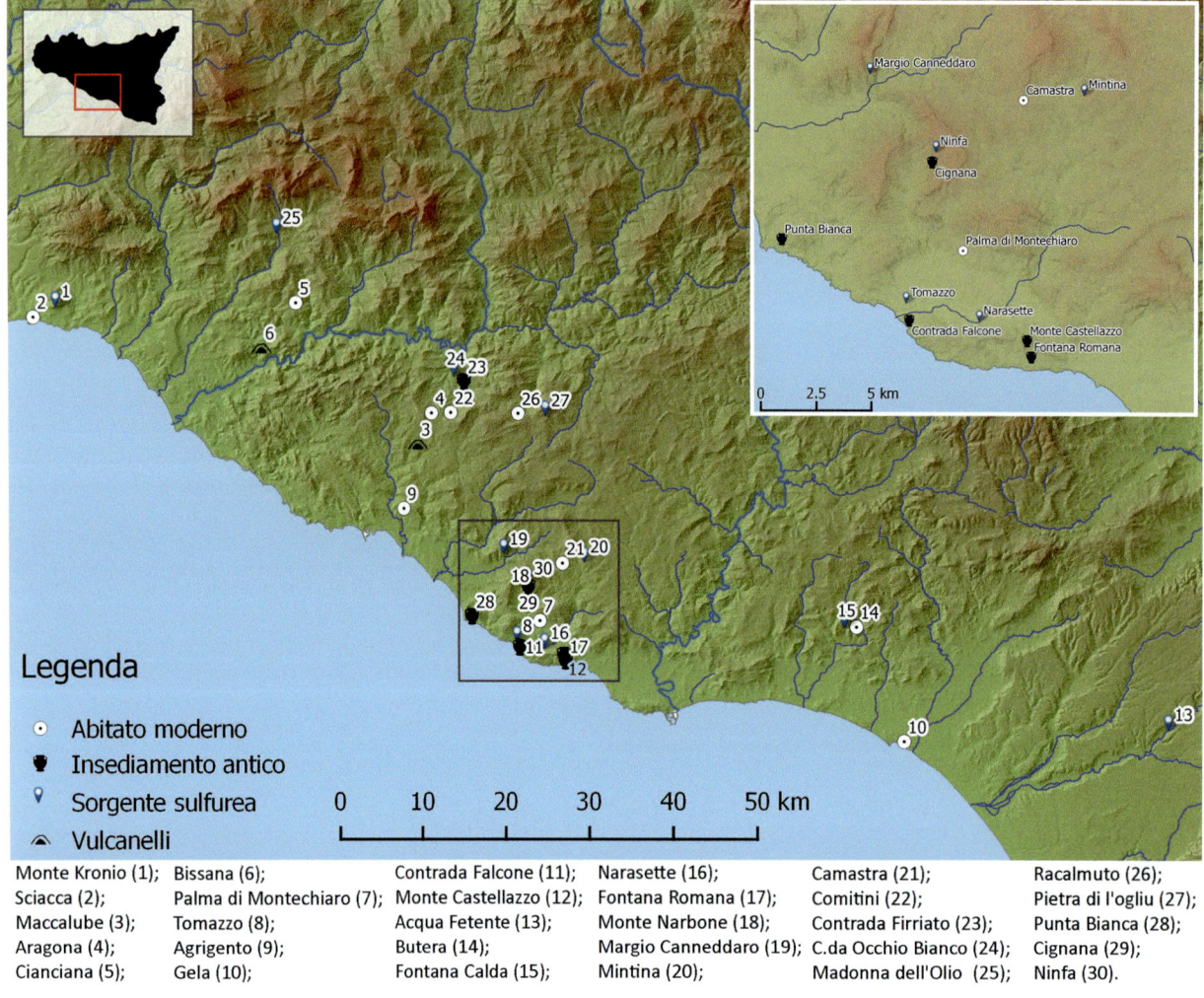

Figure 1. Carta della provincia di Agrigento con i siti citati nel testo

gli altari e non portano il fuoco a questa ramaglia: non appena vi pongono le viscere delle vittime, se compare il dio e accetta l'offerta, la sterpaglia, anche se verde, inizia ad ardere per suo conto e senza bisogno di attizzare il fuoco soffiandovi: è il nume che causa l'incendio. Il fuoco scherza con i partecipanti al sacro banchetto, poi si innalza in forma di lingue sinuose, senza bruciare coloro che sfiora, e non ha altro significato se non quello di segnale annunciante che i voti si sono compiuti in modo conforme alle prescrizioni. Lo stesso territorio di Agrigento erutta fontanazzi di fango e proprio come le scaturigini delle fonti bastano ad alimentare i fiumi, così in questa parte di Sicilia la terra vomita terra, in un perpetuo rigurgito, senza che si verifichi mai scarsità di suolo.'

Il sito di riferimento, oltre alle grotte del monte Kronio nella vicina Sciacca[4] è, senza dubbio alcuno, quello delle Maccalube, una zona prossima alla periferia di Agrigento in cui ampie sacche di gas assieme alla presenza di acqua e strati di argille, danno luogo a spettacolari esplosioni e a vulcanelli di argilla.[5]

Non mi pare superfluo annotare come Solino parli di due affioramenti di bitume: il primo in prossimità del *collis Vulcanius*, dove oggi è riconosciuto il tempio di Vulcano.[6]

Il secondo, invece, quello dei campi agrigentini, ben distinto potrebbe essere identificato proprio con il sito odierno della Macalube aragonesi.[7]

---

[4] Tiné, Torelli 2013; Caminneci 2014.

[5] Pace 1958: 525 e Castellana 1981: 238.

[6] Sul bitume affiorante in un lago nell'immediata periferia agrigentina e in prossimità di un colle dedicato a Vulcano: si veda anche Plinio, Nat. Hist. XXXV, 179.

[7] Sul toponimo Maccalube cfr. Caracausi 1993, sv. Sul culto di Vulcano in area agrigentina Castellana 1981 assieme a Cultraro 2016. Interessante anche, in Cultraro 2012: 192, il riferimento al contesto di una tomba a circolo scavata alle pendici dell'Etna, nel comune di Paternò, contrada Salinelle, in prossimità di numerose sorgenti minerali, di affioramenti bituminosi e di fenomeni di vulcanesimo secondario simili alle maccalube agrigentine.

Figure 2. Palma di Montechiaro. C.da Tomazzo. *Xoana* lignei. Museo P. Orsi, Siracusa

Nell'area delle Maccalube, in prossimità dell'odierno centro di Aragona, accumuli di metano e altri gas, assieme alla natura argillosa dell'area, danno origine a rigonfiamenti del suolo seguiti da modeste e repentine espulsioni di materiale sulla superficie. Legato alle eruzioni fangose è il deposito, in prossimità dei fori, di minerali che danno la caratteristica colorazione chiara. Di qui il toponimo 'occhio bianco' abbastanza diffuso nel circondario.

In realtà il fenomeno dell'affioramento di bitume e della presenza di vulcanelli, oltre che in territorio di Aragona, è presente anche in quello di Cianciana dove, in contrada Bissana, sono conosciuti da tempo immemorabile vulcanelli, bolle di gas e fenomeni esplosivi.

Xoana e culti salutari nel territorio palmese: il caso di Tomazzo e Narasette

La presenza di sorgenti di acqua calda o semplicemente resa sulfurea dall'attraversamento della sequenza del Messiniano nota come Serie Gessoso-solfifera, ha suscitato, come abbiamo visto, da sempre l'attenzione dei viaggiatori e degli abitanti del luogo.

In prossimità del centro urbano di Palma di Montechiaro nei primi anni del '900 G. Caputo segnala il rinvenimento in contrada Tomazzo, di un deposito archeologico in prossimità di una sorgente di acqua sulfurea.[8] Il sito è cruciale nell'ottica delle vie di comunicazione tanto in senso EW quanto da e verso l'entroterra più a nord. Tomazzo si trova alla base della corona collinare che circonda il basso corso del fiume Palma. Fin da epoca preistorica l'area per la sua fertilità e in quanto strategica nell'ottica dei rapporti tra Agrigento e Gela fu fittamente insediata, lo testimoniano i diversi nuclei di sepolture a forno e le numerose tracce di frequentazione sparse per tutta la conca.[9] Subito a SE di Tomazzo si trova la stazione preistorica di contrada Falcone,[10] mentre un importante asse viario porta alle prime pendici di Monte Castellazzo su cui si sviluppa un centro con caratteristiche urbane che dall'età arcaica arriva alla prima età ellenistica.

Nel corso di lavori per irreggimentazione delle acque, provenienti da alcune sorgenti vicine, gli operai trovarono un deposito archeologico tra cui spicca la presenza di almeno tre *xoana* in legno associati a vario

[8] Caputo 1938; Caputo 2012. Più in generale: Giontella 2012; Grifoni Cremonesi 2005; Gasperini 2006.
[9] Si veda il recente riesame dei dati e la proposta di ricostruzione del paesaggio di Schirò 2018 e 2019.
[10] Lo Vetro 2003.

Figure 3. Palma di Montechiaro. Vasca romana di c.da Narasette. Vista da Sud

materiale ceramico (Figure 2). Giacomo Caputo propone di datare almeno una delle sculture in legno al VII secolo a.C. Rinvenne anche, durante lo scavo successivo alla scoperta casuale, delle tavolette in legno e le interpretò come supporti per formule magiche che venivano lasciati andare a galleggiare sul piccolo rivolo d'acqua sgorgata dalle sorgenti, frammenti di un trono fittile, statuette femminili in trono, maschere fittili femminili.

Non mi sembra peregrino il confronto con quanto documentato in area irpina, attorno a un'area interpretata come cultuale per la dea Mefite. Anche in prossimità del laghetto della Mefite di Ansanto (Av) sono stati rinvenuti *xoana* e inoltre a tal proposito annoto, ma è una circostanza che merita un approfondimento, che P. Orsi, dopo aver esplorato forse troppo fugacemente la conca palmese acquisisce alle collezioni del Museo siracusano una collana.[11] Nel santuario di Mefite sono comuni i doni rituali di collane e altri monili come ex voto per avvenute guarigioni. E, ancora, assieme agli *xoana* sono stati rinvenuti resti di pasto e tizzoni e mascherine di terracotta.[12] Non è questo il contesto per tentare di attribuire a una precisa divinità l'area sacra di Tomazzo, tuttavia, ricordato il sito di Acqua fitenti lungo il vallone Mazzarrone a destra del Dirillo, dove è presente ancora oggi la tradizione di immergersi e di bere l'acqua minerale offrendo in voto parte dei vestiti o lasciando oggetti appesi agli alberi, relitto di più antichi riti.[13] Osservo che una componente costante in diversi miti è la presenza del serpente, iconostasi di Asclepio, in relazione alle qualità guaritrici delle acque sulfuree.

Il dato della presenza di zolfo in sospensione e delle vicine bocche di zolfara, assieme alle caratteristiche del record ceramico, non lasciano dubbi nel Caputo sull'interpretazione del sito come un luogo sacro. Mi pare particolarmente interessante il confronto, proposto dallo stesso Caputo, con la sorgente di contrada Fontana Calda, ai piedi del rilievo su cui sorge il centro moderno di Butera nel nisseno.[14] La consacrazione con piccoli edifici cultuali delle fonti e dei luoghi di raccolta delle acque era, del resto, abbastanza diffusa: un rapido confronto si trova in Frontino a proposito dell'*Aqua Vergine*.[15] L'area archeologica, così come i materiali conservati presso il museo P. Orsi di Siracusa, meritano una attenta rilettura.[16] Così come risultati importanti potrebbe dare un riesame degli *xoana* anche solo da un punto di vista stilistico-iconografico.[17] G. Caputo

---

[11] Caputo 2012: 23, e nota 2.
[12] Caiazza 2010: 247 in particolare. Loffredo 2012.
[13] Caiazza 2010: 252.
[14] Adamesteanu 1958.
[15] Front., *De Aquae ductu Urbis Romae*, I, 9: '... aedicula fontis apposita'.
[16] Orsi 1928.
[17] Caputo pensava ad Artemide, per la presenza del polos e ipotizzava

sottolineava, inoltre, come l'area fosse stata oggetto di una monumentalizzazione nel V secolo a.C.

Al sito di Tomazzo è direttamente collegabile topograficamente e dal punto di vista storico quello, importante e paradossalmente meglio noto, di contrada Narasette.[18] Qui in prossimità di un asse viario WE si conserva una grande vasca quadrangolare con un fine rivestimento in malta idraulica e un pilastrino a sezione quadrangolare al centro.[19] La vasca di Narasette era finalizzata alla raccolta di acqua sulfurea proveniente da una sorgente vicina e appare pienamente inserita in un ampio sito la cui continuità di vita va da età classica fino alle porte del basso medioevo. Interessante, nella pubblicazione di Caputo, anche la notizia che nei pressi di Tomazzo, fino al 1908 e al terremoto di Messina, sgorgava una seconda sorgente sulfurea,[20] probabilmente Caputo si riferiva proprio alla fonte di Narasette, oggi prosciugata.

A queste due emergenze bisogna aggiungere la struttura detta 'fontana romana', lungo una variante dell'asse viario che, lasciato il basso corso del fiume Palma, raggiunge le pendici meridionali di Monte Castellazzo.[21] Qui il toponimo è significativo e indica una ampia vasca circolare, ancora in uso, anch'essa come la vasca di Narasette, con un pilastro centrale, un rivestimento in fine malta idraulica e cocciopesto e un canale di adduzione ancora integro e funzionante.

La sorgente di acqua sulfurea di contrada Margio Canneddaro, tra passato e presente

A nord della Portella di Cignana, che assieme al terrazzo marino di Narbone fa da spartiacque tra il bacino idrografico del fiume Naro e quello del Palma, nella vallata del vallone Borraitotto uno dei principali affluenti di sinistra del Naro, si estende l'ampio sito archeologico di Margio Canneddaro. Le ricognizioni topografiche e una raccolta sistematica del materiale archeologico consentono di tracciare in maniera abbastanza precisa i contorni cronologici e le caratteristiche dell'insediamento. Nel sito è stata effettuata una raccolta totale per quadrati.[22] Si tratta di una ricca villa inserita nella rete produttiva dello zolfo[23] su cui, a partire dalla fine del IV secolo d.C., si sovrappone un più ampio villaggio aperto. In questa sede voglio porre l'attenzione su una caratteristica geologica che potrebbe aver determinato lo sviluppo dell'area. L'insediamento di contrada Margio Canneddaro (Figure 4) si sviluppa ai piedi di un piccolo rilievo collinare

Figure 4. Agrigento c.da Margio Canneddaro. Sorgente di acqua minerale

sfruttandone le fertili pendici argillose; subito a sud del sito, invece, sgorga una sorgente di acqua sulfurea che sembra aver mantenuto nel tempo un suo ruolo preminente nel circondario. Infatti sebbene sia ormai persa la topografia originaria, in quanto l'asse viario moderno ha apportato delle modifiche irreversibili con canalizzazioni in cemento e con probabili deviazioni del corso d'acqua, si mantiene vivo il ricordo delle acque salutari e gli abitanti del posto se ne riforniscono periodicamente in piccole bottiglie per il trattamento di affezioni della pelle e degli occhi.

Un ulteriore approfondimento bisogna dedicarlo alla scoperta, durante un intervento di emergenza nel 1964, ad Agrigento nell'immediata periferia dell'area urbana del *municipium*, ma ancora all'interno del tessuto urbano della città ellenistica, di un butto di contenitori per *Lykion*. I reperti sembrerebbero provenire da un contesto di produzione ceramica: sono, infatti, ancora non verniciati e hanno dei difetti di cottura che li rendevano inutilizzabili, tuttavia il loro alto numero lascia intendere che fossero destinati ad un laboratorio, appunto, per la produzione del medicamento.[24]

---

riti di tipo ordalico, Caputo 1938: c. 630.
[18] Castellana 1983.
[19] Castellana 1983: fig. 52.
[20] Caputo 1938: c. 683.
[21] Castellana 1994 con bibliografia precedente.
[22] Burgio 2013: 38–39.
[23] Zambito 2018: 83–85.

---

[24] Sui contenitori agrigentini Taborelli 2015 con bibliografia precedente; in particolare p. 91 e il paragrafo intitolato significativamente 'questioni aperte' in cui l'autore esclude che i contenitori potessero essere destinati al santuario extraurbano di

Non conosciamo la ricetta di questo unguento elaborato su base vegetale, tuttavia uno dei suoi eccipienti era senza'altro il bitume e il fatto di utilizzare l'acqua solforosa dava, probabilmente, una maggiore efficacia al composto.[25] Plinio dichiara chiaramente che il sulphur vivum era uno degli ingredienti dei colliri, misto ad aceto e bitume.[26]

Mintina, acqua calda, acqua fitusa. Purificazione degli armenti e sorgenti di acqua sulfurea

A pochi km da Margio Canneddaro scorre il vallone Mintina, un idronimo derivante dalla presenza di acque sulfuree presente a ovest di Monte Grande dove confluisce nel vallone di Monte Grande e poi nel fiume Palma e, a nord est rispetto al contesto di cui ci stiamo occupando, a E del comune di Camastra. Nel territorio di Camastra alle sorgenti del vallone Mintina, si trova una grotta da cui sgorga l'acqua sulfurea. I pastori del luogo, periodicamente, portano le greggi ad immergersi in questa acqua per purificarne la lana e per rimuovere i parassiti.[27]

La diffusione del toponimo 'Mintina' (e dei suoi derivati 'Mintinelli', 'Acqua Mintina' e simili) ben documenta la frequentazione e il ruolo che questa caratteristica doveva conferire ai siti interessati.[28] Accanto a questi toponimi parlanti va posto quello di Occhio Bianco, presente in territorio di Comitini, un altro modo di indicare i fenomeni di vulcanesimo secondario che caratterizzavano il territorio e derivante dal colore lasciato dai minerali in sospensione nelle acque e che venivano depositati presso le sorgenti. Contrada Occhio Bianco è lambita dalla strada ortogonale al corso del fiume Platani che proviene dal sito minerario di contrada Firrìo.[29]

Ho potuto documentare che i pastori del luogo hanno l'abitudine di portare le loro greggi, periodicamente, a fare i bagni nei pressi della sorgente sulfurea così come sono soliti fare i pastori alla periferia di Cianciana (Ag) nelle vicinanze del Santuario della Madonna dell'Olio.[30] Come a Camastra anche a Cianciana i pastori, periodicamente, fanno immergere le greggi nel vallone.[31]

Merita qui ricordare la presenza, probabilmente non casuale, nel territorio cianciancese, di un culto a *Heracles*, testimoniato da una mazza miniaturistica conservata al Museo di Gela e proveniente da Cianciana con la dedica: *Botakos toi Heraklei*.[32]

Sul culto di Eracle alla periferia della *chora* akragantina si veda la lettura di Clemente Marconi[33] che pone l'attenzione sulla presenza di sorgenti di acqua sulfurea a Colle Madore, sito in cui si sono trovati materiali indigeni assieme a materiali di importazione. Di notevole interesse risulta l'edicola di età arcaica scolpita, con figura alla fontana, interpretata come Eracle, da collocare verosimilmente su una banchina addossata alla parete orientale di uno degli ambienti scavati. Secondo Marconi, a Colle Madore sarebbe documentato un culto patriarcale in cui Eracle figura come eroe 'di frontiera', legato al processo di ellenizzazione delle comunità indigene dell'entroterra a partire dalla seconda metà del VI sec. a.C. e la cui presenza sarebbe collegata direttamente alla sorgente di acqua sulfurea. Diversa la prospettiva di F. Frisone[34] che tende a sganciare la figura di Eracle, e dei miti legati alla sua presenza siciliana, dall'univocità del rapporto tra mondo indigeno e colonie e, in ambito coloniale, pone l'accento sulle differenti chiavi di lettura nei diversi contesti sicelioti.

Come a Cianciana, anche nell'immediata periferia del centro moderno di Racalmuto, subito a NW del un sito di c.da Casalvecchio interessato da una lunga frequentazione che va almeno dal V secolo a.C. ad almeno il V d.C. e che è pienamente inserito nella filiera produttiva dello zolfo, è attestata una 'Petra di l'ogliu': si tratta di uno spuntone di calcare che emerge da un'ampia distesa di argille plioceniche e ai piedi del quale periodicamente si hanno fenomeni di affioramento di bitume come testimoniato dai contadini locali. L'acqua meteorica, infatti, in questo settore del territorio agrigentino intercetta delle vene di oli bituminosi che in estate, quando il corso del vallone è meno impetuoso, riescono ad addensarsi e vengono raccolti dagli abitanti del luogo.

Asclepio.
[25] Scribonio Largo ci informa sulla sua origine licia, cfr. Taborelli 2014: 26.
[26] *N.H.* XXXV, 177. Sull'associazione di zolfo e bitume nelle ricette di Scribonio si veda Zambito 2018: 26, con riferimenti bibliografici.
[27] Ricca la documentazione sul rapporto tra acque minerali, pastorizia e specifiche aree di culto, si vedano Santino Frizell 2004 e Veronese 2010. In questa sede ricordo che dall'area del santuario di Asclepio ad Agrigento provengono dei modellini di torelli lasciati come ex voto assieme a serpenti e ad arti umani miniaturistici. De Miro: 415. Cfr. Bassani 2010: 238, con riferimento alla notizia in Diodoro di un culto di Gerione fondato da Ercole, proprio in Sicilia, presso Agirion.
[28] Zambito 2018: 41.
[29] Zambito 2018: 54-57.
[30] Per un confronto sulla persistenza in epoca cristiana di culti di età classica si vedano: Archetti 2015; Binazzi 2012; Manselli 1982 e Spanu 2008.
[31] Caerulei fiant puro de sulphure fumi tactaque fumanti sulphure balet ouis. 'Si formano fumi cerulei di puro zolfo e, toccata dallo zolfo fumante, la pecora guarisce'. Ov. Fast. IV, 739-740. Sul passo cfr. Zambito 2018: 21. E più in generale CAM 2007.
[32] Manganaro 1996: 79, e nota 19 con riferimento a Dubois 1989: 177, nr. 159 dove, però, la località di rinvenimento è collocata a 5 km a est di Gela. In questo caso non si tratterebbe del centro agrigentino di Cianciana di cui sto parlando.
[33] Marconi 1999: 298-299 in particolare.
[34] Frisone 2017, con ampia discussione della bibliografia precedente, in particolare p. 148-149, sul rilievo da Colle Madore.

Molto più denso di testimonianze è il sito di Monte Grande alle cui pendici orientali scorre il Vallone Mintina e in cui le ricognizioni sistematiche dell'Università di Palermo hanno messo in evidenza una lunga frequentazione dell'area.[35] Probabilmente questa occorrenza di siti archeologici in prossimità di fonti di acqua sulfurea è da mettere in relazione con l'utilizzo delle aree marginali per il pascolo. Ancora in età medievale l'area alle pendici di Monte Grande era nota per l'abbondanza di mandrie e greggi: ne sono testimoni un documento databile all'epoca di Guglielmo II[36] e la frequenza con cui si trova il toponimo 'Marcato' nella zona.[37]

Impianti suntuari nelle ville dello zolfo: le fontane e le vasche

La rete insediativa e di vie di comunicazione nell'area solfifera era fortemente condizionata, come ho detto sopra, oltre che dall'orografia del territorio anche dalle attività estrattive: non era possibile, infatti, stabilirsi in prossimità delle bocche di miniera perché le attività fusorie avrebbero impedito di condurre i fondi agricoli e di vivere in maniera salubre. Si osserva, pertanto, uno stringente rapporto tra le attestazioni di aree residenziali e le zolfare. Anche perché l'esigenza di controllare direttamente e in maniera continua l'estrazione e la raffinazione, portavano i gestori e i proprietari delle licenze a costruire le loro abitazioni a non grande distanza dalle miniere stesse.

Certo, questo controllo era esercitato direttamente tramite delle iscrizioni su tegola che trasferivano sul lingotto le informazioni essenziali sui luoghi e sugli attori della produzione, tuttavia chi investiva direttamente nella estrazione ed era coinvolto nell'esportazione del minerale, aveva la necessità di gestirle direttamente.

Le *tegulae sulphuris*, le matrici per i lingotti, sono l'indicatore principale dell'esistenza di attività estrattive in un sito. La loro presenza in un determinato contesto può essere indizio di tre fenomeni: la produzione stessa delle iscrizioni, ma questo è valido solo, finora, per il centro urbano e perturbano di Agrigento e, parrebbe, per l'area di contrada Gerace presso Enna; il loro stoccaggio prima del loro utilizzo e, infine, il loro uso presso le bocche di miniera caso documentato, ad oggi, solo in due siti uno racalmutese e uno nei pressi del centro di Miniera Lucia.[38] Le caratteristiche dell'insediamento legato alle zolfare sono, ormai, abbastanza chiare e delineate: i centri di controllo dell'estrazione (ville e villaggi) devono essere abbastanza distanti dalle bocche di miniera per non essere contaminati dai fumi venefici che si sprigionavano, ma, allo stesso tempo, dovevano consentire un costante e valido controllo dei processi estrattivi. Per questo motivo, dunque, le ville e i villaggi fin qui censiti non distano più di sei Km dall'area di estrazione. Le attività di survey mi hanno consentito, quindi, di tracciare uno schema di riferimento per lo studio dell'insediamento, in chiave diacronica, nell'area a est di Agrigento. Molti dei siti che ho potuto studiare si caratterizzano come ampie *villae* che, spesso, hanno apparati suntuari che prevedono complessi sistemi idraulici e, a volte, l'impianto di piccoli *balnea* privati.

La villa di contrada Punta Bianca (Figure 5), oltre a caratterizzarsi per la sua invidiabile posizione, quasi in prossimità della linea di costa e su un'altura che consentiva ai proprietari di controllare gran parte del golfo, disponeva sicuramente di una canalizzazione e di vasche per la raccolta delle acque. Il sito, purtroppo, non è stato scavato sistematicamente, ma è stato più volte danneggiato dagli scavatori di frodo. Sul terreno si osservano ampie porzioni di blocchi rivestiti di cocciopesto e, inoltre, è stato possibile raccogliere e schedare due manufatti, uno in pietra calcarea e uno, invece, in terracotta riferibili all'uso dell'acqua. Si tratta di un grosso tubo a sezione circolare, con il bordo ingrossato e stondato e dell'orlo di un grande *labrum* in calcare, con la parete a sezione quadrangolare e la superficie interna finemente levigata (Figure 6). Poco più a nord di Punta Bianca, si sviluppa il sito di Cignana. Qui si utilizzavano le abbondanti risorse idriche dell'area per servire un complesso sistema di captazione.[39] Queste acque rifornivano anche le terme di cui sono conservati alcuni ambienti decorati in *opus tessellatum*, databili tra il I gli inizi del III secolo d.C.[40]

Purtroppo anche la villa di Cignana è stata oggetto di un importante e devastante intervento di vandalizzazione: con un mezzo meccanico sono state sconvolte le stratigrafie in gran parte dell'area e un sistema di vasche è stato completamente distrutto. Sul terreno rimangono ampi lacerti delle strutture rivestite con malta idraulica. Subito a NE della villa di Cignana, a circa 500 metri, si segnala il toponimo 'Ninfa' che, assieme a una imponente struttura idraulica in cocciopesto e malta, contribuiscono a chiarire il quadro del sistema di approvvigionamento idrico dell'area.[41]

Conclusioni e prospettive di ricerca

Un primo aspetto che merita un approfondimento è la definizione dei culti nei pressi delle sorgenti: in

---

[35] Belvedere, Burgio 2009; Burgio 2012.
[36] Schirò 2018: 10, e n. 18.
[37] Marcato può essere considerato un sinonimo di mannara e indica i circoli di pietre a secco all'interno dei quali si ricoveravano le greggi, cfr. Caracausi 1993: 954 sub voce. Dall'arabo marqad.
[38] Zambito 2018.

[39] Fiorentini 1993; Rizzo-Zambito 2008 e 2014.
[40] Paci 2006 and Riparbelli 1992.
[41] Sul toponimo Ninfa Belvedere, Burgio 2009 con bibliografia precedente.

Figure 5. Punta Bianca. Area della villa. Vista da Nord

Figure 6. Palma di Montechiaro. Località Punta Bianca. Grosso tubulo e *labrum* in calcare

questa direzione, un grande aiuto potrebbe derivare dalla revisione dei materiali provenienti dallo scavo di Tomazzo. Sembra opportuno, a tal proposito, verificare l'esistenza di una fase arcaica relativa alla frequentazione del santuario. Altra questione riguarda la pertinenza a divinità 'salutari' stricto sensu legate a fenomeni come la transumanza e la pastorizia come Eracle, oppure a figure muliebri riconducibili alla fertilità (dei campi, degli armenti o delle donne). La prosecuzione, inoltre, del censimento dei siti collegati agli affioramenti di acque sulfuree può aggiungere ulteriori tasselli alla comprensione di alcune scelte poleografiche e legate alla scelta di percorsi e tracciati viari rispetto ad altri.

Da un lato il censimento delle evidenze archeologiche legate all'affioramento di acque minerali e termali consente di contestualizzare le testimonianze di culti che hanno una lunga continuità e persistenza sul territorio, dall'altro si comprendono meglio le ragioni che portarono allo sviluppo della trama insediativa legata anche allo sfruttamento minerario come avviene per il sito di contrada Margio Canneddaro, con la sua sorgente sulfurea, che fungeva da centro catalizzatore per lo sfruttamento delle risorse del territorio. Andrà meglio chiarita, auspicabilmente con indagini archeologiche, la sua valenza cultuale. Così come importanti risultati potrà dare lo studio analitico dei reperti da Tomazzo e questo sia per quanto concerne la loro datazione, quanto per l'interpretazione complessiva dell'area. Anche in rapporto al territorio e al vicino sito di Narasette. Allo stato della documentazione, credo, non è agevole approfondire ulteriormente le modalità con cui queste aree furono abbandonate e, a volte, rifunzionalizzate. Mi riferisco, in particolare, al sito di Cignana/Ninfa dove c'è una continuità insediativa almeno dalla tarda età del rame. Subito a valle di Ninfa, in età tardo antica, arriva il *vicus* di Cignana la cui interpretazione e lettura è ancora in fieri.

Riferimenti bibliografici

Adamesteanu, D. 1958. Butera. Piano della Fiera, Consi e Fontana Calda. *Memorie Antiche dei Lincei* 44: 205–672.

Annibaletto, M., Bassani, M., Ghedini, F. 2014 (eds). *Cura preghiera e benessere. Le stazioni curative termominerali nell'Italia romana* Edizioni Quasar, Roma.

Archetti, G. 2015. La diffusione del cristianesimo lungo le vie d'acqua. Suggestioni dall'area alpina, in G. Archetti, L. Giarell, F. Zoncoron, P. Zanovello (eds) *Aquae divinae. Riti e miti nelle Alpi tra preistoria e cristianità*: 67–93 Breno (Bs). Edizioni Quasar, Roma.

Barone, G. 1989. Formazione e declino di un monopolio naturale. Per un storia sociale delle miniere di zolfo, in S. Addamo (ed.) *Zolfare di Sicilia*: 59–116. Sellerio Editore, Palermo.

Bassani, M. 2010. Greggi e mandrie fra termalismo e profezia. *Gerión* 30: 185–208.

Bassani, M., Bressan, M., Ghedini, F. 2012. *Aquae patavinae. Montegrotto terme e il termalismo in Italia aggiornamenti e nuove prospettive di valorizzazione. Atti del II convegno nazionale* (Padova, 14–15 giugno 2011). Edizioni Quasar, Roma.

Barbera, P., Vitale, M.R. 2017. *Architetti in viaggio. La Sicilia nello sguardo degli altri*. Siracusa.

Belvedere, O., Burgio, A. 2012. *Carta archeologica e sistema informativo del parco archeologico e paesaggistico della Valle dei Templi di Agrigento*. Palermo.

Binazzi, G. 2012. *Il radicamento dei culti tradizionali in Italia fra tarda antichità e alto medioevo: fonti letterarie e testimonianze archeologiche*. L'Erma di Bretschneider, Roma.

Burgio, A. 2012. Il progetto 'Cignana' (Naro-Palma di Montechiaro, Sicilia centro-meridionale). Prospezione archeologica intorno alla villa romana di Cignana. Primi dati sulle dinamiche del popolamento antico, in *Griechen in Übersee und der historische Raum, Internationales Kolloquium Universität* Göttingen: 127–139. Atti del convegno, Gottingen 13–16 Oktober 2010 = Göttinger Studien zur Mediterranean Archäologie 3. Verlag Marie Leidorf, Gottingen.

Burgio, A. 2013. Dinamiche insediative nel comprensorio di Cignana. Continuità e discontinuità tra l'età imperiale e l'età bizantina. *Sicilia Antiqua* X: 31–53.

Caiazza, D. 2010. Le «fontane dei bambini» e altri culti e tabù delle acque in Terra di Lavoro. Cenni sulla natura e motivazione delle offerte votive, in H. Di Giuseppe and M. Serlorenzi (eds) *I riti del costruire nelle acque violate*: 245–268. Atti del Convegno Internazionale Roma, Palazzo Massimo 12–14 giugno 2008. L'Erma di Bretschneider, Roma.

Cam, M.T. 2007. *La médicine vétérinaire antique. Sources écrites, archéologiques, iconographiques*. Actes du Colloque International (Brest, 9–11 septembre 2004), Presses Universitaires, Rennes.

Caminneci, V. 2014. Ad aquas. Historical sources and archaeological evidence about Sciacca's thermalism in antiquity, in N. Gullì (ed.) *From Cave to dolmen. Ritual and symbolic aspects in the prehistory between Sciacca, Sicily and the central Mediterranean*: 59–64.Bar I.S. Archaeopress, Oxford.

Caputo, G. 1938. Tre xoana e il culto di una sorgente sulfurea in territorio geloo-agrigentino. *Monumenti Antichi dei Lincei* XXXVII: 642–646, 679–682.

Caputo, G. 2012. Dal culto popolare dell'acqua sulfurea al culto autocratico del toro di Falaride. *Mare Internum* 4: 23–27.

Carlino, A. 2009. *La Sicilia e il Grand Tour. La riscoperta di Akragas, 1700-1770*. Roma.

Castellana, G. 1981. Sull'origine del culto di Efesto-Vulcano nel territorio di Agrigento. *Parola del Passato* CXCIX: 234–243.

Castellana, G. 1983. Nuove ricognizioni nel territorio di Palma di Montechiaro (Agrigento). *Sicilia Archeologica* 52 53: 119–146.

Castellana, G. 1994. Palma di Montechiaro. *Bibliografia topografica delle colonie greche e delle isole tirreniche* XIII: 300–310. Pisa–Roma.

Chellini, R. 2002. *Acque sorgive Salutari e Sacre in Etruria (Italiae Regio VII). Ricerche Archeologiche e di Topografia Antica*. British Archaeological Reports International Series 1067. Oxford.

Cultraro, M. 2012. I Siculi all'ombra del vulcano: per una proposta di definizione dell'età del Bronzo Recente e Finale nella media valle del Simeto, in M. Congiu, C. Miccichè, S. Modeo (eds) *Dal mito alla storia. Atti del VIII Convegno di Studi sulla Sicilia Antica*: 181–203. Sciascia editore, Caltanissetta-Roma.

Cultraro, M. 2014. Le fontane ardenti degli Dei: fenomeni geologici e pratiche di culto nella valle del Platani (Agrigento) nella preistoria, in A. Musco, G. Parrino (eds) *Santi, Santuari, Pellegrinaggi, Officina di Studi Medievali*: 1–14. Palermo.

Cultraro, M. 2016. Elementi di interesse paletnologico nei depositi di olio bituminoso della Sicilia centro-occidentale, in E. Mangani, A. Pellegrino (eds) Για το φίλο μας. *Scritti in ricordo di Gaetano Messineo*: 129–136. Edizioni espero Monte Compatri (Roma).

De Miro, E. 2003. *Agrigento II. I santuari extraurbani. L'Asklepieion*. Rubettino editore, Soveria Mannelli.

De Miro, E., Calì, V., Sfameni Gasparro, G. 2009. *Il culto di Asclepio nell'area mediterranea*. Gangemi editore, Roma.

Dubois, L. 1989. *Inscriptions grecques dialectales de Sicile*. Rome.

Fiorentini, G. 1993–1994. Attività di indagini archeologiche della Soprintendenza Beni Culturali e Ambientali di Agrigento. *Kokalos* 39–40: 717–733.

Frisone, F. 2017. 'Tirando il dio per la giacchetta...'. Eracle nella Sicilia antica fra Calcidesi, Dori e altri, in M. Congiu, C. Micchiché, S. Modeo (eds) *Eracle in Sicilia. Oltre il mito: arte, storia, archeologia*: 137–16. Sciascia editore, Caltanissetta.

Giontella, C. 2012. «… *Nullus enim fons non sacer* …». *Culti idrici di epoca preromana e romana (Regiones VI-VII)*. Fabrizio Serra editore, Pisa–Roma.

Grifoni Cremonesi, R. 2005. Il rapporto dell'uomo con le manifestazioni geotermiche in Italia dalla Preistoria all'alto Medioevo, in M. Ciardi and R. Cataldi (eds) *Il Calore della terra. Contributo alla Storia della Geotermia in Italia*: 10–26. Edizioni ETS, Pisa.

Loffredo, F. 2012. La dea Mefitis: dalle mofète del Sannio a Abano Terme. *I quaderni del ramo d'oro on-line* 5: 176–188.

Lo Vetro, D. 2003. L'epigravettiano finale di passo Falcone (Palma di Montechiaro, Agrigento). *Origini* 25: 47–74.

Manfredi, L.I. 2016. Le miniere, la metallurgia e il sacro nel nord Africa fenicio-punico. *Rivista di Studi Fenici* 44: 153–163.

Manganaro, G. 1996. Figurazioni e dediche religiose della Sicilia greca e romana. *Zeitschrift für Papyrologie und Epigraphik* 113: 77–81.

Manselli, R. 1982. Resistenze dei culti antichi nella pratica religiosa dei laici nelle campagne. *Cristianizzazione ed organizzazione ecclesiastica delle campagne nell'alto Medioevo: espansione e resistenze* (10-16 aprile 1980): 57–108. Spoleto.

Marconi, C. 1999. Eracle in terra indigena?, in S. Vassallo (ed.) *Colle Madore. Un caso di ellenizzazione in terra sicana*: 293–305. Regione siciliana, Assessorato dei beni culturali e ambientali e della pubblica istruzione, Palermo.

Mattias, P. and Guerra, M. 2008. *Le miniere nelle Marche. II parte. Miniere e mineralizzazioni. Giacimenti e vicende*. Roma.

Orsi, P. 1928. Miscellanea Sicula. I. Esplorazione topografica nell'Agro di Palma di Montechiaro (Girgenti). *Bullettino di Paletnologia Italiana* 48: 46–61.

Pace, B. 1958. *Arte e civiltà della Sicilia antica* I–IV. Firenze.

Paci, G. 2006. Idroterapia e religiosità alle Aquae Albulae presso Tivoli, in L. Gasperini (ed.) *Usus veneratioque fontium. Fruizione e culto delle acque salutari nell'Italia romana. Atti del Convegno internazionale di studio su Fruizione e culto delle acque salutari in Italia* (Roma-Viterbo 29–31 ottobre 1993): 255–275. Edizioni Quasar, Roma.

Paoletti, M. 2009. «Questa rovina è indicibilmente bella e pittoresca»: le antichità della Sicilia e il culto della Grecia classica nel XVIII secolo, in C. Ampolo (ed.) *Immagine e immagini della Sicilia e di altre isole del Mediterraneo antico*: 195–220, figg. 89–107. Pisa.

Riparbelli, A. 1992. Sorgenti salate e miniere di sale, in *Les eaux thermales et les cultes des eaux, Actes du Colloque* (Aix-les-Bains, 28–30 septembre 1990): 349–377. Centre de Recherches A. Piganiol, Tours.

Rizzo, M.S., Zambito, L., Giannici, F., Giarrusso, R. and Mulone, A. 2014. Anfore di tipo siciliano dal territorio di Agrigento, in N. Poulou-Papadimitriou, E. Nodarou, V. Kilikoglou (eds) *LRCW 4: late Roman coarse wares, cooking wares and amphorae in the Mediterranean: archaeology and archaeometry : the Mediterranean : a market without frontiers*: 213–223. British Archaeological Reports International Series 2616. Archaeopress, Oxford.

Santillo Frizell, B. 2004. Curing the Flock: The Use of Mineral Waters in Roman Pastoral Economy, in B. Santillo Frizell (ed.) *Pecus. Man and Animal in Antiquity. Proceedings of the Conference at the Swedish Institute in Rome* (September 9–12, 2002): 39–59. The Swedish Institute in Rome and individual authors. Roma.

Schirò, G. 2014. Ecclesia Agrigenti: note di storia e archeologia urbana. *Quaderni Digitali di Archeologia Postclassica* 3.

Schirò, G. 2018. Ecclesia Sancti Leonardi: un luogo di ospitalità sulla strada Agrigento - Licata nel XIII secolo. *Quaderni Digitali di Archeologia Postclassica*: 203–233.

Schirò, G. 2019. Alcuni esempi di strutture di ospitalità lungo le strade della diocesi di Agrigento alla luce delle fonti medievali (XII–XIII sec.). *Mare internum* 11: 111–124.

Spanu, P.G. 2008. 'Fons vivus'. Culti delle acque e santuari cristiani tra tarda antichità e alto medioevo, in *L'acqua nei secoli altomedievali* (Spoleto, 12–17 aprile 2007): 1029–1078. Spoleto.

Speciale, C. and Zambito, L. 2017. Reproducing Cato. An experimental approcah to Cato's recipe, in R. Alonso, J. Baena and D. Canales (eds) *Playing with* the time. Experimental archaeology and the study of the past (proceedings of 4th. International Experimental Archaeology Conference (8–11 May 2014, Museo de la Evolución Huma. Burgos, Spain): 289–294. Madrid.

Taborelli, L. 2014. Per l'archeologia di un farmaco. Produttori e contenitori di Lykion in epoca ellenistica. *Territori della Cultura* 14–15: 26–33.

Taborelli, L. 2015. I contenitori per il Lykion di Akragas. *Sicilia Antiqua* 12: 87–98.

Taborelli, L. 2018. Sull'isola del Gattopardo, in N. Reggiani (ed.) *Papiri, medicina antica e cultura materiale. Contributi in ricordo di Isabella Andorlini*: 75–87. Parma.

Tiné, V., Torelli, L. 2013. Il complesso speleo-termale del Monte Kronio di Sciacca (AG) tra mitologia, speleologia e archeologia, in M. Bassani, M. Bressan and F. Ghedini (eds) *Aquae salutiferae : il termalismo tra antico e contemporaneo : atti del convegno internazionale* (Montegrotto Terme, 6–8 settembre 2012): 109–118. Edizioni Quasar, RomaPadova.

Veronese, F. 2010. Appunti sul culto di Eracle e Gerione tra storia e archeologia. *Hesperìa* 26: 29–46.

Archaeology and Hydrogeology in Sicily: Solunt and Tindari[1]

Giovanni Polizzi,[1] Vincent Ollivier,[2] Olivier Bellier,[3] Edwige Pons-Branchu[4] and Michel Fontugne[5]

[1] Università degli Studi di Palermo, Freie Universität Berlin/Von Humboldt Stiftung: giovannippolizzi@live.it
[2] CNRS, Aix Marseille Univ, Minist Culture, LAMPEA, Aix-en-Provence, France: ollivier@mmsh.univ-aix.fr
[3] CEREGE – Aix Marseille Univ, CNRS, IRD, INRA, Coll. France, Aix-en-Provence, France
[4] Laboratoire des Sciences du Climat et de l'Environnement, UMR 8212, CNRS, CEA, UVSQ, Université Paris Saclay, Gif-sur-Yvette, F-91190, France: edwige.pons-branchu@lsce.ipsl.fr
[5] Aix Marseille Univ, CNRS, Minist Culture, LAMPEA, Aix-en-Provence, France: michel.fontugne@wanadoo.fr

Abstract: This contribution concerns the relationship between the hydrogeological and archaeological characteristics of Solunt and Tindari (North Sicily), within the framework of the Watertraces project. The two cities face the Tyrrhenian Sea, on the northern coast of Sicily, are characterised by a relatively recent urban history compared to other Greek or Phoenician cities, being founded in the 4th century BC. The Hellenistic-Roman city of Solunt was moved from the promontory of Solanto to Mount Catalfano after the devastation of the first in 397 BC. The new position of the habitat offered advantageous defensive conditions and allowed it to control a vast territory, from the Gulf of Palermo to the Gulf of Termini and the hinterland. The steep slopes of Mount Catalfano did not prevent the inhabitants realising a regular urban network. Considering the hydrogeological characteristics of the mountain, the rise of hydrothermal waters along fracking planes led to the formation of small surges which were used by the inhabitants and whose waters were collected through tanks and basins. In this way, the city could benefit from a double supply of water: precipitation and natural rises.

Tindari is also located on a cliff overlooking the sea. The site was founded at the beginning of the 4th century BC. The city soon became an important commercial centre due to its advantageous geographical location. As for Solunt, the orographic features provided an important natural defence of the site, reinforced by powerful ramparts embracing the contour lines. The discovery of travertine formations close to a cistern underlines, as in Solunt, the importance of the hydrogeological context, here characterised by karst exes whose waters served the city and more probably the public area.

In Solunt and Tindari, the hydrothermal and karstic springs are now dry. This could be related to a seismic event described in the ancient sources. We know in fact that the city of Tindari was shaken by several earthquakes during its history. One of them is reported to us by Pliny and dates back to the 1st century BC. Another one would have caused several devastations in the city during the late period. Ongoing analyses at the archaeological, geomorphological, hydrogeological and palaeosimological levels, coupled with geochronology and isotopic water geochemistry, little by little, we discover the history of the water supply and management of these two key cities of the Hellenistic-Roman Sicilian world.

Keywords: hydrogeology, sources, cisterns, Solunt, Tindari

Introduction

Interdisciplinary studies have been widely developed for several years. Collaborations between archaeologists and geomorphologists promote increasingly detailed reconstructions of landscape dynamics.[2] In Sicily, conversely, the study of ancient cities has focused more widely on their archaeological aspects, above all urban planning and architecture, while the geological, geomorphological and environmental aspects have been integrated there for only a few decades. Today, with the rise of recent scientific projects, this type of integrated approach seems to be better invested in by the community of archaeologists.

In Sicily, the first analyses of this type concerned Termini Imerese, during the study of the Cornelio aqueduct by Oscar Belvedere.[3] The author proposed an integrated reading of the landscape and geology of the territory of the Roman colony. His work also dealt with the topographical and structural aspects of the Himera landscape, adding paleoclimatic and geological data to clarify the causes of the wealth of the Chalcidian colony.[4] Also worth mentioning are specific isolated studies, including geological analyses, on the indigenous sites of western Sicily during the 1990s. In Colle Madore, the study carried out by Stefano Vassallo on a shrine, probably linked to the cult of local sulphurous waters, saw the participation of a geologist who helped the team understand the context.[5] For Monte Maranfusa, a monograph directed by Francesca

---

[1] The project leading to this publication has received funding from Excellence Initiative of Aix-Marseille University – A*Midex, a French 'Investissements d'Avenir' programme.
[2] https://www.topoi.org/.

[3] Belvedere 1986.
[4] Belvedere *et al.* 2002.
[5] Vassallo 1999.

Spatafora incorporates a geological analysis, treated separately from the topographical and archaeological aspects of the site.[6] Later, geomorphological studies at Tindari focused on the impacts of earthquakes on the ancient city[7] and its harbour.[8] Finally, the most recent research on Syracuse, conducted by Sophie Bouffier, has sought to determine the origin of the waters supplying the city through the Galermi aqueduct.[9]

In this contribution, our objective is to show how we have approached this recent orientation of classical archaeology, which integrates archaeology, human ecology and environmental sciences, on the study of two cities, Solunt and Tindari, founded in the 1st millennium BC on the northern coast of Sicily, and how we were able to understand part of the policy of the authorities of these cities to supply the population with drinking water. The natural element could indeed influence the choices of the inhabitants, for instance the location of the buildings or the urban organisation. According to Philippe Fraisse, the urban planning of Delos could be determined by the presence of numerous natural depressions, which could offer reserves of natural water and thus determine the location of neighbourhoods, by constituting open-air reservoirs.[10]

As part of the Watertraces programme, in October 2018 an interdisciplinary team carried out research at both sites. The aim was to understand the geomorphological and hydrogeological specificities and to detect possible impacts of the environmental context on the development of Solunt and Tindari,[11] as well as to verify through rock and water sampling the possible rise of the waters of the sub-soil within sites. This last aspect, discussed in the following chapters, is well known in the geological context of the Sicilian landscapes.[12] Sicily has rich reserves of hydrothermal or thermo-mineral water; these waters are in contact with magmatic fluxes that enriched it with minerals and the pressure caused by the heat of the subsoil allows the waters to rise and create thermal springs whose temperature is variable.[13] For the first time we reported the traces of this phenomenon in Solunt and we were able to relate it to the archaeological remains, allowing a new and innovative reading of the water supply in apparently anhydrous sites.

Solunt and Tindari do not present today clear traces of a water supply other than rainwater systems (cisterns). However, while the Mediterranean climate has always been influenced by annual and interannual irregularity, with very dry periods, the question is raised as to how a city of some eight thousand inhabitants could have been founded, develop over almost eight centuries, between the 8th century BC – 3rd century AD, with a water autonomy of almost 10,000 m^3: how were the periods of water deficit overcome? The issues of diversity and origin of water supply patterns were crucial. The field missions carried out under the Watertraces project allowed us to cross hydrogeological data and archaeological remains, and to show that the local populations were able to take advantage of resources, that have now disappeared, in order to provide abundant water to the cities, whatever its needs.

Solunt and Tindari: a similar geological and climatic framework, but with specific characteristics

Solunt and Tindari are located on the northern coast of Sicily among high mountains. Our previous research had shown that at these sites the rise in waters could be verified in geological time as well as in recent times.[14] Our research was conducted from this point of view: it was always thought that for the hydraulic supply of a city that rainwater collected in cisterns was sufficient. In view of the rather dry nature of the Sicilian climate, of a thermo-Mediterranean type (up to three consecutive months of dry season over the year), it seems necessary to consider additional contributions only to waters resulting from precipitation.

Solunt, founded by the Carthaginians c. the 4th century BC, is located on Mt Catalfano, a limestone massif 235 m a.s.l., marked by its steep slopes (average of 32%) and cliffs (average of 75%) overlooking the Tyrrhenian Sea. From the lithological point of view (Figure 1), this region belongs to the deposits of the Bacino Imerese,[15] where different types of formations are followed from top to bottom, including:

- Limestones of the Caltavuturo Formation (Eocene), the Crisanti Formation (Upper Middle Cretaceous Jurassic) and the Fanusi Formation (FUN), which corresponds mainly, in our sector, to a Mesozoic Brechic and Dolomitic limestone complex (at a 40° dip towards the south-west), whose series of the Lias (Lower Jurassic) approaches a thickness of 300 m.
- Dolomites and dolomite veins, with high fracture and karstification permeability, emerging along the coast, especially at Sant'Elia (Scillato-SCT Formation).

---

[6] Spatafora 2003.
[7] Bottari *et al.* 2008.
[8] Bottari *et al.* 2012.
[9] https://hydromed.hypotheses.org/893
[10] Fraisse 2015–2016: 864–865. See also the example of the large open cistern of 'House B' south-east of the Theatre, which, according to the excavators, had a collective function, being used by the inhabitants of the neighbouring houses. Chamonard 1933: 146–147.
[11] Polizzi *et al.* 2017; Ollivier 2020.
[12] Bonica Santamaria 2001: 22–27.
[13] Montanari *et al.* 2017.
[14] Polizzi *et al.* 2017 ; Polizzi *et al.* 2020.
[15] Catalano *et al.* 2013: 61.

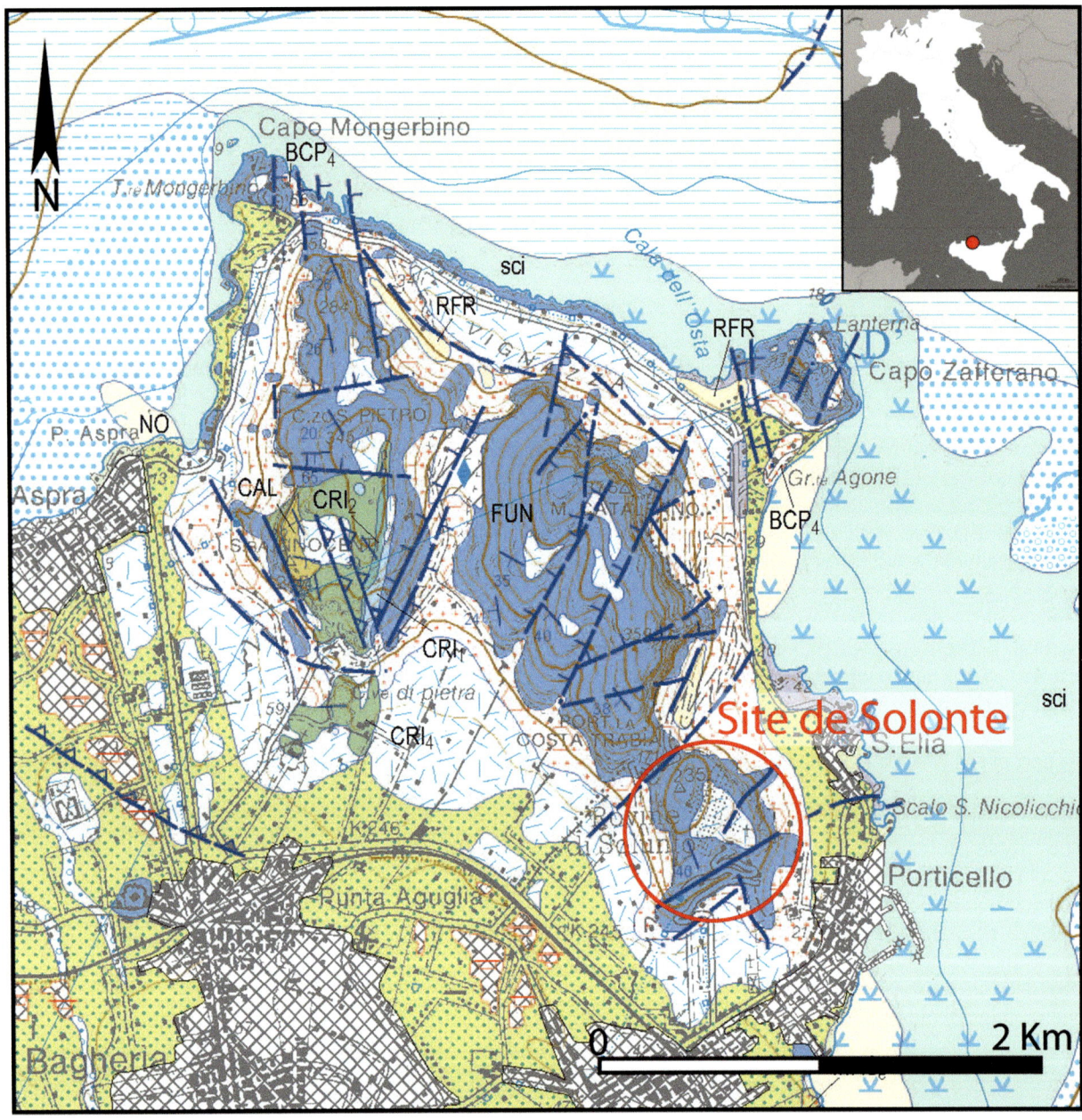

Figure 1: Geological map of Monte Catalfano (Catalano *et al.* 2013).

- Mufara Formation (MUF) clay limestone. The latter are impermeable and are not flush in this area.

This morphological unit, structurally belonging to the Palermo Mountains, is intensely fractured by relatively active tectonics. Along the main fractures widened by the karst there are caves of varying importance, the depths of which can reach -66 m, which contain important forms of concretions (speleothems).[16]

Some 160 km to the east, the town of Tindari was founded at the beginning of the 4th century BC on the promontory of the same name,[17] on the Tyrrhenian side of the Peloritain Mountains. This feature, elongated north–west/south–east, is characterised by a notable cliff on the north-east sector and by slopes more or less accentuated to the west. Tindari itself is located in the north-eastern sector of Sicily, and its lithology (Figure 2) is part of the northern range, which constitutes the structural unit of the Peloritain Mountains.[18] The latter consists of superimposed levels, the main ones being:

- Aspromonte Unit (PMP, MLE a-c): metamorphic rocks composed mostly of highly fractured grey

---

[16] Catalano *et al.* 2013: 31.
[17] Diod., XIV, 78, 5–6.
[18] Marbot 1961: 311.

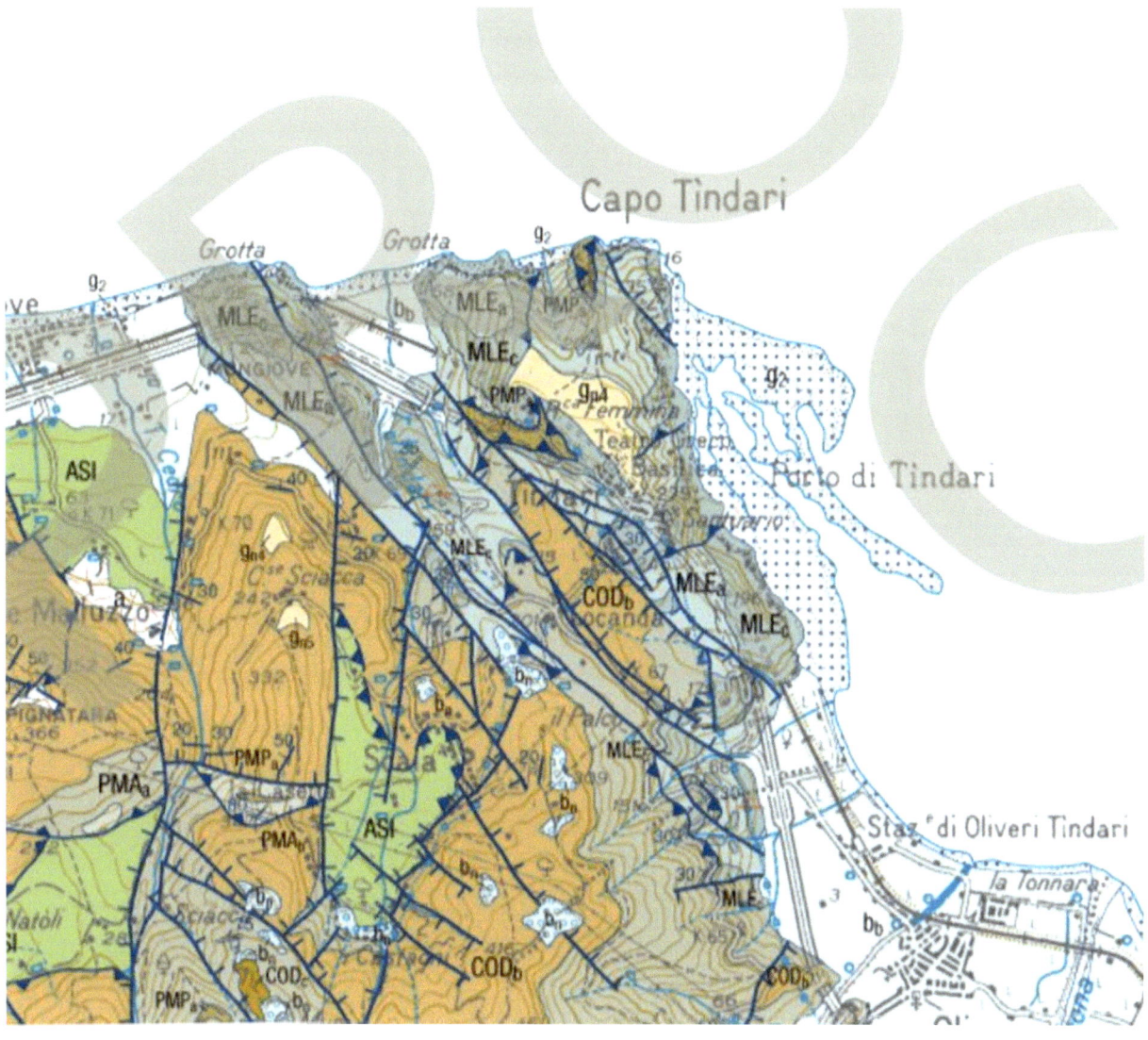

Figure 2: Tindari geological map (Catalano *et al.* 2013).

marble[19] (biotitic gneiss, muscovite-biotitic schist, micaschist, amphibolites and marbles).[20]
- Capo d'Orlando Flysch (COD): poly-genetic conglomerates with arenites and thin clay levels.[21]
- Current alluvial and coastal deposits (g2): gravel and sand originating from erosion of metamorphic rocks.[22]

From the morphostructural point of view, the sector is crossed by the Tindari Fault, oriented NNW[23] and crossing eastern Sicily from the Aeolian Islands to the northeast sector of Etna. This fault is characterised by straight-lateral and extensional movements dated from the middle and upper Pleistocene.[24] These movements have caused several earthquakes over the centuries,[25] one of the most important, probably 6° on the Richter scale,[26] caused extensive damage in Tindari shortly before the abandonment of the site, in the 8th century AD.[27]

Tectonic and seismic movements directly influence the rise of the fluids by artesian conditions by fracturing the various aquifers and reservoirs. The composition of the freshwater, hydrothermal, hypothermal or karstic, by rebalancing with atmospheric pressure, then generates carbonate precipitation and concretions that directly testify to the hydrogeological origin of the waters.

---

[19] Carbone, Messina and Lentini 2011.
[20] Catalano *et al.* 2013: 18–20, 66–71.
[21] Catalano *et al.* 2013: 71.
[22] Catalano *et al.* 2013: 85.
[23] De Astis *et al.* 2003.
[24] Catalano and Di Stefano 1997.
[25] Neri *et al.* 2003.
[26] Bottari *et al.* 2008: 221.
[27] Spigo 2005: 18.

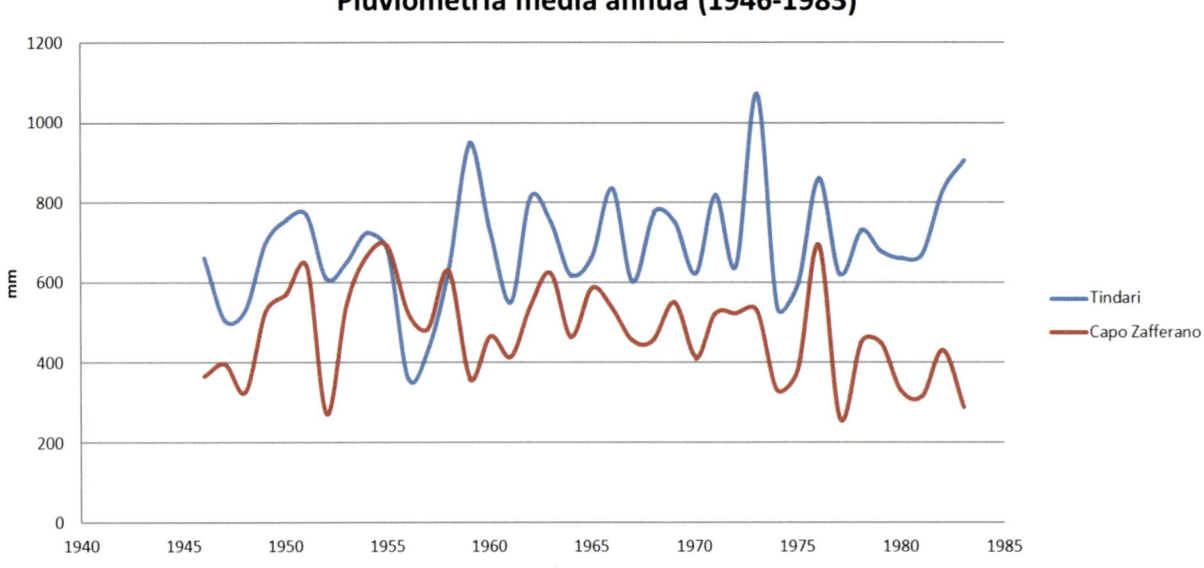

Figure 3: Mean rainfall at Solunt (Capo Zafferano) and Tindari (years 1946–1983) (by G. Polizzi).

In addition to geological data, the understanding of the ancient environment and the availability of water can be perceived through studies on climate variations. The analysis and integration of paleoclimatic data was mainly focused on its relationship to the development of societies and the fall of the great civilisations of the Bronze Age[28] and the Roman Empire.[29] This research area, however, is under development in Sicily, dependant on the enrichment of the database concerning the historical period.[30] As can be seen in the article by Pasta *et al.* (this volume), most of the data cover a chronological range that is too wide and too low a resolution for the period in question. Studies carried out at Selinunt[31] show, for example, that the period between 350 BC and the 1st century AD recorded a gradual increase in temperatures and a decrease in water reserves, while between the 6th and 4th centuries BC the rainfall was greater. These data are also confirmed for the Morgantine site in central Sicily.[32]

For a hypothesis of the water availability in Solunt and Tindari during Antiquity we therefore analysed the rainfall data of the 19th century, for a period of time between 1946 and 1983 (Figure 3). The weather station of Solunt was located in the promontory very close to Capo Zafferano at a height of 22 m a.s.l., and it operated between 1928 and 1999. For Tindari, the weather station was at the site itself, at a height of 280 m a.s.l., and it functioned between 1926 and 1983.[33] Generally, Sicilian precipitation is concentrated in the cold period of the year (autumn/winter). The isohyet maps show the distribution of rainfall at the regional level over the period 1921–2005 (Figure 4). Areas with high rainfall are concentrated in the mountain ranges, where the average rainfall is between 600 mm and 1600 mm per year, in central-southern and south-western Sicily, which are the driest areas. In the rest of Sicily, the average rainfall is between 300 mm and 800 mm per year. More than 80% of this rainfall is concentrated between October and March. The dry season can last between three and six months.

In Solunto, the trend is towards a gradual decrease in rainfall from the end of the 1970s. This trend has been recorded throughout central-southern Italy and is accompanied by a gradual increase in temperatures. The average rainfall remains between 700 and 260 mm. At the annual level, the rains are concentrated mainly in the autumn/winter months (between October and February). During the spring a reduction occurs, sometimes to zero during the summer months. Over 73 years of measurements, 45 months of July were dry. In the summer months, and especially in August, intense rains can be recorded, the value of which is between 30 and 50 mm, in the form of heavy showers. The climatic conditions of Tindari are not very different, although it is possible to count on slightly more abundant rains. Here the average rainfall is 300 to 1000 mm/year.

---

[28] Cline 2014.
[29] Harper 2019. For the ongoing debate on the relationship between the fall of Rome and the climate see Leveau *in press*.
[30] Sadori *et al.* 2016.
[31] Ortolani, Pagliuca 2007.
[32] Sadori *et al.* 2013: 1981.

[33] The data used in this chapter have been collected on the website http://www.osservatorioacque.it, where they can be consulted free of charge.

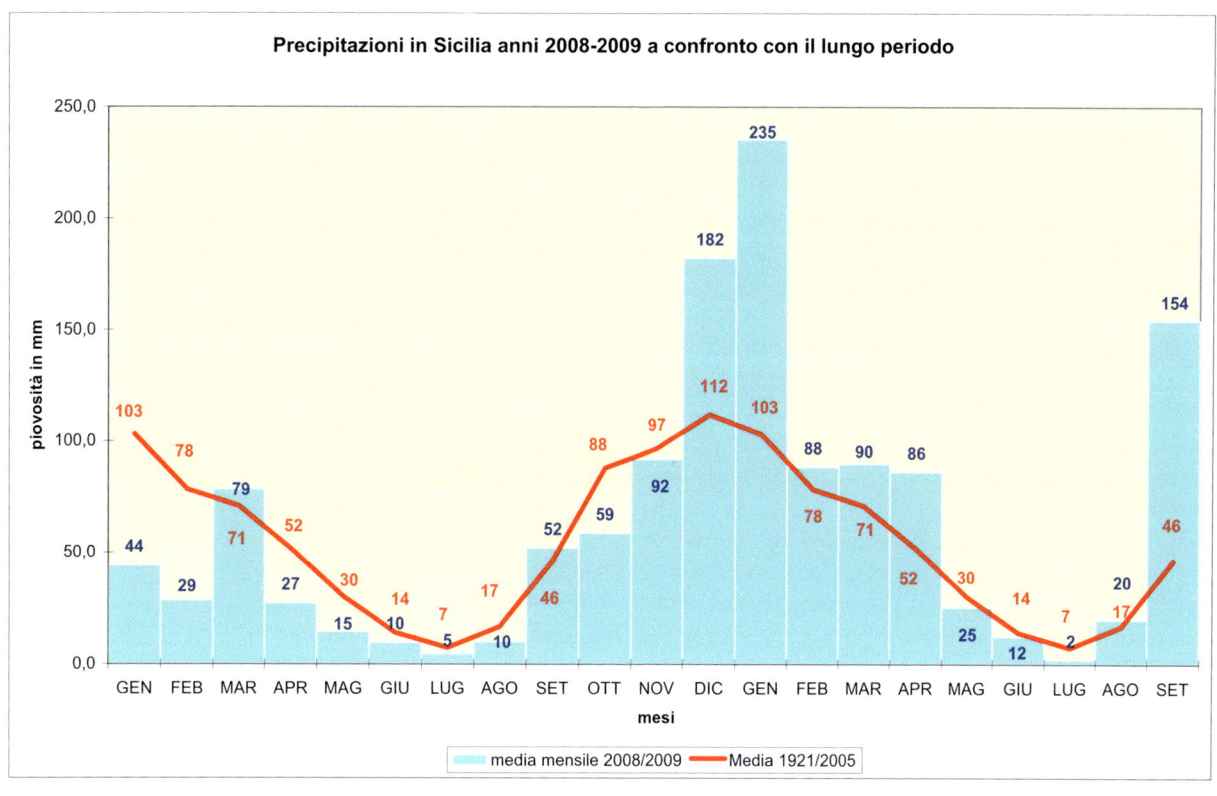

Figure 4: Average rainfall in Sicily (2008/2009/long period ratio).

For both sites, the differences in precipitation in the year are significant and the minimal curve (Figure 4) shows us the current difficulties in water supply if adequate storage is not foreseen. One might wonder, for example, how the inhabitants of the city reacted during the driest periods if they had to rely only on rainwater. If we transpose similar climatic conditions during Antiquity, especially considering the length of dry periods, the sites in question could be affected by significant water crises. In addition, in Solunt, the calculation of the rainfall flow rate, cross-referenced with an estimate of water consumption per person, shows that there was a need for a greater quantity of water than just stored rainwater.[34]

On the basis of the rainfall data, and considering the demographic values and a need for water of 7 litres/day/person for the winter and 13 litres/day/person for the summer,[35] the quantity of water is sufficient, in Solunt and Tindari, for a population of 3,000 and 4,000. If, on the other hand, a value of 25 litres/person/day is considered[36] then it can be estimated that rainwater alone was insufficient to meet population needs, since these would increase exponentially, by estimating a maximum number of inhabitants. So, we had to look elsewhere for other resources. This is demonstrated, for example, by the fact that the most important Hellenistic-Roman cities of Sicily had, in addition to the cisterns, additional water supplies, consisting of internal or nearby sources, such as Morgantina[37] and Palermo,[38] or aqueducts that brought in water from further away, such as Halaesa,[39] Kalé Akté,[40] Taormina,[41] and Termini Imerese.[42]

Our research was thus to find out how the two cities under discussion were able to supplement the meteoric resources, and we searched locally for former traces of water that have now disappeared. In this way, we propose the hypothesis that this additional water came from underground lifts, was captured in tanks, and/or used in collective buildings.

The phenomena of artesian rises of water along fault planes

The Monte Catalfano massif, where Solunt is located, has a significant specific permeability and a large capacity for storage and recharge of meteoric waters.

---

[34] Statistics on the water requirements of the inhabitants of Solunt are based on research conducted in Pompeii (Dessales 2008) and Pergamon (AvP I.4: 37–39). They will be published in a forthcoming monograph.
[35] Data drawn by AvP I.4.
[36] Derived from data FAO. Dessales 2008: 37.
[37] Bruno 2015.
[38] Agnesi 2019; Polizzi 2020a.
[39] Burgio in this volume.
[40] Collura 2019: 257.
[41] Muscolino 2020.
[42] Belvedere 1986.

The data collected during field missions made it possible to identify a series of main tectonic accidents generally oriented NE–SW. But more importantly, several secondary beam fracturing stalling from W to NW have calcitic laminations filling the fault planes.[43] These mineralisations indicate a control of fracturing on the geometry of fluid circulations and highlight the local carbo-gas potential. In Solunt, these upward fluid flows seem to have worked on a geological scale but some were active during the occupation of Monte Catalfano. Indeed, several veins and archaeological structures show laminations and concretions on the walls. The possible hydrothermal character of the waters of Solunt would be represented today, in the nearby bay, by paravolcanic phenomena[44] with deep waters rich in minerals.[45] In addition, near Pizzo Cannita and Villabate, less than 5 km from Monte Catalfano, hydrothermal activity is also reported in a text dating from the 16th century.[46] All these indicators suggest that the artesian rise of the waters worked during the occupation of Solunt.

The present urban layout of Solunt dates back to c. the 2nd century BC. At that time a programme of monumentalisation involved both the collective and private spaces.[47] The excavations have restored a well-integrated water network in the urban setting, where every drop of water was valued. The city, in fact, has many cisterns, individual or collective, to store rainwater and subsoil. These presented two construction techniques: either dug into the rock, and of irregular plan, or masonry and regular plan.[48] The roof received rainwater that was brought into the tanks through vertical or horizontal pipes. The cisterns dug into the rock still have an irregular shape and a niche in the south-west corner. The walls are waterproofed with a mortar based on lime and sand, often restored over the centuries, and their capacity is between 20 – 30 m^3. The dating of these tanks is estimated to the 4th – 3rd centuries BC. What we found during our field missions, is that the location of these cisterns is linked to the limestone substrate and conditioned by the need to exploit the groundwater veins, through the presence/absence of the fractures of the mountain. The presence of spring water left traces through limestone concretions visible at the south-west corners of the cisterns. The plan in Figure 5 shows that several cisterns were dug at the level of the natural fractures of the mountain that we noticed on the ground (the red dotted line), and that several cisterns often exploit the same natural fracture.

We observe the same phenomenon in the construction of community structures, as at the South Baths (Figure 6), located in the south-east sector of the city, near the old urban gate. The building occupies an area of almost 300 m^2, organised on two terraces.[49] A cistern to the west of the *frigidarium* Nord (Room H) received rainwater, but also captured the hydrothermal rising waters via a fracture in the mountain. The tank has a capacity of almost 20 m^3, has an L-shaped plan, and is on a slope towards the north. The walls are waterproofed with a lime and sand coating. The same natural fracture probably fed a small fountain inside the room. Of the latter, there only remains an empty space for the housing of a block, with a drain hole on the west wall of the room (Figure 7).

The discovery of the practice of recovering groundwater in the installation of the cisterns allowed us to look again at some poorly understood archaeological structures, i.e. within the sacred sector of the city (Figure 8). The importance of water in the cults of the Phoenician-Punic world has recently been well highlighted.[50] It is likely that in Solunt, too, as in other Punic sites on the island (i.e. Monte Adranone),[51] water played an important role in the ritual sphere. The sanctuary of Solunt, currently being studied via Elisa Chiara Portale's archaeological excavations, is organised on two terraces and equipped with three main sacred buildings (Figure 8, AB, C, DE). A fourth building further south (Figure 8, F), probably to be read as a service space for the sanctuary, has a large irregular cistern, the only one for the entire religious sector, installed on a fracture of the mount. It is likely that this tank exploited the resurgent waters, as suggested by the presence of a niche in the south-west corner and the presence of calcareous concretions (Figure 10).

Traces of concretions were also found on a sporadic architectural block from the 'Edificio sacro a due navate',[52] just north of the service building (Figure 9). The block in question (44 x 29 x 19 cm) has a rectangular shape and a bevelled side with a hole in a circular section of 6 cm in diameter, inside which are the concretions. We cannot say anything about its exact function, but the traces of stratified concretions suggest the passage of a spring water that was in some way collected and used. The same reduced diameter suggests a water supply function, while all the pipes linked to the passage of rainwater or runoff have a maximum width of 19 cm, probably related to the need to clean the pipes, but this would not be the case for the block in question.

---

[43] Catalano *et al.* 2013: 175–176.
[44] Sulli *et al.* 2011.
[45] Grassa *et al.* 2006.
[46] Amico 1856; Di Matteo 1986.
[47] Portale 2006.
[48] Polizzi 2019; Polizzi 2020b.

[49] For a general description of the building, see Trümper 2020: 368–376.
[50] Usai 2014; Fumadó Ortega 2019.
[51] Trombi 2020.
[52] Portale 2020.

Figure 5: Plan of Solonte with indication of cisterns and natural fractures (elaboration: G.I.S., G. Polizzi).

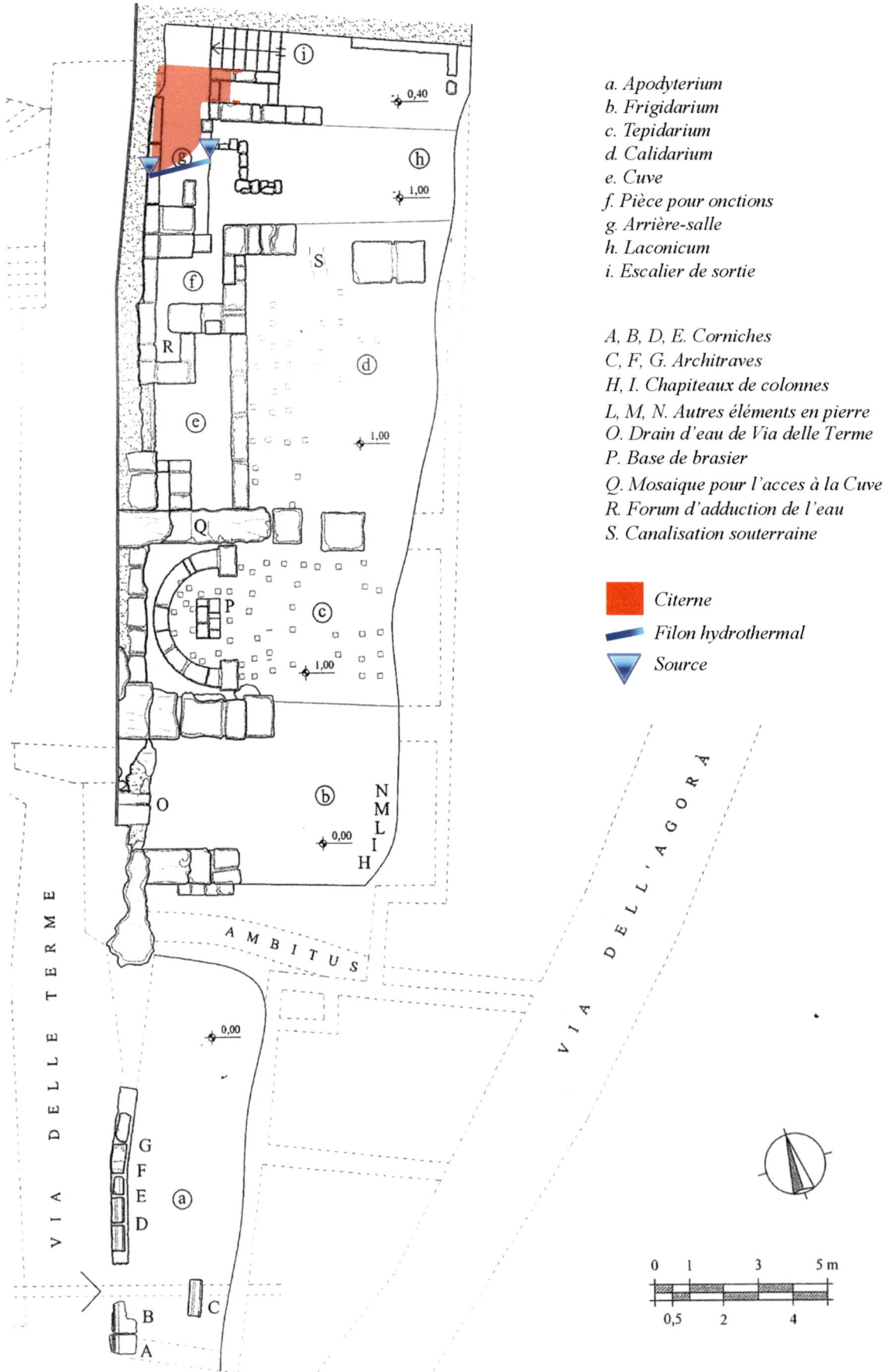

Figure 6: Solunt. Plan of South Baths (from Polizzi *et al.* 2017).

1. Source alimentant la citerne
2. Source et prise d'eau creusée dans la roche

Figure 7: Solunt. Orthophotography of the west wall of room 'h' of the South Baths (photo: G. Polizzi, elaboration V. Dumas).

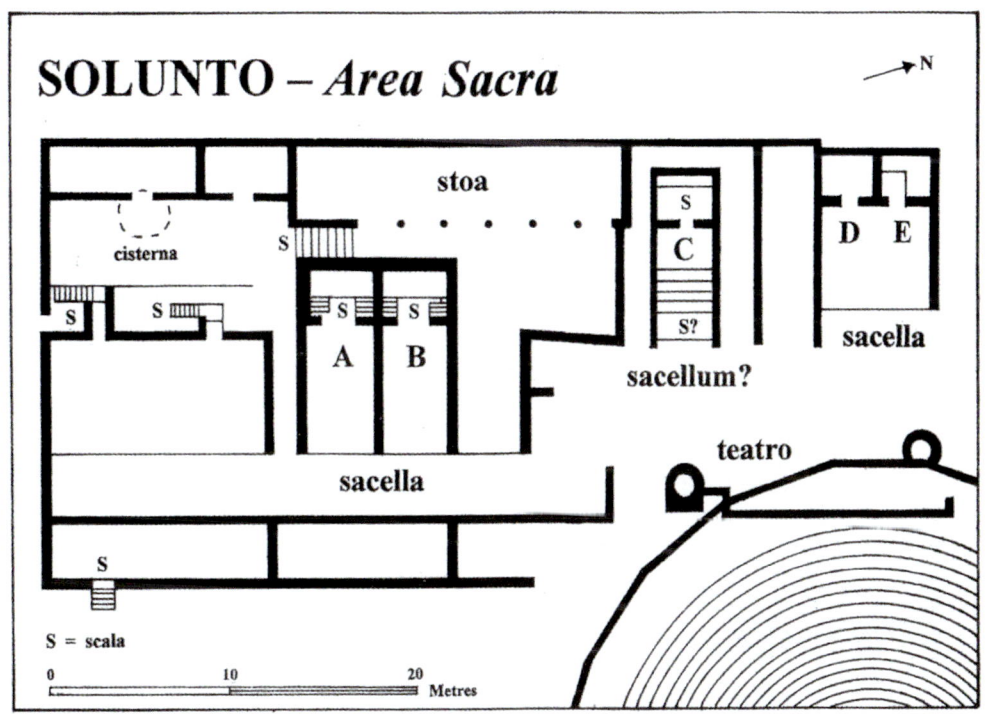

Figure 8: Solunt. Plan of the sacred terrace.

Table 1: Results of U/Th dating performed on calcitic laminations of substratum fracturations and archaeological structures in Solunt.

Labcode	Sample description	[²³⁸U] ppm		[²³²Th] ppb		d²³⁴U$_M$ (‰)		(²³⁰Th/²³⁸U)		(²³⁰Th/²³²Th)		Age (ka)-before 2017
7251-	SOL 16-01 A	0.093	0.001	6.834	0.055	9.90	1.00	1.071	0.003	44.47	0.12	sup 500 kans
7252-	SOL 16-01 B	0.167	0.001	0.633	0.005	16.89	1.00	1.021	0.003	820.74	2.18	sup 500 kans
7253-	SOL 16-03	0.134	0.001	0.160	0.001	25.56	0.71	1.052	0.002	2694.49	4.14	sup 500 kans
7435-	Solunt-17-02	0.194	0.002	2.138	0.017	-2.64	1.31	0.991	0.001	275.56	0.39	sup 500 kans
7436-	Solunt - 4A	0.120	0.001	0.481	0.004	2.84	0.86	1.005	0.002	770.78	1.73	sup 500 kans
7437-	Solunt - 4B	0.138	0.001	1.828	0.015	5.70	0.97	1.001	0.002	123.37	0.19	sup 500 kans

Dating of calcification and concretion

In all surface-usable hydrogeological indices, absolute dating has been undertaken. In order to better characterise over the long term fluid transfers within the massif, U/Th (MC-ICPMS Neptune +) have been made[53] on the series of laminations lodged in the fracturations of the geological formations and then of the archaeological structures.

The development of calcitic laminations of fracturations in the Brechic limestones yielded results all greater than 500 Kans (Table 1) indicating artesian upwelling conditions in place prior to the Middle Pleistocene, probably during the regional Plio-Pleistocene tectonic activity phase. The samples taken at the small baths, although carried out on concretions affecting ancient structures, gave the same type of dating. It will be necessary to determine why such a result could have been obtained. Another architectural block, coming from the 'Edificio sacro a due navate' (Figures 9 and 11), and even more clearly affected by laminations developed during its operating period, will be analysed using the same methods to provide further clarification on this point.

Elementary analyses on laminations of substratum fracturations and archaeological structures

In parallel with the U/Th dating, geochemical analyses by X-ray fluorescence on the laminations and concretions of the site were undertaken. The spectrum of X-rays emitted by matter is characteristic of the composition of the sample, and by analysing this spectrum we can deduce the elementary composition, i.e. the mass concentrations in elements.

Samples were previously dried at 60°C for 48 hours. The elemental analysis was performed using an Olympus Vanta C series portable x-ray fluorescence unit (pXRF), using energy dispersive x-ray fluorescence (ED-XRF)

Figure 9: Solunt. Sporadic architectural block from the 'Edificio sacro a due navate' (photo: G. Polizzi).

with two-beam ground-based geochemical calibration, measuring Mg, Al, Si, Ca, S, P, Ti, V, Cr, Mn, Fe, Co, Ni, Cu, W, Zn, Hg, As, Pb, Bi, Se, Th, U, Rb, Sr, Y, Zr, Nb, Mo, Ag, Cd, Sn, and Sb. This is particularly in order to define the distinct elementary signature of each of the elements that can characterise the deep or karst origin of the reservoirs whose resurgences are at the origin of the concretions studied. All elements with an atomic number less than 11 (H, He, Li, Be, B, C, N, O, F, Ne, Na) are not detected individually by the pXRF technology used and are classified as Light Element (LE) by the Olympus Vanta C geochemical software.

XRF analysis shows that the concretions/laminations of the substratum fractures (Figures 11–14, Table 2) are mainly composed of calcium (especially CaCO3 calcium carbonate in its compound form). Other elements present such as Silica (Si) Aluminium (Al) or Iron (Fe) are also relatively common in carbonate geological contexts. The absence of elements Na, SO4, Cl, Pb, Zn, Au, Ag... (alkaline, sulphate or metalliferous mineralisation) indicates that these are weakly mineralised waters of karstic origin and not hydrothermal (in the hydrogeological and geological sense). On the other

---

[53] By Edwige Pons Branchu, LSCE, UMR 8212, Paris Saclay, France.

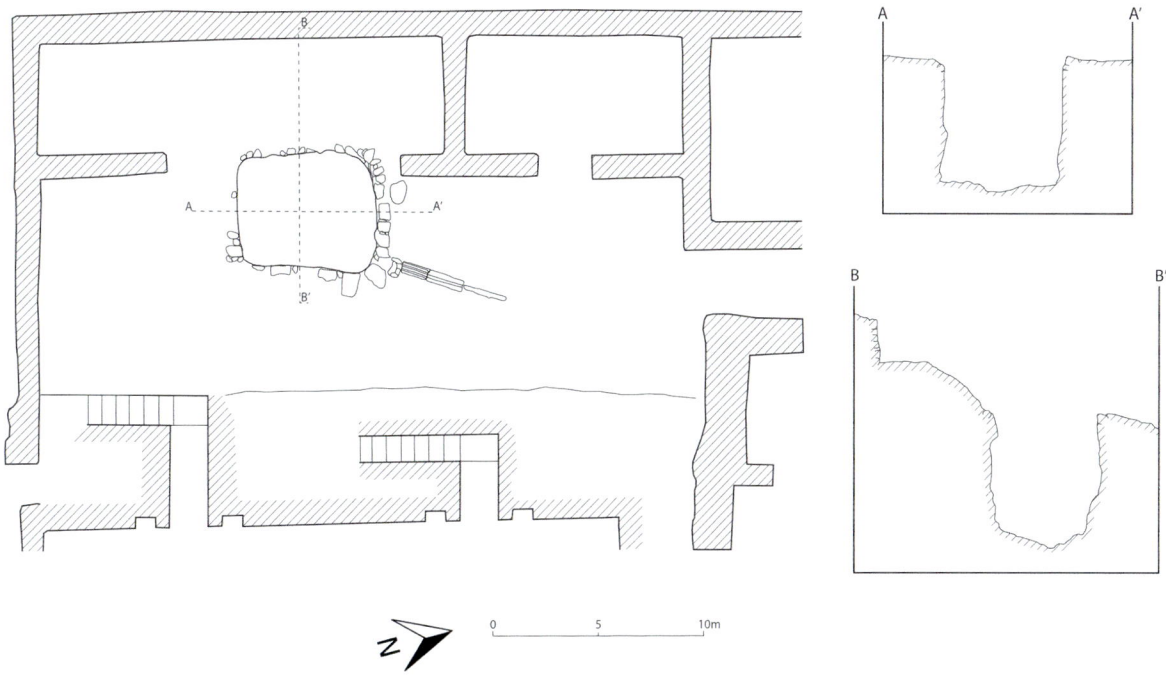

Figure 10: Solunt. Plan and sections of the South-West building of the sacred terrace (elaboration: G. Polizzi).

Figure 11: Solunt. Digital microscope details of the laminations of the SOL 18/19 sample at the architectural block of the 'Edificio sacro a due navate' (photo: V. Ollivier).

Figure 12: Solunt. General view of the calcitic lamination of the SOL 18/10 sample within substratum fracturing (photo: V. Ollivier).

hand, the geochemical results of the concretions affecting certain architectural structures of the baths and of the sacred building with two naves (Figures 15, 16, Tables 3, 4) indicate a much richer elementary composition. Calcium (Ca) is present in much smaller amounts, as is Silica (Si) Aluminium (Al) or Iron (Fe) and there are many other elements, including Sulphur (S), Manganese (Mn), Copper (Cu), Zinc (Zn) or Lead (Pb)

Figure 13: Solunt. Detailed digital microscope view of the calcitic laminations of the SOL 18/10 sample within substratum fracturing (photo: V. Ollivier).

which, although in modest quantities, underline the more youthful character of the encrusting waters. This point of hydrothermalism is a marker of the change in the origin of the waters whose engines remain to be determined. The relative improvement in water quality prompted by this change is also a parameter to be explored in relation to the human occupations of the site which have directly benefited from it.

These first data on Solunt, combining U/Th dating and elementary analyses, bring new arguments in terms of the existence, from the Pliocene/Pleistocene, then, after an indefinite period of time, up to Antiquity, of a specific hydrogeology in the Mont Catalfano massif. The morphostructural and tectonic context characteristic of this sector, where a strong porosity and tensions at the origin of intense fracturing dominate. The hydrogeological consequence was the development of

Figure 14: Spectrum of the elements composing the SOL 18/10 sample at the level of the Solonte substratum fracturations (elaboration: V. Ollivier).

Table 2: List of elements composing the SOL 18/10 sample at the level of the Solonte substratum fracturations.

El	PPM	+/- 3σ
LE	60.52%	0.83
Ca	38.03%	0.79
Si	9550	620
Al	4300	2000
Fe	600	110
Sr	32	6
Nb	7	7

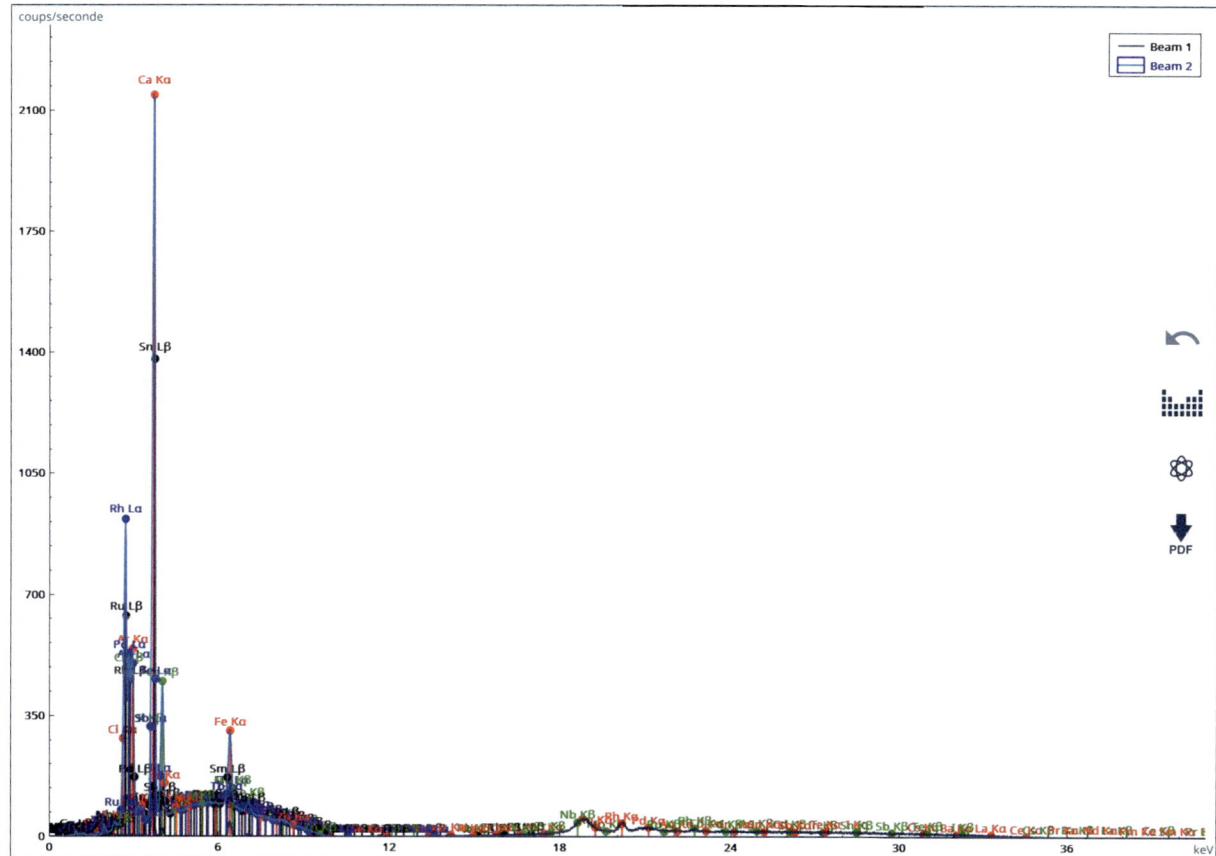

Figure 15: Spectrum of elements composing the SOL 18/08 sample at the level of the Solonte South Baths (elaboration: V. Ollivier).

emergences with hydrothermal potential that allowed an efficient water supply to the city of Solunt, this, in turn, augmenting the water stored in the tank devices harvesting meteoric waters.[54]

The same hydraulic problems facing Solunt are found at Tindari (Figure 17). The city was founded in 396 BC by Denys the Elder (Diod. XIV, 78, 5-6), and the first occupation of the summit took place in the south-west sector, now occupied by the sanctuary of the Madonna del Tindaro.[55]

Since the first excavations, cisterns were found dug into the rock, leading to the supposition that the city could have been supplied exclusively by rainwater.[56] This hypothesis has been maintained, even in recent years, since the discovery of tanks in new excavations;[57] the discovery of a public, large vaulted cistern has helped maintain this hypothesis.[58] This cistern, of rectangular shape and large dimensions (16 m x14 m; depth not determined), probably had a public function

---

[54] Polizzi et al. 2017.
[55] Leone and Spigo 2008: 102.
[56] Lamboglia 1953: 73.
[57] Ravesi 2018: fig. 15; Leone and Spigo 2008: 44, 48, 77.
[58] Gulletta 2012: 300-301.

From Hydrology to Hydroarchaeology

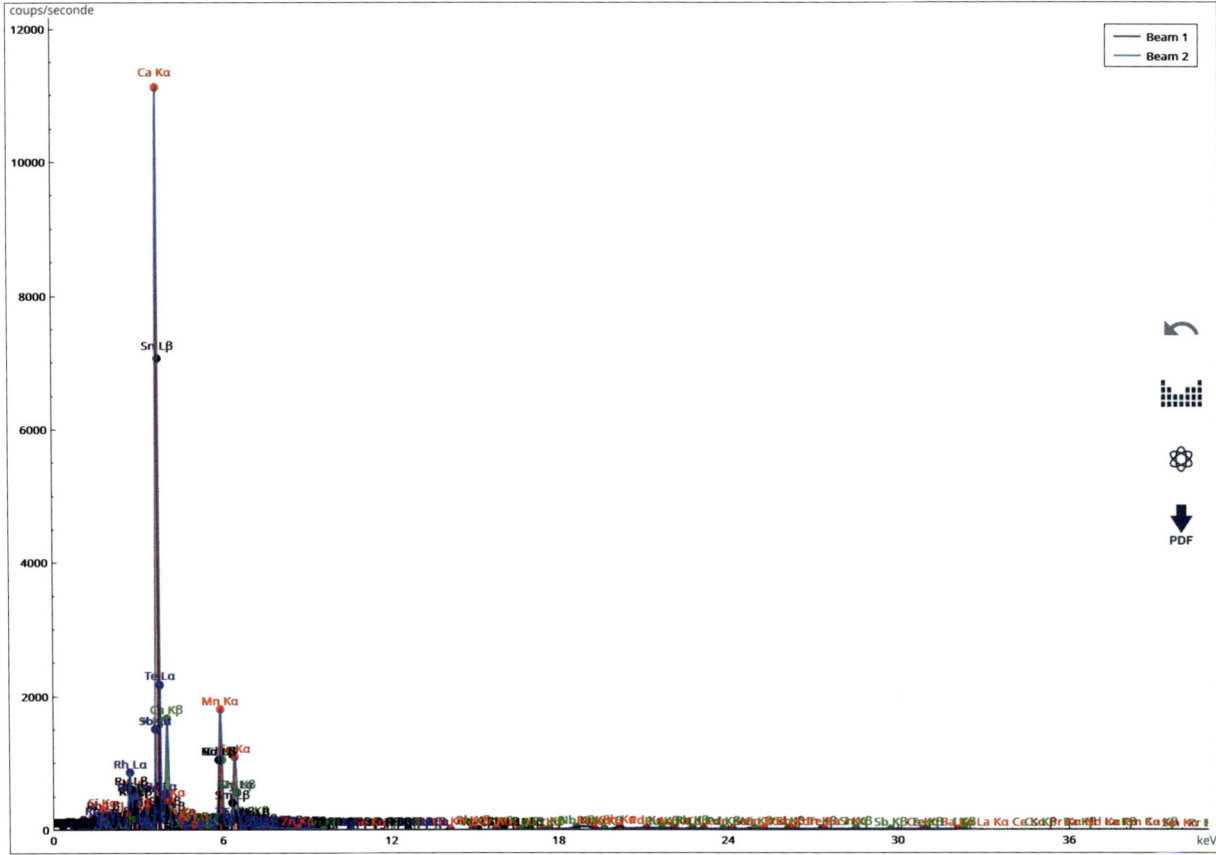

Figure 16: Solunt. Spectrum of elements composing the SOL 18/19 sample at the architectural block of the 'Edificio sacro a due navate' (elaboration: V. Ollivier).

Table 3: List of elements composing the SOL 18/08 sample at the small thermal baths of Solonte.

El	PPM	+/- 3σ
LE	90.43%	0.49
Ca	7.42%	0.26
Fe	6170	410
Si	5990	950
Al	4600	3800
K	2760	230
S	1030	250
Ti	680	590
Mn	130	100
Th	42	28
Sr	35	7
Cu	27	26
Zn	24	16
Mo	18	12
Pb	16	12
Rb	10	5
Zr	9	8

Table 4: List of elements composing the SOL 18/19 sample at the architectural block of the 'Edificio sacro a due navate', Solonte.

El	PPM	+/- 3σ
LE	59.53%	0.92
Ca	22.76%	0.51
Si	7.16%	0.21
Al	3.62%	0.30
Mn	3.42%	0.12
Fe	2.037%	0.075
K	5700	210
Ti	3060	690
S	3030	180
P	1250	270
Pb	286	31
V	270	95
Cr	210	150
Y	202	14
Ni	164	46
Cu	113	35
Sr	113	10
Zn	102	24
Zr	92	11
As	60	23

and collected the waters of the theatre, as in a similar example at Solunt, dated to the 2nd century BC.[59] However, its architecture, which uses bricks extensively, suggests that it was restored during the Imperial age (1st century AD).

The hypothesis of an exclusive supply by rainwater, however, has to be ruled out because of the discovery of an ancient aqueduct in the territory of the ancient city. The survey conducted by Michele Fasolo in the nearby countryside, found pipes of an aqueduct that led the waters towards the city from perennial springs located upstream of the village of Scala, 3 km south of Tindari.[60] The poor state of conservation of the aqueduct means that a precise chronology cannot be put forward, nor its complete route, however the latter would seem to cross along its route various sites of Late Republican and Imperial times. The pipes of this aqueduct, of conical shape, and with openings for inspection not always present, have dimensions inferior to the pipes of the aqueducts known elsewhere in Sicily[61] (length: 42.5 cm; max mouth diameter: 9.5 cm; wall thickness: 5 cm), and find no parallels among similar ancient works (Figure 18). The presence of mortar at the ends of the conserved segments suggests that the water passing through them was under pressure.

We should therefore ask ourselves when, and in what environmental context, this aqueduct was built. The presence of an aqueduct, in fact, would require a significant commitment by the city, especially if it had internal sources within the city. One of these sources was detected during the Watertraces missions. Indeed, in addition to the aqueduct and cisterns, the ancient city could be supplied by a natural source from within the city – located along Teatro Greco Street, in an area of the city that, according to hypotheses still under study, housed the Tindari agora/forum.[62] This ancient source, now dry, is located between the public space marked by the presence of the so-called 'Basilica' and the fortifications, more precisely between towers VI and VII[63] (Figure 17, n. 18).

We have proposed the existence of a natural source in this sector after explorations in 2017, during which we were able to recognise concretions characteristic of the presence of an exsurgence near a tank. A more targeted exploration carried out in 2018 confirmed this hypothesis and recognised a calcareous tufa bank formed by the presence of karstic or hydrothermal water. Finally, the concretion was cleaned in April 2019, making it possible to recognise a niche with traces of tile concrete. A second niche appeared on the northwest side. Further south, the tank was cleaned. The proximity of the source to the cistern had suggested

---

[59] Polizzi, Torre 2018.
[60] Fasolo 2013: 97, fig. 109.
[61] See the contribution by Aurelio Burgio in this volume.
[62] Ravesi 2018.
[63] For the numbering of the towers, see Leone 2020: fig. 6.

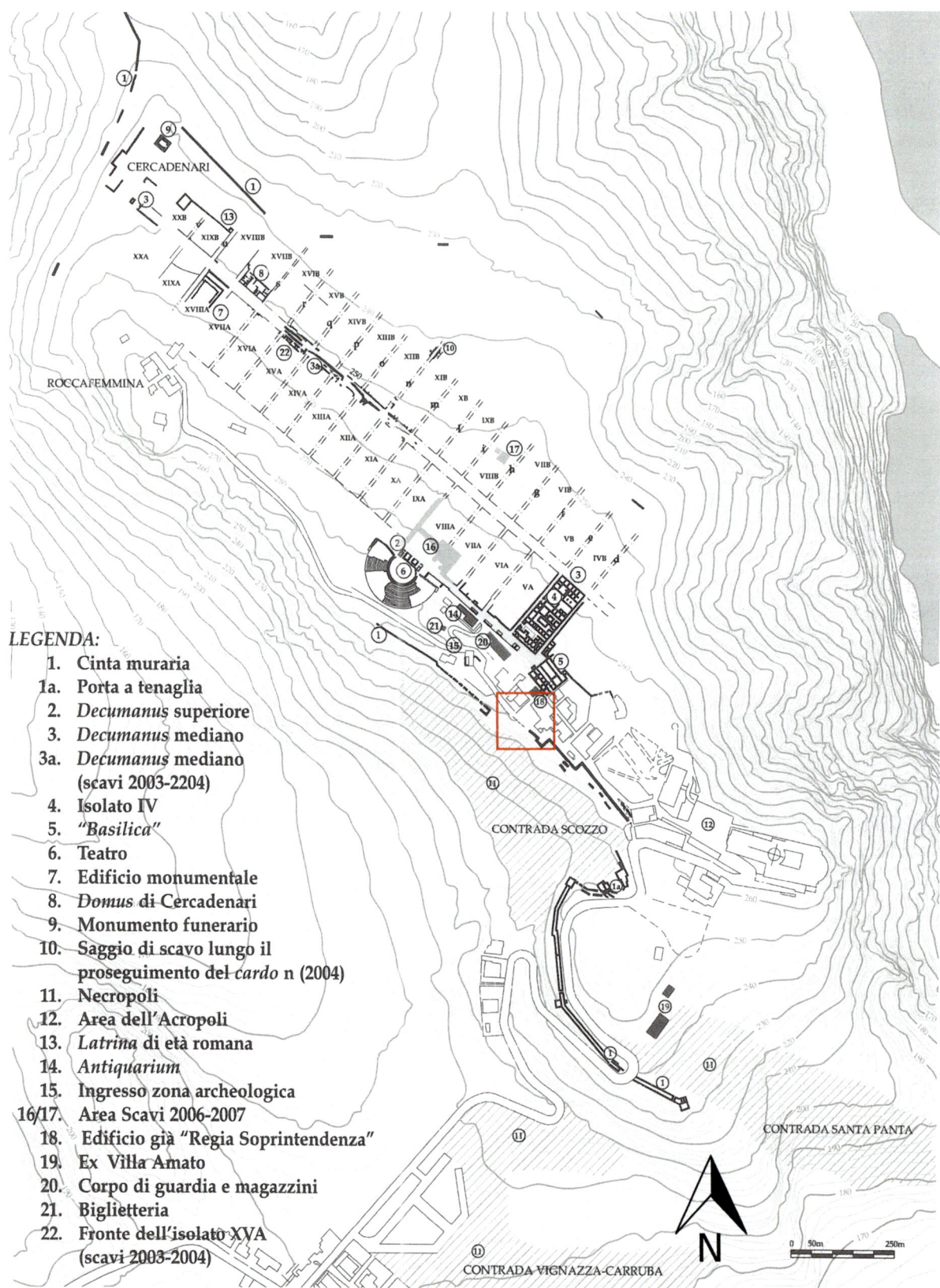

Figure 17: Map of Tindari. In red, the area of the spring (from Spigo 2005).

Figure 18: Tindari. Aqueduct pipe (photo: G. Polizzi).

a possible functional connection between the two, with an interpretation of the latter as the source water reservoir.

It is important to note that the area of the ramparts near the source appears to be characterised by access to the city, as suggested by a map of 1814[64] (Figure 19). Due to modern works, it is no longer possible to verify the exact position of this access, however the strategic position of the source should be noted – close both to the public space and an access to the city through the ramparts.

A new mission in February 2021, under the Watertraces project, has allowed further examination of the remains of the source. After cleaning the limestone tuff and the cistern, we were able to better understand the architecture and functioning of the hydraulic system (Figure 20), organised in three sectors from north to south: in the north it is possible to see a normal fault, which should correspond to the rising point of the water; further south (2 m), a limestone tuff formation arranged with three opposite niches is partially hidden to the east by a modern reinforced concrete platform that prevents a complete reading of the monument; finally, further south, and at a level c. 60 cm lower, a tank/tank was installed.

The limestone tuff is arranged with three niches, artificially dug. The main niche (niche 1) has a width of 0.89 m and a preserved height of 0.71 cm. It has walls covered by irregular stones glued by concretionary effect. To the north of this niche is another one of oval shape, 1.35 m wide and 1 m deep, preserved to a height of 0.68 m. The internal walls of the niche have been artificially regularised. Further north, a modern concrete staircase covers the space between the fault and the tuff. To the south of the central niche is the local metamorphic rock, partially covered by limestone tuff arranged in the shape of a niche, now almost completely destroyed. Only the bottom part is preserved, to a width of 50 cm. Here the geological substrate was cut vertically, sheltering stone blocks of regular shape (Figure 21).

To the south of the calcareous tufa is the cistern/reservoir. It is partially covered by the modern platform, which does not allow its development to be perceived towards the north. In addition, parallel to the southern wall a modern sewer was installed at the bottom of the tank, which caused damage to its east and west walls as well as to its bottom. The southern part of the tank has a regular shape, with a straight wall, 3.90 m long, oriented east–west. The hollow walls in the rock are coated with waterproofing concrete in three layers, the first with a lime and sand base, the second with coarse tile concrete, and the third with finer tile concrete. The bottom of the tank is made of coarse tile concrete, laid on a layer of lime-based and medium-sized pebbles (0.05/0.10 m). The height is preserved, near the eastern angle, to 0.80 cm. On the south wall, 30 cm from the bottom, is a horizontal layer of concretions, 1 cm thick:

---
[64] Ferrara 1814.

From Hydrology to Hydroarchaeology

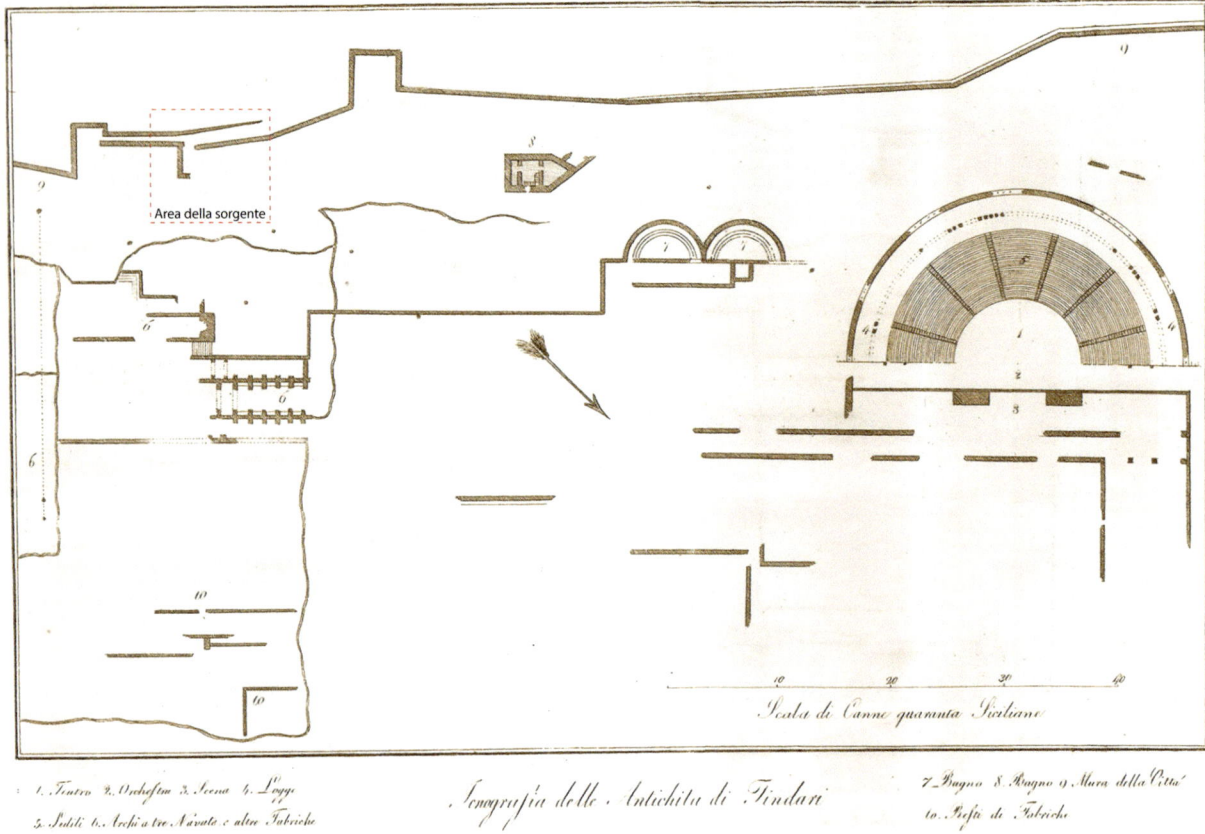

Figure 19: Old plan of Tindari (from Ferrara 1814).

Figure 20: Tindari. General view of the source. South-west view (elaboration: G. Polizzi).

Figure 21: Tindari. Detail of the artificially cut geological substrate (photo: G. Polizzi).

Figure 22: Tindari. Detail of the horizontal layer of concretions (G. Polizzi).

this stratum is the result of the prolonged deposition of spring waters (Figure 22). This form of carbonatation indicates the presence of spring water inside the tank. Through analyses focused on laminations of this type, it is also possible to define the carbonation rhythms in terms of their intensity and chronology; these will have to wait for a future field mission, but it is now possible to define the very nature and origin of the waters responsible for these carbonation phenomena.

Calcareous concretions (travertine, tufa, caving, calcareous crusts) generally provide valuable information on: the origin and quality of the waters (karstic, hydrothermal, hypothermal, meteoric), the hydrodynamics of the environment (depending on the concretion morphotype), the state of the vegetation cover (participating in the dissolution of limestones and the remobilisation of carbonates), the climatic context over time (stimulated expression or constraint of carbonation), anthropogenic activities (sensitivity of carbonatogenesis processes to detritism and pollutants). As at Solunt, one of the approaches adapted to the study and characterisation of these different components is elemental X-ray fluorescence analysis.

Elementary analyses on the encrustations and laminations of the archaeological structures of Tindari

Considering the facies, crystallisation modes, and morphologies observed, the carbonate formation of the tufa type of the via Teatro Greco in Tindari is directly related to the supply of water from karstic or thermo-mineral sources. Biochemical construction phenomena related to the activity of algae and bacteria also played a fundamental role in the genesis of these encrustations. The type of inlays encountered is comparable to the tufa and travertines of river, source, or cascade that can be encountered in natural contexts. However, it was necessary to ensure that it was not a hydrothermal/hypothermal component in order that we could better understand the hydrogeological functioning of the massif and determine the origin of the waters, a parameter directly influencing the modalities of their exploitation during the occupation phases of the site of Tindari.

The X-ray fluorescence analyses carried out on the travertine formation (Figure 23, Table 5) confirm the origin and part of the composition of the waters responsible for the concretions. The dominant element is Calcium (Ca = 27.14%), which underlines, if necessary, the supersaturation in carbonates of the waters. The second element present in greater proportion in the measured elemental cortege is Silica (Si= 6.54%), a common element in this type of substratum. The other elements remain as traces and indicate only a very discreet influence of processes of circulation of fluids from deep contexts. The low mineralisation of concretions refers to contexts related rather to karstic or pseudo-karstic domains.

In the morphostructural context of the Tindari headland, the hydrogeological mechanisms of emergent systems can be complex. The first results of the isotopic analyses of Strontium in the current waters (an element also present in the concretions), collected at the source of Casa della Vita, a few hundred metres from the formations studied, allow us to estimate the location of the reservoirs by delivering a $^{87}Sr/^{86}Sr$ ratio of 0.707970 characteristic of the lower Oligocene/Miocene lithologies to which the local Capo Orlando Flysch Formation belongs.

The waters of the via Teatro Greco concretions appear to have karstic or pseudo-karstic origins, as indicated by the X-ray fluorescence data, and therefore do not fall under hydrothermal mechanisms. In detail, these rocks probably underwent, during the Neogene (Lower Miocene?), one or more phases of alteration at the origin of the development of alterations. The transition from bedrock to other forms is usually done through a karstified horizon, characterised by the presence of karst-like conduits, in which sometimes the other form still remains in place. This horizon often develops between a saprolitisation front (in contact with the other form) and an alteration front (in contact with the healthy mother rock). To this must be added the presence of karst-type drains, taking advantage of the vacuum generated by the presence of faults and fracturing. Karstification or pseudo karstification of flyschs develops in the core fault (breach) and decreases in the damaged area. The 'karstic' conduits connected in the unsaturated and saturated zone of the massif pass the pressure wave and form a surface resurgence.

Conclusions

From the integrated analysis of the Solunt and Tindari sites certain initial findings can be made. Data from hydrogeological observations, isotopic analysis (Sr) of waters and XRF measurements indicate to us relatively clearly that the ancient cities were well supplied with hydrothermal/hypothermal (Solunt) or karstic (Tindari) waters, in probable support of meteoric waters. New data are still being processed but all seem to support this first finding of a meteor/deep water model[65] that can be validated both in geochemistry and in terms of the resulting surface formations (concretions and travertine buildings). The question of a connection between the location of ancient cities, hydrogeology and water management is now clearer.

From the hydrogeological point of view, and with the help of the first geochemical measurements obtained, the observation of reservoirs, appearing more and more mineralised, of the Pliocene/ Pleistocene, in ancient times at the city of Solunt makes it necessary to search for the origins of such a phenomenon, as well as its validation. Such research involves new reflections on the link between the evolution of the origins of these water sources, their quality, their availability, and the modalities of human occupations, including the management of such resources over the long term.

---

[65] Polizzi *et al.* 2017.

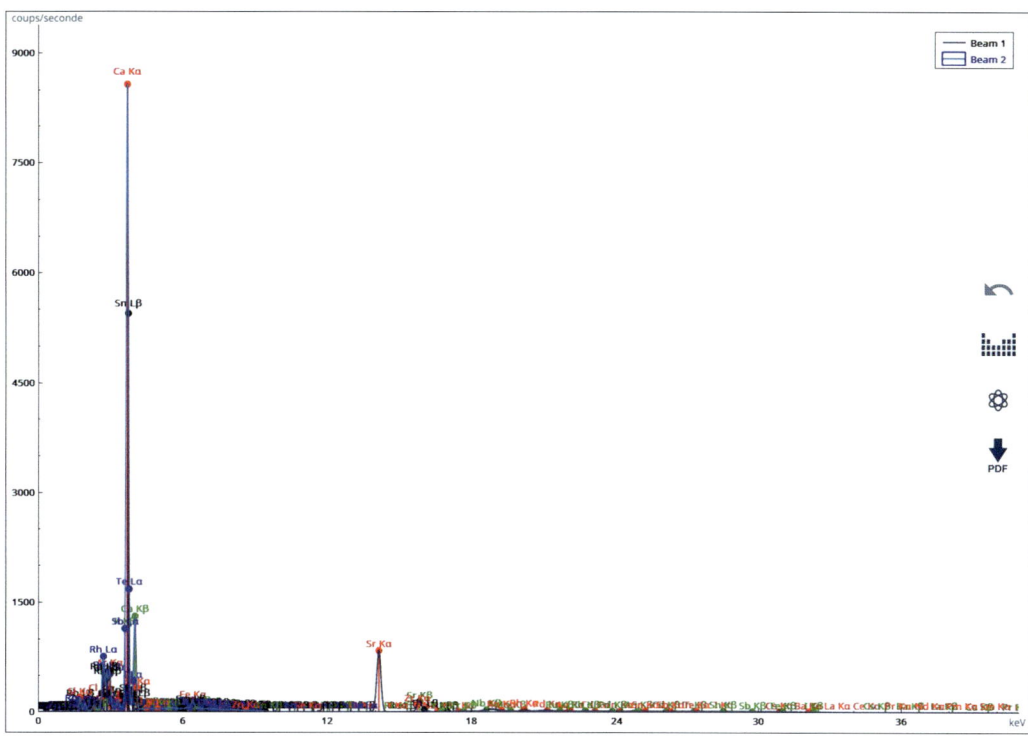

Figure 23: Tindari. Spectrum of elements composing the sample TIN 18/01 at the calcareous tuff of via Teatro Greco (elaboration: V. Ollivier).

Table 5: List of elements composing sample TIN 18/01 at the calcareous tuff of via Teatro Greco in Tindari.

El	PPM	+/- 3σ
LE	63.4%	1.1
Ca	27.14%	0.80
Si	6.54%	0.24
Sr	9350	290
Al	7900	2800
K	3440	220
Fe	3090	300
S	2830	240
P	1160	350
Ti	820	640
Mn	260	140
Zr	119	36
Th	88	58
Cu	35	29
Zn	29	19
Mo	16	16

At Solunt, the urban organisation, as well as the location of the sacred spaces, seem determined by the points of spill of the deep waters; a sacred building, and several houses, have cisterns for collecting them, in addition to the rainwater. The flow of these waters was then guaranteed by a system of pipes and various tanks that directed excess water out of the city. The 'South Baths' were built over a spring, which, in addition to its strategic position at the entrance of the city, must have determined the siting of the building in this area.

Tindari's water source suggests specific attention to the urban organisation of the city, given its proximity to the public sector of the 'Basilica' and the ramparts. It is likely that the opportunities offered by the source could also have encouraged the creation of a gateway in this sector.

Even if the dating currently available does not provide a coherent chronology to the phases of ancient occupation of the sites, the link between sources/water spill and archaeological structures remains clear: in Solunt and Tindari there is a very close relationship between the environment, water infrastructure and urban organisation, which confirms the importance of a study integrating hydrogeology and archaeology in this case.

Bibliography

Agnesi, V. 2019. *Palermo e l'acqua perduta.* Palermo: Plumelia edizioni.

Amico, V. 1856. *Dizionario topografico della Sicilia di Vito Amico tradotto dal latino ed annotato da Gioacchino Dimarzo.* Palermo: Tipografia di Pietro Morvillo.

AvP I.4. (Garbrecht, G.) 2001. *Altertümer von Pergamon I.4. Stadt und Landschaft 4. Die Wasserversorgung von Pergamon.* Berlin: De Gruyter.

Belvedere, O. 1986. *L'Acquedotto Cornelio di Termini Imerese.* Roma: L'Erma di Bretschneider.

Belvedere, O. and E. Termine 2005. L'urbanizzazione della costa nord-orientale della Sicilia e la struttura urbana di Tindari. *Omni paede stare. Saggi architettonici e circumvesuviani in memoriam Jos De Waele:* 85–92. Napoli: Electa Napoli.

Belvedere, O., Bertini, A., Boschian, G., Burgio, A., Contino, A., Cucco, R.M. and Lauro, D. 2002. *Himera III.2.* Roma: l'Erma di Bretschneider.

Bernabò Brea, L. 1952. Tyndaris, Tindari (Sicilia, Messina). *Fasti Archeologici* VII: n. 2107.

Bernabò Brea, L. 1965. Due secoli di studi, scavi e restauri del teatro greco di Tindari. *Rivista dell'Istituto nazionale d'archeologia e storia dell'arte* XIV: 99–114.

Bernabò Brea, L., Cavalier, M. 1965. Scavi in Sicilia. Tindari. Area Urbana. L'Insula IV e le strade che la circondano. *Bollettino d'Arte* III–IV: 205–209.

Bonfiglio, L., Mangano, G., Pino, P. 2010. The contribution of mammal-bearing deposits to timing late Pleistocene Tectonics of Cape Tindari (North-Eastern Sicily). *Rivista italiana di Paleontologia e Stratigrafia* 116: 103–118.

Bonica Santamaria M.L. 2001. Il termalismo in Sicilia. *Archivio Storico Messinese* 82: 21–103.

Bottari, C., Bottari, A., Carveni, P., Saccà, C., Spigo, U. and Teramo, A. 2008. Evidence of seismic deformation of the paved floor of the decumanus at Tindari (NE, Sicily). *Geophysical Journal International* 174: 213–222; doi.org/10.1111/j.1365-246X.2008.03772.

Bottari C., Urbini, S., Bianca, M., D'Amico, M., Marchetti, M. and Pizzolo, F. 2012. Buried archaeological remains connected to the Greek-Roman harbor at Tindari (north-east Sicily): results from geomorphological and geophysical investigations. *Annals of Geophysics* 55: 223–234.

Bruno, G. 2015. La drastica riduzione della risorsa idrica a Morgantina nel periodo greco-romano, in Maniscalco, L. (ed) *Morgantina duemilaquindici. La ricerca archeologica a sessant'anni dall'avvio degli scavi.* Palermo: Regione Siciliana.

Carbone, S., Messina, A., Lentini, F. 2011. *Note illustrative della Carta Geologica d'Italia alla scala 1:50.000 'Milazzo - Barcellona P.G.'.* Palermo: Istituto Poligrafico dello Stato.

Catalano R., Avellone, G., Basilone, L., Contino, A., Agate, M. 2013. *Note illustrative della Carta Geologica d'Italia alla scala 1: 50000 Foglio 595 'Palermo'.* Palermo: Istituto Poligrafico dello Stato.

Catalano, S., Di Stefano, A. 1997. Sollevamenti e tettonogenesi pleistocenica lungo il margine tirrenico dei Monti Peloritani: integrazione dei dati geomorfologici, strutturali e biostratigrafici. *Il Quaternario* 10: 337–342.

Chamonard, J. 1933. Fouilles de Délos. *Bulletin de Correspondance Hellénique* 57: 98–169.

Cline, E. 2014. *177 B.C.: The Year Civilization Collapsed.* Princeton: Princeton University Press.

Collura, F. 2019. *I Nebrodi nell'antichità: Città Culture Paesaggio.* Oxford: Archaeopress.

De Astis, G., Ventura, G. and Vilardo, G. 2003. Geodynamic significance of the Aeolian volcanism (Southern Tyrrhenian Sea, Italy) in light of structural, seismological and geochemical data. *Tectonics* 22(4): 1040–1057.

Dessales, H. 2008. Des usages de l'eau aux évaluations démographiques. L'exemple de Pompéi. *Histoire Urbaine* 2(22): 27–41.

Di Matteo, S. 1986. *Opuscoli del marchese di Villabianca. La fontanagrafia oretea : le acque di Palermo.* Palermo: Giada Editore.

Fasolo, M. 2013. *Tyndaris e il suo territorio. Volume II. Carta archeologica del territorio di Tindari e materiali.* Roma: mediaGEO.

Ferrara, F. 1814. *Antichi edificj ed altri monumenti di belle arti ancora esistenti in sicilia disegnati e descritti dall'Abate Francesco Ferrara.* Palermo: Tipografia Reale di Guerra.

Fraisse, Ph. 2015–2016. Délos. Morphologie urbaine. *Bulletin de Correspondance Hellénique* 139–140: 862–864.

Fumadó Ortega, I. 2019. L'apparition et la diffusion des citernes en Méditerranée phénico-punique. Un bilan, in S. Bouffier, O. Belvedere and S. Vassallo (eds) *Gérer l'eau en Méditerranée au premier millénaire avant J.-C.:* 169-183. Aix en Provence: Presses universitaires de Provence.

Grassa, F., Capasso, G. and Favara, R. 2006. Chemical and Isotopic Composition of Waters and Dissolved Gases in Some Thermal Springs of Sicily and Adjacent Volcanic Islands, Italy. *Pure and Applied Geophysics* 163: 781–807; doi.org/10.1007/s00024-006-0043-0.

Gulletta, M.I.P. 2012. Tyndaris: per uno status quaestionis sulle ipotesi di ubicazione dell'agora-foro, in C. Ampolo (ed.) *Sicilia occidentale. Studi, rassegne, ricerche:* 297–303. Atti delle Settime

Giornate Internazionali di Studi sull'area elima e la Sicilia Occidentale. Pisa: Edizioni della Normale.

Harper, K. 2017. *The Fate of Rome. Climate, Disease, and the End of an Empire.* Princeton: Princeton University Press.

Lamboglia, N. 1953. Gli scavi di Tindari (1950–1952). *La Giara* 2: 70–81.

La Torre, G.F. 2004. Il processo di 'romanizzazione' della Sicilia: il caso di Tindari. *Sicilia Antiqua* I: 111–146.

La Torre, G.F. 2006. Urbanistica e architettura ellenistica a Tindari, Eraclea Minoa e Finziade: nuovi dati e prospettive di ricerca, in M. Osanna and M. Torelli (eds) *Sicilia Ellenistica, consuetudo italica. Alle origini dell'architettura ellenistica d'occidente*: 140–151. Atti dell'Incontro di Studio (Spoleto, 5–7 novembre 2004). Rome: Edizioni dell'Ateneo.

Leone, R. 2020. L'archeologia a Tindari tra 1814 e 1845, in C. Malacrino, A. Quattrocchi, R. Di Cesare (eds) *L'antichità nel regno. Archeologia, tutela e restauri nel Mezzogiorno preunitario*: 311–316. Reggio Calabria: Kore s.r.l.

Leone, R. and Spigo, U. (eds) 2008. *Tyndaris 1. Ricerche nel settore occidentale: campagne di scavo 1993-2004.* Palermo: Regione Siciliana, Assessorato dei Beni Culturali, Ambientali e della Pubblica Istruzione.

Leveau, Ph., Le destin de l'Empire romain dans le temps long de l'environnement à la lumière des géosciences et des biosciences, in *Annales. Histoire, Sciences Sociales, in press.*.

Marbot. V. 1961. Mise au point sur la structure de la Sicile. *Annales de Géographie* 70(379): 311–319.

Montanari D., Minissale, A., Doveri, M., Gola, G., Trumpy, E., Santilano, A. and Manzella, A. 2017. Geothermal resources within carbonate reservoirs in western Sicily (Italy): A review. *Earth-Science Reviews* 169: 180–201.

Muscolino, F. 2020. Le cisterne di Taormina in età ellenistico-romana, in V. Caminneci, M.C. Parello and M.S. Rizzo (eds) *Le forme dell'acqua. Approvvigionamento, raccolta e smaltimento nella città antica*: 51–62. Atti delle Giornate Gregoriane XII Edizione (Agrigento 1–2 dicembre 2018). Bologna: Ante Quem.

Neri, G., Barberi, G., Orecchio, B. and Mostaccio, A. 2003. Seismic strain and seismogenic stress regimes in the crust of the southern Tyrrhenian region. *Earth Planetary Science*: 97–112.

Ollivier, V. 2020. Géomorphologie, archéologie et gestion de l'eau « Identités remarquables » des relations plurimillénaires entre environnements et sociétés?, in S. Bouffier and I. Fumadó Ortega (eds) *L'eau dans tous ses états. Perceptions Antiques*: 13–28. Aix en Provence: Presses Universitaires de Provence.

Ortolani, F. and Pagliuca, S. 2007, Le variazioni climatiche in Italia Centrale negli ultimi 10.000 anni. *Quaderni della Società Geologica Italiana* 1: 14–18.

Polizzi, G. 2019. Gestione e uso dell'acqua a Solunto: le cisterne, in S. Bouffier, O. Belvedere, S. Vassallo (eds) *Gérer l'eau en Méditerranée au premier millénaire avant J.-C.: 193-209.* Aix en Provence. Presses universitaires de Provence.

Polizzi, G. 2020a. Il sistema idrico urbano di Solunto fra IV e I secolo a. C., in S. Celestino and E. Rodríguez (eds) *X Congreso Internacional de Estudios Fenicios y Púnicos / International Congress of Phoenician and Punic Studies: 1903-1919.* Mérida: CSIC - Junta de Extremadura - Instituto de Arqueología (IAM).

Polizzi, G. 2020b. La gestione delle acque nelle città fenicio-puniche della Sicilia: riflessioni e prospettive a partire dalle evidenze di Mozia, Solunto e Palermo, in E. Bianchi and M. D'Acunto (eds) *Opere di regimentazione delle acque in età arcaica. Roma, Grecia e Magna Grecia, Etruria e mondo italico*: 161–184. Roma: Quasar.

Polizzi, G., Ollivier, V., Fumadó Ortega, I., Bouffier, S. 2017. Archéologie et hydrogéologie. *Chronique des activités archéologiques de l'École française de Rome. Sicile.* http://cefr.revues.org/1705; doi.org/10.4000/cefr.1705.

Polizzi, G., Ollivier, V., Bellier, O., Pons-Branchu, E., Fontugne, M. 2020. Evidenze di un sistema idraulico alternativo in Sicilia e in Grecia, in V. Caminneci, M.C. Parello and M.S. Rizzo (eds) *Le forme dell'acqua. Approvvigionamento, raccolta e smaltimento nella città antica*: 305–315. Atti delle Giornate Gregoriane XII Edizione (Agrigento 1–2 dicembre 2018). Bologna: Ante Quem.

Polizzi, G., Torre, R. 2018. Il balaneion dell'agorà di Solunto. *Mare Internum* 10: 59–72.

Portale, E.C. 2006. Problemi dell'archeologia della Sicilia Ellenistico-romana: il caso di Solunto, *Archeologia Classica* LVII: 49–114.

Portale, E.C. 2020. Scultura ellenistica e paesaggio urbano: i casi di Tindari e Solunto, in M. Trümper, G. Adornato and Th. Lappi (eds) *Cityscapes of Hellenistic Sicily*: 239–261. Roma: Quasar.

Ravesi, M. 2018. Agora/foro di Tindari: considerazioni alla luce dei recenti scavi in via Omero, in M. Bernabò Brea, M. Cultraro, M. Gras, M.C. Martinelli, C. Pouzadoux and U. Spigo (eds) *A Madeleine Cavalier*: 393-404. Collection du Centre Jean Bérard 49. Napoli: Centre Jean Bérard.

Sadori, L., Ortu, E., Peyron, O., Zanchetta, G., Vannière, B., Desmet, M. and Magny, M. 2013. The last 7 millennia of vegetation and climate changes at Lago di Pergusa (central Sicily, Italy). *Climate of the Past* 9: 1969-1984.

Sadori, L., Giraudi, C., Masi, A., Magny, M., Ortu, E., Zanchetta, G. and Izdebski, A. 2016. Climate, environment and society in southern Italy during the last 2000 years. A review of the environmental, historical and archaeological evidence. *Quaternary Science Reviews* 136: 173–188.

Spatafora, F. 2003. *Monte Maranfusa. Un insediamento nella media valle del Belice. L'abitato indigeno.* Palermo: Regione Siciliana.

Spigo, U. 2005. *Tindari. L'area archeologica e l'antiquarium*. Milazzo: Rebus Edizioni.

Spigo, U. 2006. Tindari. Considerazioni sull'impianto urbano e notizie preliminari sulle recenti campagne di scavo nel settore occidentale, in M. Osanna and M. Torelli (eds) *Sicilia Ellenistica, consuetudo italica. Alle origini dell'architettura ellenistica d'occidente*: 97–105. Atti dell'Incontro di Studio (Spoleto, 5–7 novembre 2004). Roma: Edizioni dell'Ateneo.

Sulli, A., Pepe, F., Pennino, V., Lo Iacono, C., and Agate, M. 2009. Evidences of mud volcanoes in the Palermo and Termini Gulf (N Sicily offshore). *Geoitalia, VII Forum Italiano di Scienze della Terra*.

Trombi, C. 2020. L'acqua a Monte Adranone: approvvigionamento idrico e uso cultuale, in V. Caminneci, M.C. Parello and M.S. Rizzo (eds) *Le forme dell'acqua. Approvvigionamento, raccolta e smaltimento nella città antica*: 231–244. Atti delle Giornate Gregoriane XII Edizione (Agrigento 1–2 dicembre 2018). Bologna: Ante Quem.

Trümper, M. 2020. Development of Bathing Culture in Hellenistic Sicily, in M. Trümper, G. Adornato and Th. Lappi (eds) *Cityscapes of Hellenistic Sicily*: 347–390. Roma: Quasar.

Usai, E. 2014. Dall'archeologia dell'acqua: canali, vasche, piscine, pozzi…Alle implicazioni cultuali nei santuari fenici e punici di Sardegna, in *Antike und moderne Wasserspeicherung*: 158– 180. Internationaler Workshop vom 11.–14.05.2011 in Pantelleria (Italien). Rahden: Verlag Marie Leidorf.

Vassallo, S. 1999. *Colle Madore. Un caso di ellenizzazione in terra sicana*. Palermo: Assessorato Regionale dei Beni Culturali e della Pubblica Istruzione.

Wilson, R.J.A. 2018. Archaeology and earthquakes in late Roman Sicily: unpacking the myth of the *terrae motus per totum orbem* of AD 365, in M. Bernabò Brea, M. Cultraro, M. Gras, M.C. Martinelli, C. Pouzadoux and U. Spigo (eds) *A Madeleine Cavalier*: 445-466. Collection du Centre Jean Bérard 49. Napoli: Centre Jean Bérard.

Baia (Bacoli-NA):
l'acqua e il suo utilizzo nel complesso delle Terme romane

Daniele De Simone[1]

[1] Aix Marseille Univ, CNRS, Minist Culture, CCJ, Aix-en-Provence, France – Ales S.p.A. (danieledesimone@email.it)

Abstract: The Terme di Baia are one of the most magnificent and complex examples of Roman architecture and engineering, which occupies, thanks to an impressive terracing system, the entire eastern side of the Gulf of Pozzuoli and that, thanks to the geomorphological characteristics of the area (the caldera of the Campi Flegrei), is characterised by the presence of numerous hydrothermal springs whose exploitation dates back at least to the 3rd/2nd century BC. The richness of the hydrothermal springs, mainly exploited for the 'sudationes', is contrasted by the absence of sources of drinking water and water for domestic use, a factor that makes the solutions adopted here particularly significant in the development of water management systems of the Roman world. The lack of sources of drinking water is, in fact, effectively resolved through the creation of an extensive and complex system of collection, storage and distribution of rainwater, the dimensions of which are still largely unknown and unstudied. This system was to strengthened and reorganised over the following centuries, and made more functional to the changing needs of the area of Baia, thanks to its connection, at the end of the 1st century BC, with the 'Acqua Augusta' pipeline, also known as the Serino Aqueduct (from the name of the locality, in the province of Avellino, where the springs were identified). In this contribution we will try to show what relationships exist between the geomorphology of the site, linked to volcanism, the presence of hydrothermal springs, and the expedients used by the Roman designers, from the Late Republican era through to the entire 4th century AD, to exploit this great wealth. The solutions adopted allowed this favoured residential district, a destination for Roman elites, Republican and Imperial, to prosper for a long time, and it was only geomorphological changes that led first to a slow decline and then to total abandonment.

Keywords: roman engineering, hydrothermal springs, water management systems,

Introduzione

Il complesso delle Terme di Baia rappresenta uno dei più grandiosi esempi dell'architettura e dell'ingegneria romana. *Baiae* è un ricco quartiere residenziale situato a nord di Pozzuoli[1] che, grazie alle sue caratteristiche geomorfologiche, all'intero del settore Nord della Caldera dei Campi Flegrei, vede la presenza di numerose sorgenti idrotermali il cui sfruttamento risale almeno alla metà/fine del III – Inizi II sec. a.C.[2] A questa ricchezza di acque idrotermali (se ne contano almeno sei) è da segnalare la totale assenza di sorgenti di acque potabili, utili per l'alimentazione umana, in opposizione alla presenza di numerose sorgenti di acque potabili abbondanti nelle vicine Miseno e Cuma. Un problema, quest'ultimo, certamente presente presso i progettisti ed i ricchi proprietari che in epoca repubblicana fanno realizzare le prime ville. La soluzione adottata fu quella di realizzare un complesso sistema di gestione, stoccaggio e scarico delle acque estremamente complesso e al giorno d'oggi ancora non pienamente conosciuto.[3] Nonostante la complessità di questa infrastruttura che innerva la totalità dei complessi riportati in luce nel secolo scorso ad oggi non esiste uno studio complessivo volto ad individuarne le reali dimensioni.[4]

In questo contributo si cercherà di ricostruire, sulla base dei dati a nostra disposizione, frutto di ricognizioni sul campo e delle ricerche speleologiche intraprese da chi scrive in collaborazione con il Parco Archologico dei Campi Flegrei e l'Associazione Cocceius,[5] le fasi costitutive di questo complesso architettonico e di come, nei secoli si sia passati da un sistema basato esclusivamente sulla raccolta e stoccaggio delle acque pluviali, ad un sistema misto, alimentato dalle acque qui trasportate grazie da una delle più grandi opere ingegneristiche del mondo romano: l'Acquedotto Augusteo. Un aggiornamento, quest'ultimo, alla base

---

[1] Amministrativamente legato a Cuma e non alla vicina *Misenum*; vedi Camodeca 1997: 289–306.
[2] Da ultimo Medri 2013: 121–144, al quale si rimanda alla bibliografia.
[3] Da alcuni anni però il gruppo di speleologi guidati da Graziano Ferrari con le loro esplorazioni hanno permesso di individuare e ricostruire interi settori di questo vasto sistema. Ferrari and Lamagna, 2013: 387–397; Ferrari-Lamagna 2017: 151–157; Ferrari and Lamagna 2018: 59–75; Ferrari and De Simone 2020: 15–153.
[4] Per una raccolta esaustiva vedi Medri 2013: 120; Nieberle and Broisch 2016: 91–115.
[5] Si tratta di un'associazione formata da tecnici del sottosuolo e da archeologi fondata nel 2018.

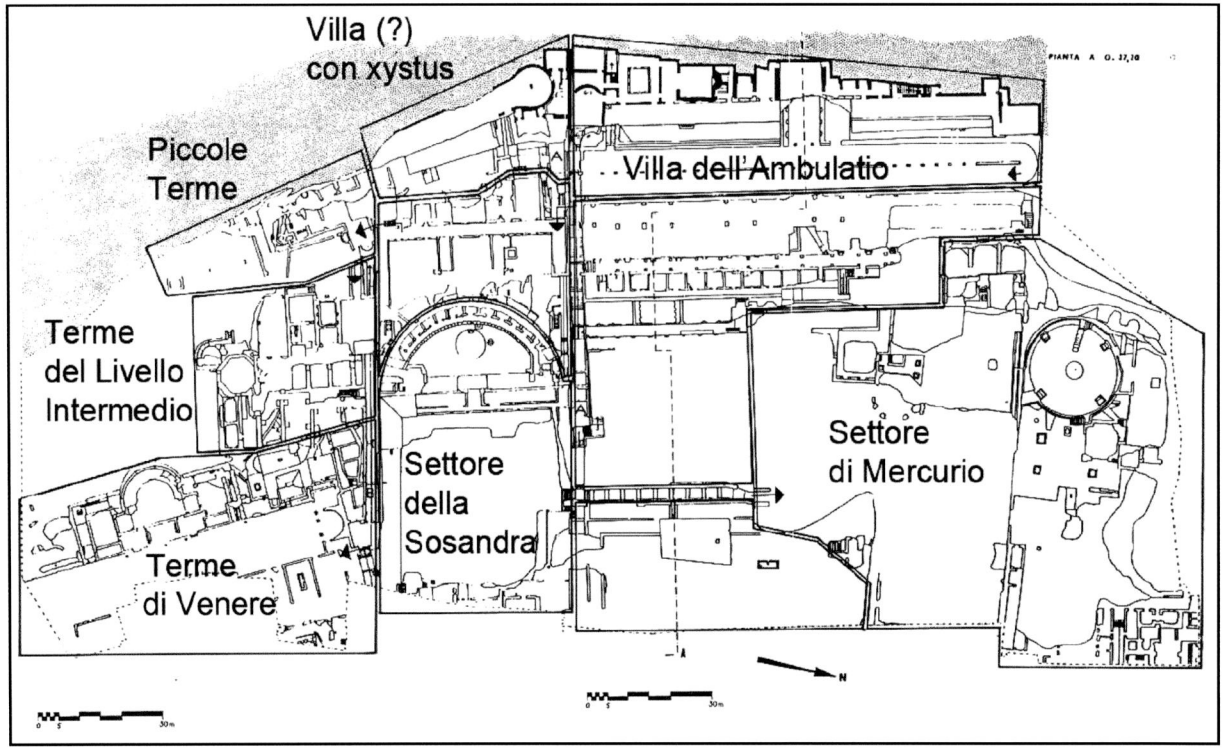

Figure 1: Planimetria delle Terme di Baia.

della enorme fortuna che Baia avrà presso le elite romane almeno nei primi tre secoli di vita dell'impero.

L'attuale suddivisione del complesso delle Terme di Baia[6] in quattro settori principali[7]è ormai ampiamente accettata dagli studiosi, e si basa su di una suddivisione operata analizzando l'organizzazione degli assi viari principali, certamente pedonali, utilizzabili ancora oggi per visitare l'area archeologica (Figure 1). Tali settori sono a loro volta organizzati in diversi terrazzamenti che discendono verso il mare,[8] in direzione Ovest/Est.[9] Questi settori sono: il Settore di Mercurio, localizzato a Nord del Parco delle terme nell'area delimitata dalla linea ferroviaria della Cumana (oggi dismessa) e dalla cd Villa dell'Ambulatio,[10] il Settore della Sosandra, che deve il suo nome al rinvenimento di una copia dell'Afrodite Sosandra di Kalamis,[11] e infine, vi è il settore di Venere, il cui nome è dovuto al rinvenimento, nel XVIII secolo di alcuni mosaici che avevano ad oggetto motivi legati a quella divinità. Questo settore è a sua volta suddiviso i tre complessi differenti: Il primo è quello delle cd Piccole Terme,[12] situate quasi alla sommità del versante, seguono poi le Terme del Settore Intermedio,[13] caratterizzate da imponenti opere di sostruzione ancora oggi in buona parte conservate e caratterizzanti quest'area del sito. Al di sotto delle Terme Intermedie vi sono le cd 'Terme di Venere'.[14]

L'area di Baia alla fine dell'epoca Repubblicana

Anche se è ipotizzabile un uso delle risorse idrotermali nell'area di Baia almeno a partire dal IV – III sec. a.C.,[15] la prima fase edilizia riconoscibile nell'area del Complesso delle Terme Romane di Baia è databiletra alla fine del II sec. a.C.. Nel corso delle fasi terminali della Repubblicail fenomeno delle villae, del tipo 'arroccate' come ci riportano le fonti, in cima al cratere di Baia,[16]

---

[6] Medri 1990: 184–237; Maiuri 1958; Sgobbo 1934: 294–309.
[7] All'interno di quali è possibile effettuare ulteriori suddivisioni interne come avviene nel caso del cd 'Settore di Venere' che va a sua volta suddiviso in: Terme Piccole, Settore Intermedio e Terme di Venere.
[8] Per una sintesi vedi: Cairoli Giuliani 1976: 369–375; Campi Flegrei 1990; Miniero 2000.
[9] Lo studio della viabilità antica è da sempre un argomento di difficile risoluzione a causa degli enormi cambiamenti che l'area in questione, a causa anche dell'abusivismo edilizio, ha subito negli ultimi cinquant'anni. Si rimanda a Borriello and D'Ambrosio 1979: 34, e Döering 2012: 50-134 per un inquadramento della problematica.
[10] Maiuri 1951: 359–364; De Angelis d'Ossat 1977: 230–234; Rakob 1988: 257–301; Medri 1990: 199–209; Di Luca 2009: 151–156.
[11] A monte del quale è sono individuabili i resti della cd 'villa con xystus', ancora oggi un edificio di difficile interpretazione; vedi De Angelis d'Ossat 1977: 238–243; Medri 1990: 209–217; Di Luca 2009: 156–161.
[12] Medri 2013: 121–144 con l'ampia bibliografia di riferimento; Di Luca 2009: 161–162.
[13] Auberson 1974: 167–178; De Angelis d'Ossat 1977: 243–254; Ling 1979: 77–140; Medri 1990: 222–227.
[14] Medri 2013: 121–125 con l'ampia bibliografia in nota.
[15] Come potrebbero far ipotizzare sia alcune tracce nella trattatistica di IV-III sec. a.C., nonchè i resti di opera in blocchi, posizionati allo sbocco del cd. *Great Anthrum*, nell'area delle Terme Piccole, vedi Medri 2013: 123–125, e Medri 2018: 549–558.
[16] D'Arms 2003.

Figure 2: Panoramica Villa dell'Ambulatio (foto : D. De Simone).

diventa una realtà consolidata.[17] Sicuramentela cd *Villa dell'Ambulatio*,[18] la ricca dimorache ancora oggi occupa una gran parte dell'area ricadente nel Parco delle Terme Romane di Baia, appartiene alle villae di questo primo periodo (Figure 2). Questa residenza, disposta già all'epoca su sei terrazze, si estendeva, con molta probabilità, anche verso Sud, includendo sia la parte alta del settore Settore della Sosandra, sia l'area occupata dalle cd *Piccole Terme*.[19] L'assenza nelle immediate vicinanze di sorgenti di acqua potabile impone ai progettisti antichi la realizzazione di un primo ed efficiente sistema di raccolta e conservazione delle acque meteoriche. Tale sistema è basato su un'ossatura formata da un complesso sistema di canalizzazioni, dotate di un complesso sistema di pozzetti e chiusini di recupero e pulizia, che innervano i terrazzamenti della villa e evitano anche il più piccolo spreco della preziosa risorsa. Elementi centrali erano le pluviali, **P.1** e **P.2** (Figure 3) che, partendo da quelli che paiono essere gli ambienti propriamente abitativi del complesso edilizio (terrazze B e C, Figure 3) garantivano la raccolta dell'acqua piovana.[20] Lo stoccaggio è la conservazione delle acque meteoriche raccolte in questi settori della villa era garantito dalla grossa cisterna **C.1,** posizionata al di sotto della zona dei cd *hospitalia* (terrazza D). Questa grande cisterna,[21] composta da cunicoli intersecantisi tra loro a formare una struttura quadrangolare, con i due bracci maggiori, che hanno direzione N/S,[22] lunghi circa 32,5 m, collegati tra loro da ulteriori quattro cunicoli intermedi perpendicolari, dei quali tre sono lunghi circa 16 m, mentre un quarto risulta lungo circa 8 m a causa di una tompagnatura antica, pare essere il punto terminale di questo primo sistema di raccolta e stoccaggio delle acque della Villa dell'Ambulatio (Figures 3, 4). Sulla base dei dati in nostro possesso possiamo stimare in circa 260 m³ di

---

[17] D'Arms 2003; Miniero 2017: 795–810.
[18] Maiuri 1958: 75–76; De Angelis d'Ossat 1977: 242; Borriello and D'Ambrosio 1979: 69; Di Luca 2009: 161; Miniero 2017: 795–800.
[19] De Angelis d'Ossat 1977: 274; Medri 2013: 121.
[20] Ulteriori dati circa la raccolta delle acque meteoriche sono di recente stati raccolti a monte di tali terrazze, dove sono emerse tracce di una serie di ambienti concamerati, rivestiti da uno spesso strato di intonaco idraulico, che paiono essere il primo elemento di un complesso di raccolta sul quale doveva basarsi il corretto funzionamento della villa in questo periodo.
[21] Bodon *et al.* 1994: 323–325.
[22] Largh. 2,20 x 2,00 m di altezza. La volta si imposta a circa 1,30 m dal piano pavimentale.

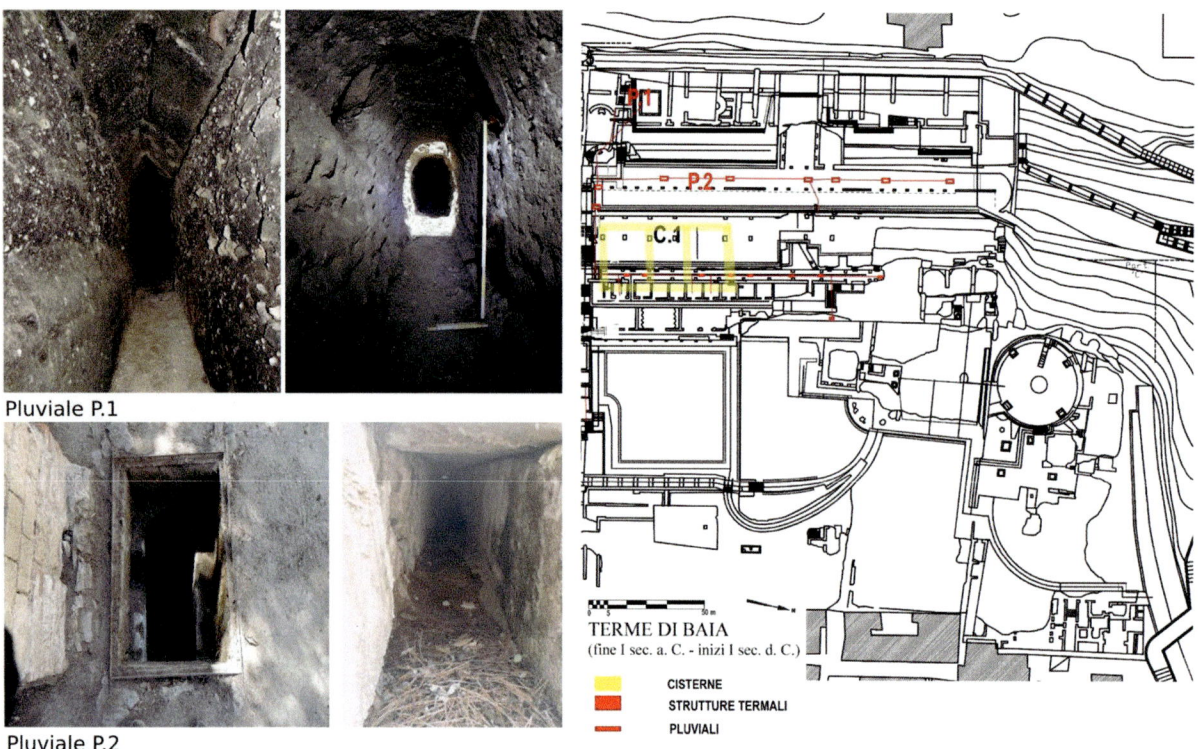

Figure 3: Particolari delle pluviali della Villa dell'Ambulatio e planimetria delle strutture ipogee.

Figure 4: Particolari della cisterna C2 della Villa dell'Ambulatio.

capienza circa per la cisterna C.1.[23] Parallelamente alla cisterna **C.1**, per questo periodo, erano attivo anche il sistema di cisterne posizionate al di sotto del II livello nel Settore della Sosandra. Questo sistema, denominato nel suo complesso **C.2, è composto** da una serie di cisterne concamerate simili, strutturalmente,[24] alla Cisterna Inferiore delle *Centocamerelle* e a quella visibile a monte della Sella di Baia.[25] Altre strutture riferibili a questo periodo e utilizzate per la raccolta delle acque meteoriche vanno individuate nel gruppo di cisterne definito **C.5**, a monte del settore della Sosandra (**Figure**

---

[23] Le misurazioni sono state effettuate dal dott. Graziano Ferrari, speleologo e profondo conoscitore delle cavità baiane, al quale va il mio più sentito ringraziamento per la collaborazione. Inoltre la stima della capienza risulta del tutto approssimativa e non definitiva poiché la cisterna è ancora in parte ingombra.

[24] Le camere sono realizzate facendo uso di opera reticolata 'antica', cosi come definita in Di Luca 2009: 156–161.

[25] Ruggiero 1759: 166–167; Mingazzini 1931: 353–355; Mingazzini 1932: 293–303; Doering 2012: 109–111.

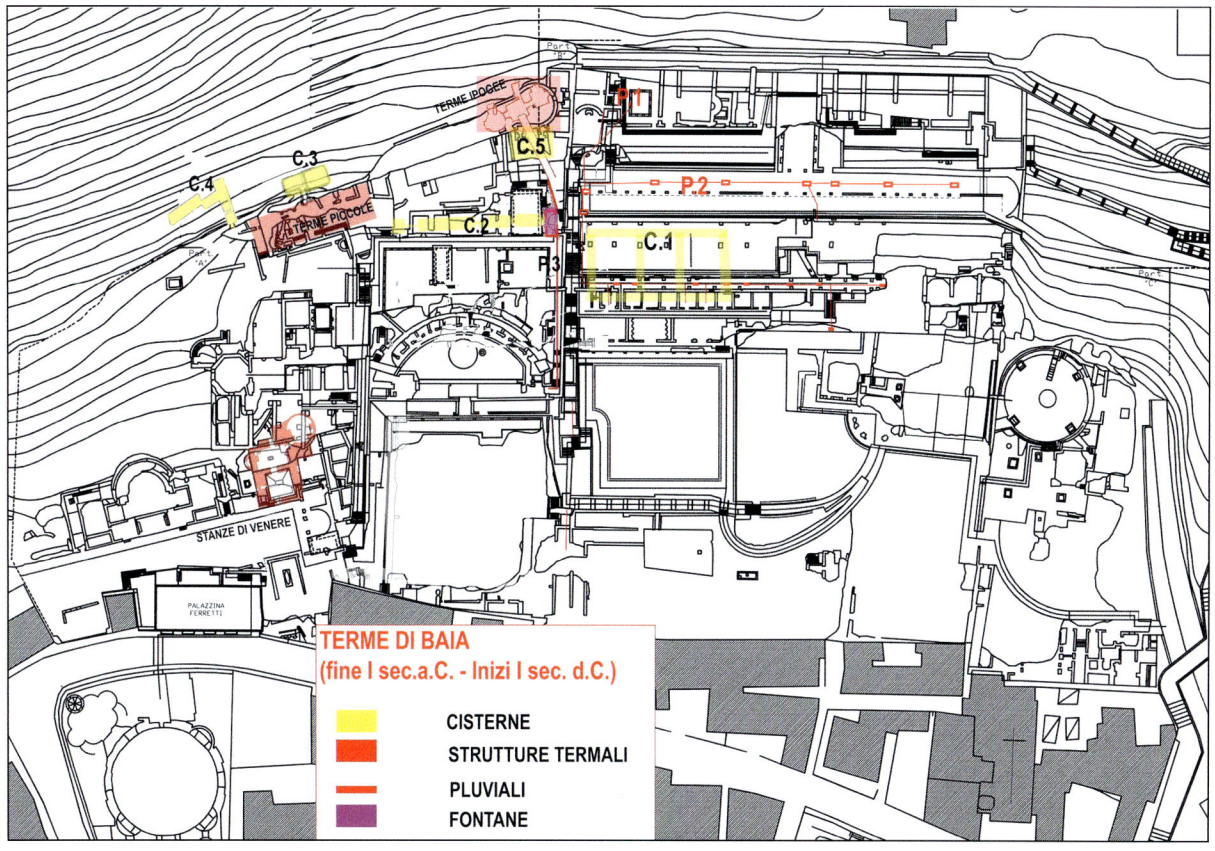

Figure 5: Planimetria delle strutture ipogee attive tra I sec. a.C. e inizi I sec.d.C.

5).[26] Procedendo verso Sud altre cisterne sono visibili in corrispondenza del livello superiore delle Piccole Terme (definite **C.3** e **C.4** – Figure 5).[27]

Questa organizzazione e distribuzione delle cisterne lungo il pendio, in diverse posizioni e a diversa altezza, permetteva una raccolta efficace delle acque piovane rispecchiando anche quanto espresso dalle fonti antiche, in particolare Seneca, in merito alle prime *villae* baiane che erano posizionate presso la sommità dei pendii piuttosto isolate, in posizione arroccata, a controllo del territorio circostante.[28] Un'organizzazione che corrisponde anche alla posizione della Villa *dell'Ambulatio* in questo periodo almeno fino alla fine del I sec. a.C. Le sorgenti idrotermali, certamente sfruttate a partire dal II sec. a.C., seguono questa organizzazione. Difatti gli ambienti termali più antichi sono posizionati a mezza costa: nei pressi della rampa Nord della Villa dell'Ambulatio, presso le cd Terme Ipogee del settore della Sosandra e presso le Piccole Terme; di poco successivo è il piccolo complesso termale delle cd Stanze di Venere generalmente datato alla seconda metà del I sec. a.C.[29] (Figure 5). Lo sfruttamento intensivo di queste sorgenti a partire dalla seconda metà del I sec. a.C., ma ancor più dall'epoca augustea, comporta nuovi e significativi cambiamenti ai complessi baiani. Sintomo di questi cambiamenti è la risistemazione dapprima delle cd *Terme Ipogee* della Sosandra e quasi contemporaneamente, a valle, delle cd *Terme di Venere*. Si tratta di due impianti termali strutturalmente molto simili, entrambi dotati di un *laconicum*, con cupola in opera cementizia, che anticipa nelle forme e nelle regole costruttive quella del vicino Tempio di Mercurio, utilizzati per le *sudationes*.

Verso la fine del I sec. a.C., in piena età augustea, si registra nel *baiano sinu* una vera e propria 'esplosione' di cantieri legati all'edilizia residenziale di lusso.[30] Le

---

[26] Queste due strutture risultano oggi irraggiungibili.
[27] Di queste cisterne, ma in generale di questo settore al di sopra delle Terme Piccole ad oggi non esiste uno studio complessivo nè un'analisi delle stesse e delle relazioni con gli impianti sottostanti; questa situazione limita molto la comprensione delle dinamiche insediative di questo settore dell'area archeologica.
[28] Molto chiaro in tal senso è l'epistola di Seneca riguardo alle prime ville fatte costruire da Pompeo, Mario e Cesare: '*in regione Baiana summis iugis montium: videbatur hoc magis militari ex edito speculari late longeque subiecta...scies non villas esse sed castro*' – Sen., *ad Luc.*, 51,11; su questa problematica vedi anche Miniero 2017: 795–810, con la relativa bibliografia ma soprattutto D'Arms 2003.

[29] Se per le Terme Piccole il loro sfruttamento è certo, per le altre due sorgenti individuate è possibile ipotizzarne un utilizzo in questo periodo anche perchè sono tra i luoghi che più di tutti, in epoca successiva vengono coinvolte in estesi e profondi lavori di rifunzionalizzazione.
[30] Hor. *Carm.* II, 18, 18–22; Hor. *Ep.* I, 1, 83–88; Cicala and Illiano 2017: 358–362; Miniero 2017: 802–803.

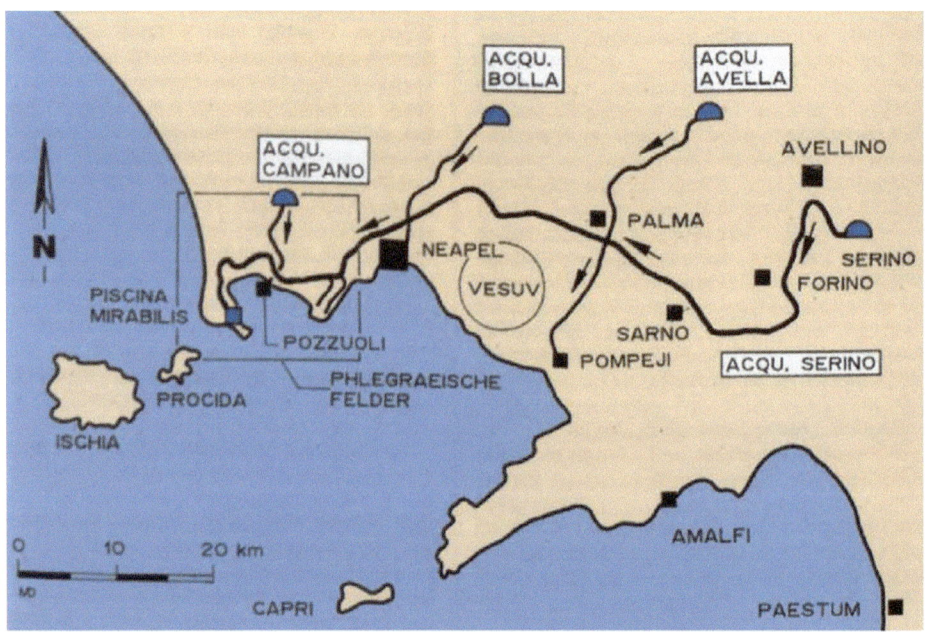

Figure 6: Ricostruzione del Percorso dell'Acquedotto del Serino (da Döering 2012) e immagine dell'epigrafe dello Scalandrone (da Camodeca 1997).

nuove esigenze, pertanto, spingono verso un sostanziale aggiornamento e ampliamento dei vari complessi edilizi che compongono le Terme di Baia. Si avvia, in particolare, il processo di separazione definitiva tra il settore della Villa dell'*Ambulatio* e il Settore della Sosandra e la separazione quest'ultima dalle cd *Piccole Terme*.[31] Una suddivisione ben sottolineata dai due assi viari pedonali che attraversano in direzione Ovest-Est i due complessi, ma ora ben organizzati e distinti, proprio in funzione della nuova organizzazione.[32]

Grazie all'epigrafe recuperata in località Scalandrone[33] sappiamo, inoltre, che l'Acquedotto Augusteo è attivo almeno dall'ultimo quarto del I sec. a.C., tanto da avere necessità di una prima risistemazione e manutenzione agli inizi del I sec. a.C.[34]. La presenza della condotta influenzerà per i secoli successivi l'evoluzione dei complessi termali fino a caratterizzarne l'organizzazione interna e i rapporti tra i vari settori (Figure 6). La disponibilità continua di acqua potabile

---

[31] Medri 2013: 119–144;
[32] Sulla presenza di assi viari pedonali che suddividono l'area si veda Giuliani 1976: 365–375, che raccoglie le prime riflessioni in merito e che al momento rimane il solo studio dedicato ai sistemi viari all'interno del complesso delle Terme di Baia.
[33] Camodeca 1997: 289–306; Ferrari and Lamagna 2010: 387–397.
[34] Medri and Soricelli 1999: 207–2014; Medri 2013: 121–144.

Figure 7: Fontana posizionata all'ingresso delle Terme Ipogee (foto: D. De Simone).

garantita da questa grande infrastruttura antica,[35] che corre nel bancone tufaceo, alle spalle del Complesso delle Terme, velocizza una serie di cambiamenti nel primordiale sistema di *Water Management* delle ville di epoca repubblicana. I 'vecchi', ma funzionali, sistemi di raccolta delle acque piovane, sono rifunzionalizzati e destinati a garantire, nella maniera più efficace e sicura possibile, lo smaltimento delle acque superficiali e di scarto verso il mare. Le pluviali **P.1** e **P.2** della *Villa dell'Ambulatio*, la pluviale del peristilio del II livello della Sosandra, **P.3**, ma in generale tutte le condotte del periodo precedente non alimentano più le sottostanti cisterne (**C.1** e **C.2**) o alcune delle fontane, come quella posta all'ingresso delle *Terme Ipogee* della Sosandra.[36]

Parallelamente al sistema dedicato allo scarico delle *acque grigie*[37] viene organizzato e reso operativo un nuovo sistema di distribuzione dell'acqua, alimentato dall'Acquedotto Augusteo, molto più funzionale alle necessità dell'area che nel corso del I sec. d.C. diviene sempre più specializzata nelle cure termali e aperte ad un pubblico ben più numeroso rispetto al passato.[38]

In questo periodo il complesso delle cisterne **C.2** è con molta probabilità a servizio delle retrostanti *Terme Ipogee*, ma più in generale sono utilizzate dalle strutture di questo settore alimentando una fontana posizionata proprio all'ingresso delle *Terme Ipogee* (Figure 7). Stessa funzione dovevano avere anche le Cisterne **C.3** e **C.4** presso il complesso delle Piccole Terme.

Il comprensorio baiano in epoca Imperiale

Il passaggio dell'Acquedotto Augusteo segna una svolta significativa nello sviluppo dei quartieri residenziali

---

[35] l'acquedotto augusteo era lungo circa 100 km e prendeva avvio dalle sorgenti Acquaro e Pelosi localizzate presso l'odierno comune di Serino (AV).
[36] Si tratta di una pluviale dotata di copertura alla cappuccina (dim. Speco 0,80 x 0,60 cm) che corre al di sotto della rampa che separa il Settore della Sosanda dal Settore di Venere, vedi Ferrari - Lamagna 2017: 151–157.
[37] A tal proposito va segnalato che scarse sono invece dati riconducibili ai sistemi di gestione delle acque nere, poiché almeno in questo periodo non sono da segnalare presenza di latrine (pubbliche o private) attive. Si tratta di un'assenza che solo più estese e puntuali indagini potranno chiarire.
[38] In questo periodo vengono individuate nuove sorgenti idrotermali tipo quelle del Settore di Mercurio e quelle del IV livello della Sosandra che comporteranno una nuova e più profonda riorganizzazione degli edifici esistenti, vedi D'Angelis D'Ossat 1976: 234-235.

Figure 8: Panoramica del Settore della Sosandra con al centro le fontane monumentali (foto: D. De Simone).

baiani ma, più in generale, di tutto il territorio flegreo. Nel corso del I sec. d.C. si assiste alla definita suddivisione dell'area in ben determinati e distinti settori oltre che all'apertura di nuovi impianti in prossimità di nuove sorgenti termominerali, forse già conosciute in precedenza ma a partire da questo periodo pienamente sfruttate: la prima è individuata a valle del Settore della Sosandra (nota in epoca moderna con il nome di *acqua della rogna*), mentre l'altra fu individuata a valle della Villa dell'Ambulatio, alle spalle del Tempio di Mercurio (Figure 1). La scoperta di queste due ulteriori sorgenti spinge ad una profonda riorganizzazione delle strutture ricadenti nel Settore della Sosandra. Dai pochi dati ricostruiti grazie all'analisi delle opere murarie verso la fine del I sec. d.C. va individuato il momento di definitiva separazione di questo complesso dalla vicina Villa dell'Ambulato.[39] Contemporaneamente si assiste alla costruzione di un nuovo grande complesso termale, quello di Mercurio, che comporta, molto probabilmente, la parziale demolizione e ricostruzione di alcune delle terrazze dell'Ambulatio,[40] caratterizzato dalla grande cupola in opera cementizia, frutto di una sperimentazione edilizia avviata già alcuni anni prima presso le Piccole Terme e nelle Stanze di Venere.[41] La realizzazione, infine di un sistema di fontane, posizionate lungo il centro prospettico sia della Villa dell'*Ambulatio* e successivamente presso il Settore della Sosandra contribuisce all'aumento esponenziale della richiesta e del consumo di acqua che grazie proprio alla presenza dell'Acquedotto Augusteo permette lo sviluppo di tutte le strutture connesse alle cure termali (Figures 8a, b).

A servizio di tutto questo enorme apparato viene impiantato un nuovo sistema di *Water Management*. Quest'ultimo è dotato di un'estesa rete di canalizzazioni e di strutture di stoccaggio delle acque (serbatoi e cisterne) ora posizionati in punti strategici del pendio, che ne necessitano di acqua in maniera

---

[39] E' possibile che questo primo nucleo fosse organizzato in maniera del tutto simile a quanto avviene per la vicina Villa dell'Ambulato con i primi terrazzi (I e II livello) utilizzati come ambienti abitativi mentre le terrazze inferiori erano usate come aree di svago, Di Luca 2009: 160–161.

[40] Non è molto chiara la motivazione alla base di queste demolizione; varie sono le ipotesi tra le quali in evento naturale disastroso, come ipotizzato in D'Angelis D'Ossat 1976: 234–235.

[41] Ling, 1979: 77–140; Rakob 1992: 229–258.

Figure 9: Sbocco del diversorium del Serino (foto: D. De Simone).

continua e costante. Punto di partenza è ormai la condotta dell'Acquedotto Augusteo, che grazie alla presenza di alcuni *diversoria*, permetteva di alimentare in maniera distinta i diversi complessi edilizi. Ad oggi sono stati individuati almeno tre canali di derivazione della condotta dell'Acquedotto Augusteo, che alimentano, partendo da Nord, rispettivamente: la *Villa dell'Ambulatio*, il complesso della Sosandra e le Piccole Terme (Figure 9). Tali canalizzazioni sono di forma rettangolare (0,80 x 1,00 m), dotate di una copertura in gettata cementizia e impermeabilizzate grazie ad uno spesso rivestimento in cocciopesto (8–10 cm).

Il primo di questi canali di derivazione alimenta il grande serbatoio **S.2** della Villa dell'Ambulatio posto al di sotto della terrazza B (Figure 15). **S.2** è composto da una serie di camere quadrangolari realizzate, probabilmente agli inizi I sec. d.C., modificando le originarie sostruzioni della villa repubblicana.[42]

L'acqua di questo serbatoio era poi immessa in una canaletta, i cui resti sono ancora visibili a Ovest dell'esedra della terrazza C (realizzata nello stesso periodo), che alimentava la fontana centrale della Terrazza D.[43] Da questa terrazza, una seconda canalizzazione, che si ricongiungeva con i canali della fontana centrale, permetteva l'afflusso di acqua presso la terrazza E giungendo nel lungo serbatoio **S.2**. Nello specifico il serbatoio **S.2** è composto da dieci camere rettangolari tra loro collegate da una serie di stretti passaggi alimentato tramite un canale di carico, individuato presso l'ottavo ambiente Nord dei soprastanti *hospitalia* della Terrazza E.[44] A

---

[42] Questo *diversorium*, recentemente individuato ha permesso di raggiungere la retrostante condotta. Le indagini speleo-archeologiche condotte in collaborazione con i tecnici dell'Associazione Cocceius, hanno permesso di esplorare circa 80 metri dell'acquedotto antico. A questo periodo, inoltre, va forse datato anche il lavoro di apertura delle sostruzioni che reggono la terrazza D. Qui il piano originario viene vistosamente abbassato creando una serie di camere accessibili tramite un corridoio. Successivamente, probabilmente nel III sec. d.C., vengono realizzati gli antistanti *hospitalia* che caratterizzano la terrazza E.

[43] Ancora non è ben chiaro come venisse alimentata la fontana dell'esedra centrale della Terrazza C che usufruire di qualche altra diramazione dell'Acquedotto ancora non identificata.

[44] In realtà un'ulteriore ingresso per le acque sembra essere quello visibile all'estremità Nord della terrazza, dove le ultime due cisterne paiono essere parte di un sistema di decantazione delle acque. Purtroppo gli importanti interventi di restauro, che hanno comportato la ricostruzione di interi tratti di muratura rendono di difficile comprensione la lettura delle strutture.

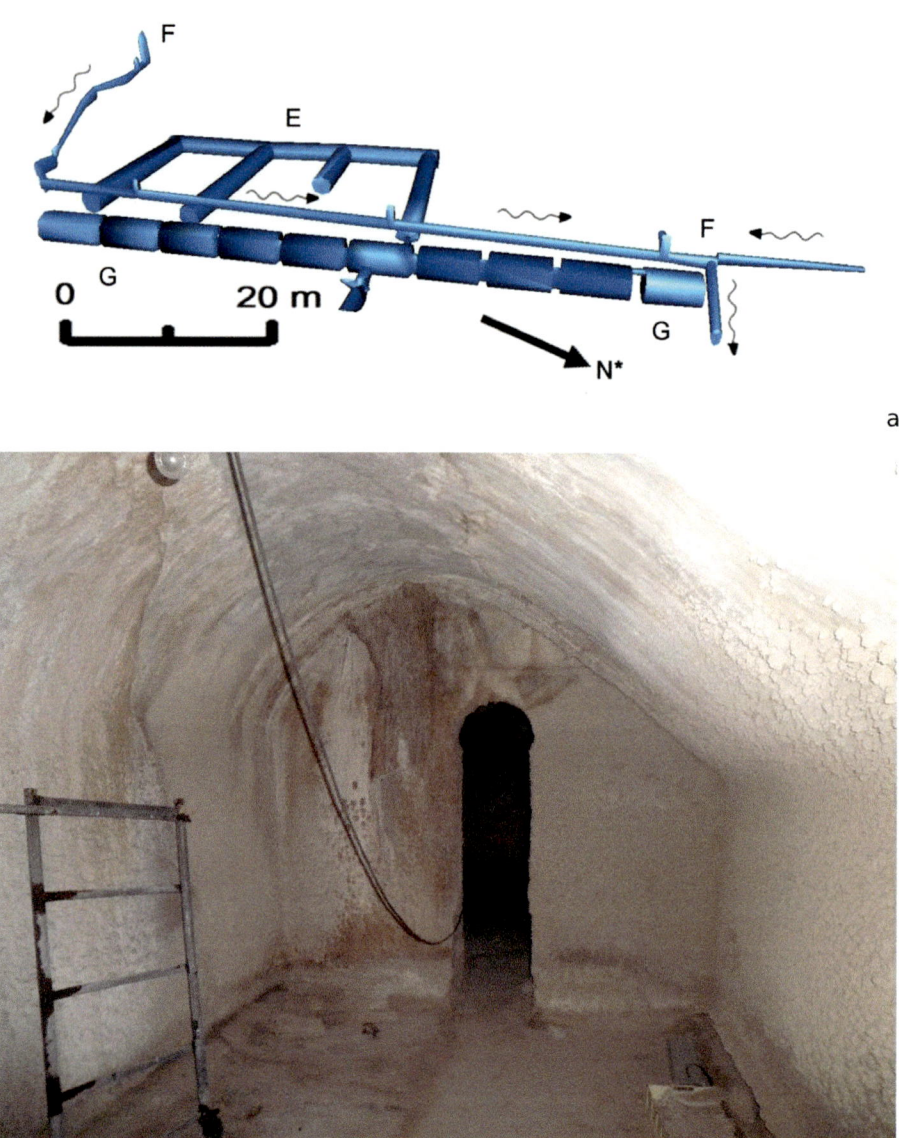

Figure 10: Ricostruzione tridimensionale delle cisterne C.2 e del Serbatoio S.1 e interno di una delle dieci camere del serbatoio S1.

testimonianza del passaggio dell'acqua vi è una vistosa concrezione calcarea che oblitera quasi del tutto il canale di immissione dell'acqua. Oltre questo punto non è più possibile ipotizzare il percorso dell'acqua a causa dei numerosi crolli che hanno portato alla perdita di altri dati utili alla ricostruzione dei percorsi nel settore inferiore della villa. Restano solo scarsi resti di quella che pare essere un'ulteriore fontana allineata a quelle visibili nei livelli superiori (Figures 10 a, b).

Segue, verso Sud, è la derivazione che alimenta il Settore della Sosandra. Questo complesso edilizio realizzato a partire dalla metà del I sec. d.C., è al centro di un'importante opera di ristrutturazione che porta alla realizzazione della grande fontana-ninfeo che occupa il III Livello e la realizzazione del Settore Inferiore del IV Livello, dove furono irreggimentate le 'nuove' sorgenti idrotermali.[45] Similmente a quanto accade alla vicina Villa dell'*Ambulatio* anche qui il centro prospettico del complesso viene arricchito da una serie di fontane monumentali alimentate grazie all'afflusso di acqua dell'Acquedotto Augusteo. Dal I livello, immediatamente a Sud il serbatoio S.3,[46] doveva esserci lo sbocco di questa seconda derivazione[47].

---

[45] La complessità strutturale di questo settore non ne permette una sicura attribuzione anche se negli anni si è pensato di identificarlo con l'*ebeterion* ricordato da Cassio Dione che Nerone fece costruire per i militari della *Classis Misenatis* di stanza nella vicina a Misenum, vedi *Campi Flegrei* 1990: 209–210.

[46] Definito comunemente 'cellaio' o 'masseria', il serbatoio S.3 è composto da quattro camere rettangolari utilizzate oggi come deposito di materiali vari; anche in questo caso risulta molto difficile effettuare misurazioni o analisi di dettaglio.

[47] Questa diramazione è del tutto identica per dimensioni e tecnica

Figure 11: Sbocco del *diversorium* del Serino presso le Terme Piccole (foto: D. De Simone).

L'acqua era poi immessa in una serie di cisterne collocate al centro del II Livello,[48] a loro volta collegate al sottostante sistema delle cisterne tardo repubblicane **C.2**. Questo sistema da un lato continuava ad alimentare la fontana posta all'ingresso delle *Terme Ipogee* (Figure 7), a Nord, mentre dal lato Est partiva una diramazione che alimentava le tre fontane principali del complesso terminando in basso, nel IV livello.[49]

L'ultima diramazione che si diparte dalla condotta del Serino è posizionata a monte delle *Terme Piccole*. Tale canale alimentava a sua volta una serie di cisterne ipogee (Figure 9), sempre a monte del piccolo edificio termale, mentre una parte dell'acqua andava ad alimentare il serbatoio plumbeo che era posizionato allo sbocco del cd

Great Anthrum[50] (Figure 11). Lo smaltimento delle acque in eccesso in questo settore del Complesso era garantito da una serie di canalizzazioniche incanalavano l'acqua nella lunga pluviale **P.3** che procedendoal di sotto della rampa Sud (che divide il Settore della Sosandra dal Settore di Venere) similmente alla condotta **P.1** terminava verosimilmente in mare.[51] Similmente alle pluviali **P.1** e **P.2** della *Villa dell'Ambulatio*, **P.3** ha sezione rettangolare, una platea costituita da elementi laterizie i piedritti con ricorsi di blocchi tufacei rettangolari rivestiti da intonaco. Il canale aveva una copertura, come in **P.1**, a doppio spiovente, in bipedali.[52] Il cunicolo segue la pendenza della scalinata per uno sviluppo di 82 m e supera 21 m. di dislivello e nel suo percorso viene alimentato da almeno altre due pluviali provenienti dal I e II livello delle Terme del Settore Intermedio, terminando nei pressi del livello inferiore del Settore della Sosandra (Figure 12).

---

costruttiva a quella che alimenta le cisterne della vicina *Villa dell'Ambulatio*. È infatti di forma rettangolare e coperta con una volta a botte realizzata con conci radiali, alta circa 1,50 m per 80 cm di larghezza.
[48] Di fianco all'ambiente decorato con il famoso mosaico policromo con scene e maschere teatrali; vedi *Campi Flegrei* 1990: 209–214.
[49] L'acqua di questa fontana finiva in una canaletta (dim. 20 x 30 cm) che correva al di sotto della rampa Nord, e terminava presso il IV livello di questo Settore. Anche qui, a causa dei crolli avvenuti nel corso dei secoli, non è chiaro il prosieguo del percorso della canaletta.

[50] Medri 2013: 121–144.
[51] Ferrari and Lamagna 2017: 151–157.
[52] Le dimensioni sono le seguenti: H. tot. 1,02 m, Largh. 0,57 m; altezza dei piedritti 0,56 m circa.

Figure 12: Ricostruzione dell'andamento della pluviale P.3 (da Ferrari 2017).

Con il II sec. d.C. il Complesso delle Terme di Baia si arricchisce di nuove strutture termali, sintomo del grande successo che ancora in questo periodo ha il comprensorio flegreo. É in questo periodo che vengono costruiti due nuovi edifici termali: le terme del Settore Intermedio, costruite colmando il dislivello tra le Piccole Terme e le Stanze di Venere, e il 'Complesso Adrianeo', che sorge a Sud del Settore della Sosandra, e del quale doveva far parte il cd Tempio di Venere (Figure 13). La realizzazione di questi due nuovi edifici crea le condizioni per nuove e profonde modifiche alle strutture precedenti, legate principalmente allo sfruttamento delle acque portate qui grazie all'Acquedotto Augusteo. Le Terme del Settore Intermedio sono tutte disposte a Est rispetto alle Piccole Terme, con tutte le varie parti che compongono il percorso termale che affacciano sul mare. Il calore proveniente dal sottosuolo è portato ai vari ambienti grazie al prolungamento del cd *Great Anthrum*.[53] Tutti questi ambienti sono sostenuti grazie ad un poderoso sistema di sostruzioni necessarie per recuperare il forte dislivello di questo tratto del pendio del cratere di Baia (circa 10 m). Al di sotto del I Livello delle Terme del Settore Intermedio, vi è una cisterna di forma irregolare,[54] alimentata dall'acqua proveniente dall'acquedotto e dalla quale prende avvio una lunga canalizzazione che attraversa, in direzione N/E – S/O le sostruzioni del II livello, terminando nella cupola del *laconicum* del complesso augusteo delle *Stanze di Venere*. E' infatti in questo periodo (II sec. d.C.) che questo piccolo complesso termale viene riconvertito in un serbatoio a servizio dei livelli inferiori. Questo comporta una rifoderatura interna delle pareti perimetrali con una spessa (circa 60 cm) parete in laterizio che copre del tutto anche gli apparati decorativi di epoca precedente lasciando in vista solo i bei stucchi visibili presso le coperture (Figure 14). Per permettere l'ingresso dell'acqua viene sfruttata un'apertura (lucernaio) nella cupola del *laconicum*; cupola che viene poi rinforzata con un grosso arco in laterizio che sorregge l'intera copertura.[55] Il secondo edificio realizzato in questo periodo è il grande complesso noto come 'Terme Adrianee'. Questo edificio, sviluppato lungo l'asse centrale (direzione SO – N/E) è segnato dalla grande esedra centrale.[56] Purtroppo da questo edificio in poi non si riesce più a seguire il percorso delle canalizzazioni a causa della presenza delle strutture del moderno borgo di Baia che hanno obliterato questo settore dell'abitato antico. Dopo un periodo di relativa stasi alla fine del II sec. d.C. si assiste

---

[53] Medri 2013; Paget 1967a; Paget 1967b: 102–113.
[54] Oggi questa cisterna è inaccessibile ed è nota solo grazie ad una serie di rilievi effettuati dal Consorzio Pinakos negli anni Ottanta del secolo scorso.
[55] *Campi Flegrei* 1990: 227–229; Ling 1979: 77–140.
[56] *Campi Flegrei* 1990: 229–231.

Figure 13: Panoramica Settore intermedio e cd Stanze di Venere (foto: D. De Simone).

Figure 14: Panoramica dell'interno di una delle Cisterne delle cd Stanze di Venere con ben visibile il muro di rivestimento funzionale alla riconversione in serbatoi di questi ambienti (foto: D. De Simone).

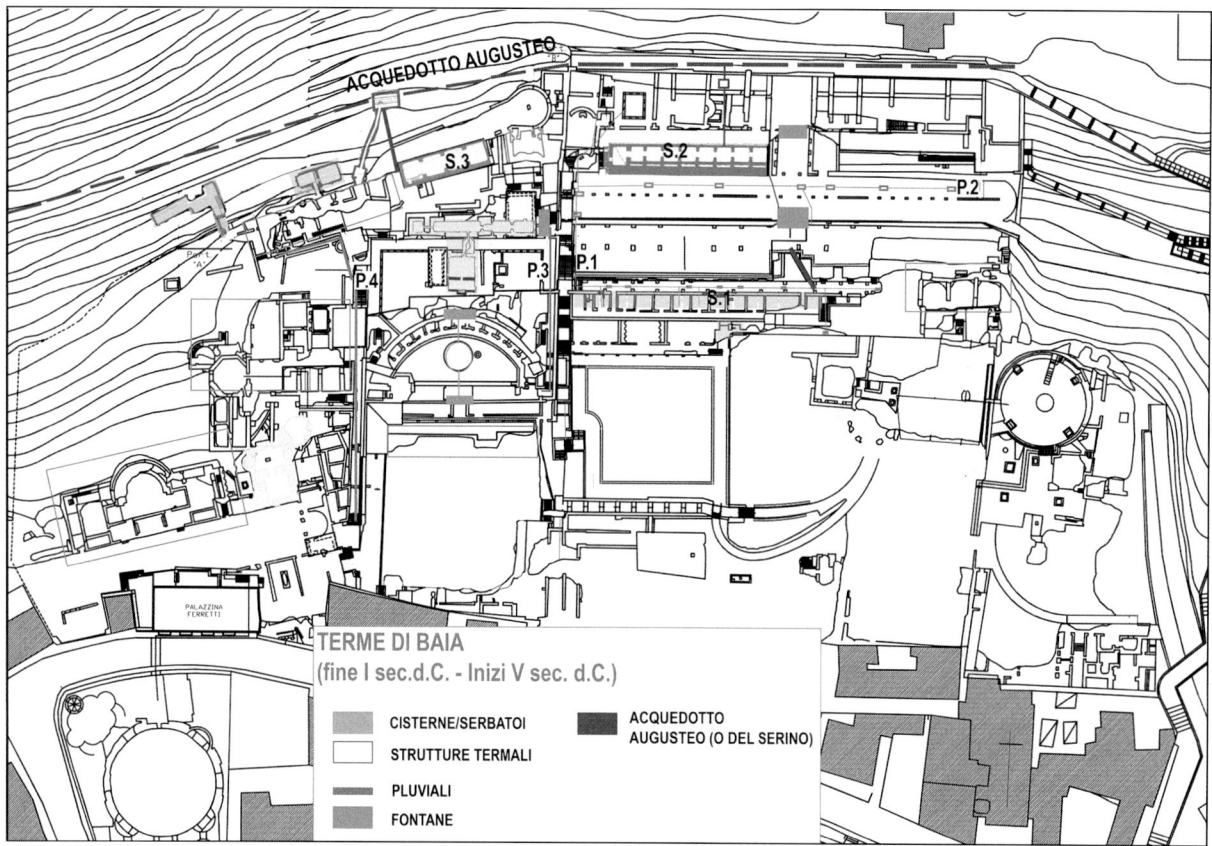

Figure 15: Planimetria delle strutture ipogee attive tra I sec. d.C. e inizi V sec. d.C.

ad una certa ripresa edilizia nel corso del III sec. d.C. Con la dinastia dei Severi, infatti, ci fu un periodo di ripresa per quest'area, grazie alla costruzione del *Palatium* Severiano e di una serie di strutture tra le quali vi dovevano essere un ninfeo e altri edifici che facevano dell'acqua il loro principale ornamento.[57]

Conclusioni. Per una prima analisi delle strutture idrauliche baiane

L' analisi delle diverse fasi evolutive dei sistemi idraulici del complesso delle Terme di Baia e della loro stretta relazione con il contesto territoriale, caratterizzato dalla presenza di importanti sorgenti idrotermali frutto della particolare cornice ambientale (la caldera dei Campi Flegrei), ci permette di osservare quali sono le dinamiche che investono questo ricchissimo quartiere residenziale. Un'area, quella del '*sinus baianus*' che grazie alle sue peculiari caratteristiche ambientali ha rappresentato uno dei luoghi di formazione sviluppo dell'ingegneria e dell'architettura romana. Uno sviluppo che appare poderoso già a partire dall'epoca tardo repubblicana, con l'impianto delle grandi ville patrizie[58] che, posizionate in sommità del pendio del cratere di Baia (e delle colline circostanti), necessitano di grandi quantità di acqua, in questo periodo esclusivamente piovana, e quindi di sistemi dedicati alla raccolta e allo stoccaggio di questa preziosa risorsa.[59] Le lunghe pluviali (**P.1** e **P.2** per la Villa dell'Ambulatio, **P.3** per la Sosandra)che innervano i vari terrazzamenti di questi complessi edilizi permettono il trasporto delle acque in posizioni adatte alla conservazione. E' questo il caso delle cisterne **C.1 e C.2**., posizionate in punti strategicamente a 'mezza costa' del pendio della collina baiana.[60] Una necessità, quella dello stoccaggio delle acque piovane, confermata dalla presenza di altre camere ipogee più a Sud delle cd Piccole Terme (**C.3** e **C.4**).[61] E' da sottolineare, inoltre, che in questo periodo fossero già attive le cd *Piccole Terme*, forse parte della più grande villa dell'*Ambulatio* e, probabilmente, le cd Terme Ipogee delle Sosandra, entrambe utilizzate principalmente per le '*sudationes*' e collegate, grazie a dei lunghi cunicoli ricavati dalla roccia a delle sorgenti

---

[57] Maniscalco 1997; Maniscalco 1995: 257–271.
[58] Miniero 2017: 795–810; Cicala and Illiano, 2017: 358–362; Lafon 2001.
[59] Approvvigionamento che andrebbe legato al fatto che tali ville altro non erano che la *pars maritima* di più grandi complessi che prevedevano anche una *pars rustica* sul versante Ovest del cratere di Baia. Al momento però gli studi sono in fase del tutto preliminare e non ci è possibile avanzare alcuna ipotesi in merito.
[60] C.1 al di sotto della terrazza E della Villa dell'*Ambulatio* mentre C.2 è posizionata al di sotto il II livello della Sosandra.
[61] Modalità simili si riscontrano anche lungo il versante Ovest del Cratere ma in generale in tutto il territorio flegreo, vedi a tal proposito Borriello and D'ambrosio 1979.

di acqua idrotermale i cui vapori permettevano il loro funzionamento.

Con il passaggio al periodo del Principato Augusteo e ancor più al Periodo Imperiale, quando l'area diviene luogo prediletto dalle nuove elité imperiali, si avvia quel processo di separazione tra il Settore dell'*Ambulatio* e il Settore della Sosandra, e di quest'ultima dalle cd *Piccole Terme*. A partire dalla fine del I sec. a.C. e gli inizi del secolo successivo la necessità di fornire acqua potabile e di buona qualità alla flotta di stanza nell'area flegrea, presso il porto di *Misenum,* spinge Augusto a promuovere la realizzazione della grande condotta dell'*Acqua Augusta* che, dalle sorgenti di Serino (AV)[62] permette di alimentare numerosi centri della Campania, ma soprattutto, permette di rifornire d'acqua quest'area della regione ora sempre più al centro dell'attenzione delle elité imperiali e dell'imperatore stesso. L'epigrafe dello Scalandrone ci permette di ipotizzare anche una certa precocità nella realizzazione (forse già tra il 30 e il 20 a.C. l'acquedotto doveva essere attivo o in fase di realizzazione) di questa grande infrastruttura[63] che ha come riflesso principale l'esplosione edilizia nell'area flegrea, in particolare l'edilizia di lusso, in continuità con quanto accadeva nel periodo tardo repubblicano.

Tornando alle Terme di Baia il passaggio della condotta dell'*Acqua Augusta*, favorisce lo sviluppo di un complesso sistema di stoccaggio delle acque basato essenzialmente sul posizionamento di una serie di serbatoi (**S.1, S.2 e S.3**) creati rifunzionalizzando le vecchie cisterne repubblicane, se non addirittura modificando parte delle sostruzioni dei diversi terrazzamenti, col duplice intento da un lato di controllo e gestione delle acque, mentre dall'altro permetterne lo sfruttamento capillare anche per motivi 'diversi' da quelli esclusivamente alimentari, come il funzionamento delle numerose fontane con i loro giochi d'acqua, i ninfei e vasche, che caratterizzano l'asse mediano degli edifici principali, come accade nella cd *Villa dell'Ambulatio* o nel complesso della Sosandra.[64] Questi aggiornamenti si distribuiscono in un lasso di tempo piuttosto lungo poiché le modifiche più imponenti vanno collocate tra l'età giulio-claudia e almeno la fine del II sec. d.C. quando, vengono completamente ristrutturate le Piccole Terme, viene costruito il Settore di Mercurio e avviene la riconversione delle 'Stanze di Venere' in serbatoio per l'acqua, a servizio probabilmente del Complesso Termale Adrianeo. Tale attività edilizia altro non è che l'adattamento alle diverse esigenze espresse di volta in volta dall'utenza baiana tra il I e gli inizi del IV sec. d.C., molto rappresentativi della fortuna del sito (Figure 15). Si tratta di lavori estremamente impegnativi e costosi, frutto di una committenza di altissimo livello la cui resa è talmente elevata che l'uso delle terme baiane e la loro fortuna travalica i limiti dell'età antica andandosi ad incardinare nel corso del periodo medievale, quando le Terme di Baia, seppur viste come luogo di perdizione o come luogo (similmente a quanto avveniva in antico) per collegarsi al mondo ultraterreno, risultano ancora ben conosciute e le sue sorgenti ancora utilizzate per le cure e il benessere del corpo. Tutta questa congerie di tradizioni troverà coronamento nel *De Balneis Puteolani* di Pietro da Eboli, testo cardine per la ripresa delle investigazioni dedicate all'area flegrea in epoca moderna e alla base delle prime ricerche archeologiche.[65]

Bibliografia

Auberson, E.P. 1964. Etudes sur les Thermes de Venus. RAAN 39: 167–178.

Borriello, M.R. and D'Ambrosio, A. 1979. *Baiae-Misenum*. Forma Italiae I, XIV. Firenze: Olschki.

Camodeca, G. 1997, Una ignorata galleria stradale d'età augustea fra Lucrinum e Baiae e la più antica iscrizione di un curator aquaeaugustae. *Atti del Convegno in omaggio a Raimondo Annecchino* (Pozzuoli, 17 gennaio 1997): 289–306. Napoli: Cartopoli Pozzuoli.

Cairoli Giuliani, F. 1976. Note sull'architettura dei Campi flegrei. *I campi flegrei nella storia e nell'archeologia*, Convegno Internazionale (Roma 4–7 maggio 1976): 145–175. Roma: Accademia Nazionale dei Lincei.

Campi Flegrei 1976. I Campi Flegrei nell'archeologia e nella storia. Atti dei Convegni Lincei, 33 (Roma 4–7 maggio 1976). Roma: Accademia Nazionale dei Lincei.

I Campi Flegrei 1990. Amalfitano, P., Camodeca, G. and Medri, M. (eds) *I Campi Flegrei. Un itinerario archeologico*. Venezia: Marsilio.

Cicala, L. and Illiano, G. 2017. Paesaggio archeologico e paesaggi urbani tra Bacoli e Miseno, in A. Aveta, B.G. Marino and R. Amore (eds) *La Baia di Napoli. Strategie integrate per la conservazione e la fruizione del paesaggio culturale*, Vol. I: 358–362. Napoli: Artstudio Paparo.

D'Arms, J.H. 2003. *Romans on the bay of Naples and other Essays of Roman Campania*. Bari.

De Franciscis, A. 1970. Baia in *EAA*, Suppl.: 133–134.

Di Luca, G. 2009. Nullus in orbe sinus Baianis praeluceta moenis, Riflessioni sull'architettura dei complessi cd 'dell'Ambulatio', della Sosandra e delle Piccole Terme a Baia. *BABesh* 84: 149–168.

Döering, M. 2012. *In der Wundersamsten gegend der welt*. Adenstedt: Parmenio.

Ferrari, G. and Lamagna, R. 2010. Il bimillenario dell'acquedotto augusteo di Serino, in F. Cucchi and P. Guidi (eds), *Atti del XXI congresso di Speleologia*, Trieste 2–5 giugno 2011: 387–397. Trieste: Edizioni Università di Trieste.

---

[62] Sgobbo 1938: 76–90.
[63] Camodeca 1997: 289–306; Ferrari and Lamagna 2010: 387–397.
[64] Medri 2012: 134–135, si tratta di tipologia termale poco rappresentata a Baia e forse riconoscibile solo nelle strutture del Settore Adrianeo e forse nelle Terme del Livello Intermedio.

[65] Sulla fortuna delle terme in epoca medievale vedi Kauffmann 1959; Russo Mailler 1979: 141–153; Annecchino 1985; I Campi Flegrei 1990: 234–235, e in ultimo Medri 2012: 130–144.

Ferrari, G. and Lamagna, R. 2017. Un nuovo cunicolo pluviale nel Parco Archeologico di Baia (Napoli), in *Campania Speleologica 2017, Atti del III Convegno Regionale di speleologia* (2–4 giugno 2017): 151–157. Napoli: Società Speleologica Italiana.

Ferrari, G., Lamagna, R. 2018. Il Parco Archeologico di Baia (Bacoli): note preliminari sulle opere idrauliche di età romana nel Settore dell'Ambulatio. *Opera Ipogea* 1: 59–75.

Ferrari, G., Guidone, I., Lamagna, R. 2015. Il sistema di vapore delle Piccole Terme di Baia, in L. De Nitto, F. Maurano and M. Parise (eds) *Atti 22° Congresso nazionale di speleologia 'Condividere i dati'*: 429–434. Petrosa Auletta : Società Speleologica Italiana.

Gullini, G. 1973. Il Criptoportico nell'architettura romana, in *Les Cryptoportiques dans l'architecture romaine. Colloque international du Centre National de la Recherche Scientifique* (École française de Rome, 19–23 avril 1973): 137–142. Rome: École Française de Rome.

Kauffmann, C. 1959. *The Bath of Pozzuoli, a Study of the Medieval Illuminations of Peter of Eboli's Poem*. Oxford: Bruno Cassirer.

Lafon, X. 2001. *Villa Maritima. Recherches sur les villas littorales de l'Italie romaine (IIIe siècle av. J.-C. / IIIe siècle ap. J.-C.)*. Roma: Bibliothèque des Écoles françaises d'Athènes et de Rome.

Levi, A. 1922. Ruderi di terme romane trovati a Baja. *MonAnt* 28: 129–154.

Ling, R. 1979. The Stanze di Venere at Baia. *Archaeologia* 106: 77–140.

Maiuri, A. 1958. *I Campi Flegrei*. Roma: Ist. poligrafico dello Stato.

Maniscalco, F. 1995. Un ninfeo severiano dalle acque del porto di Baia. *Ostraka* 4.2: 257–271.

Maniscalco, F. 1997. *Ninfei ed edifici marittimi severiani del Palatium imperiale di Baia*. Napoli: Massa.

Medri, M. 1990. Baia, in P. Amalfitano, G. Camodeca and M. Medri (eds) *I Campi Flegrei. Un itinerario archeologico*: 184–237. Venezia: Marsilio.

Medri, M. 2001. La diffusione dell'opera reticolata: considerazioni a partire dal caso di Olimpia, in *Constructions publiques et programmes éditilaires en Grèce du IIe s. av. J.C. au Ier s. ap. J.C.* (Athènes, 14–17 mai 1995). *BCH*, Supp. 39: 15–40.

Medri, M. 2012. In Baiano Sinu: il Vapor, le Aquae e le Piccole Terme di Baia, in M. Bassani, M. Bressan and F. Ghedini (eds) *Aquae Salutiferae. Il Termalismo tra antico e contemporaneo, Atti del Convegno internazionale* (Montegrotto Terme 6–8 settembre 2012): 121–144. Roma: Quasar.

Medri, M., Soricelli, G. and Benini, A. 1999. In baiano sinu: le Piccole Terme di Baia, in J. De Laine J. and D.E. Johnston (eds) *Roman Baths and Bathing, Proceedings of the First International Conference on Roman Baths held at Bath* (Bath, 30 March – 2 April). *JRA*, Supp. 37(2): 207–219.

Miniero, P. 2000. *Baia. Il castello, il museo, l'area archeologica*. Napoli: Electa.

Miniero, P. 2017. Ville romane in Baiano sinu: recenti rinvenimenti e riflessioni, in L. Cicala and B. Ferrara (eds) *Kithon Lydios. Studi di storia e archeologia con Giovanna Greco*: 795–810. Pozzuoli: Naus Editoria Archeologica.

Mingazzini, P. 1977. Le terme di Baia. *I Campi Flegrei nell'archeologia e nella storia*: 275–281. Roma: Bardi.

Napoli, M. 1958. Architettura di Baia, in F. Franco and F. Reggiani (eds) *Le meraviglie del passato*: 519–534. Milano–Verona: Mondadori.

Nieberle, M., Broisch, M., Geiermann, S. and Broser, J. 2016, Neue Forschungen zu mantiken Baiae. *Kölner und Bonner Archaeologica* 6: 91–115.

Paget, R.F. 1967a. *In the Footsteps of Orpheus*. London: Roy Publishers.

Paget, R.F. 1967b. The 'Great Antrum' at Baiae: A Preliminary Report. *BSR* 35: 102–113.

Rakob, F. 1992. Le cupole di Baia, in M. Gigante (ed.) *Civiltà dei Campi Flegrei. Atti del Convegno Internazionale*: 229–258. Napoli: Giannini editore.

Russo Mailler, C. 1979. La tradizione medievale dei bagni flegrei. *Puteoli* 3: 99–127.

Sgobbo, I. 1934. I nuclei monumentali delle Terme Romane di Baia per la prima volta riconosciuti. *Atti del III Convegno Nazionale di Studi Romani*: 294–309. Bologna: Capelli.

Sgobbo, I. 1938. L' acquedotto romano della Campania :Fontis Augustei Aquaeductus. *NSc*: 76–97.

Yegül, F.K. 1996. The Thermo-Mineral Complex at Baiae and the Balneis Puteolanis. *The Art Bulletin* 78(1): 137–161.

Natural Risks and Water Management in Delphi

Amélie Perrier,[1] Isabelle Moretti[2] and Luigi Piccardi[3]

[1] Loire Valley University, IRAMAT/ CNRS UMR 5060, Associate Professor in Ancient Greek History, former Director of Ancient and Byzantine Studies at the French School at Athens (amelie.perrier@univ-orleans.fr)
[2] Sorbonne University, ISTeP/ CNRS UMR 7193, Associate researcher in Geological Sciences (isabelle.moretti@sorbonne-universite.fr)
[3] CNR Istituto di Geoscienze e Georisorse, Via La Pira 4, 50121 Firenze, Italy, Researcher in Geological Sciences (luigi.piccardi@cnr.it)

Abstract: The archaeological site of Delphi is subject to several natural hazards including earthquakes, gravity gliding, and slope instabilities. The architectural remains at the site bear the marks of the disasters that occurred in antiquity but also show how ingeniously the ancient Greeks dealt with risks and major events. The greatest danger seems to be landslides and rock falls, which are caused by surface and underground water runoff. The period of reconstruction that followed the collapse of the Temple of Apollo in the 4th century BC seems to mark an important turning point in how ancient designers considered and responded to water-related risks. Water is appropriately channelled and later exploited for thermal baths.

Keywords: earthquakes, rockfall, underground waters, architecture, thermal baths

Introduction

Delphi, one of the historically most relevant sites of antiquity, is located in central Greece, on the slope of Mount Parnassus, at the foot of the Phaedriades cliffs (called Phlemboukos, to the east of Kastalia Spring, and Rhodini, to the west of it) (Figure 1), about 600 m a.s.l. and 10 km inland from the northern coast of the Gulf of Corinth (Figure 2). The topographical positioning of the ancient site of Delphi is indeed exceptionally scenographic and suggestive. Gustave Flaubert wrote that the oracular Sanctuary of Apollo was located in a 'landscape of religious terror' and that to have put Pythia there was 'genius'.[1]

The Sanctuary of Apollo, which housed the famous oracle and was in function between the 8th century BC and the 4th century AD, was located in the heart of the ancient city of Delphi. The city of Late Antiquity developed until the last quarter of the 6th century, when the site was abandoned, which is generally attributed to the wave of Slavic invasions that occurred c. 580–590 AD. The sanctuary was mainly excavated from 1892 to 1903 by French archaeologists. In addition to the monuments of Apollo's sanctuary and the districts of the city, the remains include the Sanctuary of Athena Pronaia, necropolises, the Castalian Fountain, the gymnasium and the stadium.

From a geological point of view, the site lies on the downthrown block of one of the major seismic faults that form the northern margin of the Corinth Rift: the Delphi Fault, which runs from Arachova to Amphissa at the foot of the southern slope of the limestone massif of Mount Parnassus. Because of it position and of the local conditions of this geologically active environment, whose processes have shaped the landscape, the archaeological site is subject to several natural hazards, the most relevant ones linked to seismicity and to the action of water, and archaeological traces of past catastrophes are preserved in the monuments. It seems that from the earliest period, already in the Mycenaean period, various strategies have been implemented, more or less successfully, to manage these risks.

In this contribution we offer a first joint analysis of the geological context and the archaeological remains in order to measure in particular the risks related to water and the relevance of the technical responses implemented to cope with them.

Geological setting

The geodynamic history of the Mediterranean region has been dominated by the convergence of Africa and Eurasia, starting since the Jurassic (200–150 Ma), inducing subduction, collision, and obduction processes, but also back-arc extension such as in the Aegean and Thyrenian zones.[2] The 2,457-meter-high Mount Parnassus that towers above Delphi formed during the Alpine-Hellenic orogeny, which has affected the Mesozoic limestone platforms of the Thetys margin from the Late Cretaceous–Eocene (100–30 Ma).[3]

Due to the clockwise rotation of the region between 15 and 8 Ma,[4] the Hellenic orogenic belt is now

---
[1] G. Flaubert, *Lettres de Grèce* (ed. 1948): 52–53, quoted in Hellmann 1992: 47.
[2] Dercourt *et al.* 1986; Menant *et al.* 2016; Ricou *et al.* 1986; Ricou 1994; Şengör and Yilmaz 1981.
[3] E.g. Nirta *et al.* 2015; Nirta *et al.* 2018, and references therein.
[4] Brun and Sokoutis 2007; Jolivet *et al.* 2015.

Figure 1: Panorama of Delphi archaeological area and the Phedriades cliffs (called Phlemboukos, to the east of the Kastalia Spring, and Rhodini, to the west of it).

oriented nearly north–south. In reaction to the south Mediterranean subduction, back-arc extension took place from the Oligocene in the Aegean Sea, and the slab quickly migrated to the south, and the trench is now located south of Crete. Since the Late Miocene, the Hellenides have been affected by N–S extension localised in the active rift zones of Corinth, Evia, and within the Central Hellenic Shear Zone.[5] The relative roles of the southward migration of the subduction-related extension and of the westward propagation of the North Anatolian fault[6] remain open to discussion.

The Corinth Rift, the tectonic depression that hosts the Gulf of Corinth (Figure 2), is characterised today by particularly high rates of deformation, which are among the highest in the Mediterranean area (c. 1.5 cm/year).[7] The extension started at the end of the Miocene[8] and the current phase of extension began c. 1 Ma ago,[9] and has an average rate of about 10 mm/yr.[10] The downthrow of the central portion is realised by slip on faults on its sides, dipping toward the axis of the depression (Figure 2 section). On the basis of geologic data and uplifted topography, vertical slip rates of the main faults, which are currently along the southern Gulf margin, appear to range between 6–7 and 11 mm/yr at Xilocastro[11] and between 3 and 8 mm/yr at Helice.[12] Paleoseismological trenching indicated a slip rate between 0.7 and 2.5 mm/yr for the Skinos fault alone (Figure 2).[13]

The evolution of the rift is well constrained into three phases:[14] (1) an initiation phase dominated by small-extension continental and lacustrine sedimentation; (2) an increase of the fault activity associated with a connection of the basins, leading from a deep-water lacustrine to marine environment; and (3) the uplift of the Peloponnesus since 1 Ma, with the exhumation

---

[5] Armijo *et al.* 1996; Moretti *et al.* 2003; Papanikolaou and Royden 2007.
[6] Sakellariou *et al.* 2017, and references therein.
[7] Moretti *et al.* 2003, and references therein.
[8] E.g. Rohais and Moretti 2017, and references therein.
[9] Armijo *et al.* 1996; Jackson *et al.* 1982; Lyon-Caen *et al.* 1988; Mouyaris *et al.* 1992; Ori 1989.
[10] Ambraseys and Jackson 1990; Billiris *et al.* 1991.
[11] Armijo *et al.* 1996.
[12] Armijo *et al.* 1996; Mouyaris *et al.* 1992.
[13] Collier *et al.* 1998.
[14] Gawthorpe *et al.* 2018; Ori 1989; Rohais and Moretti 2017.

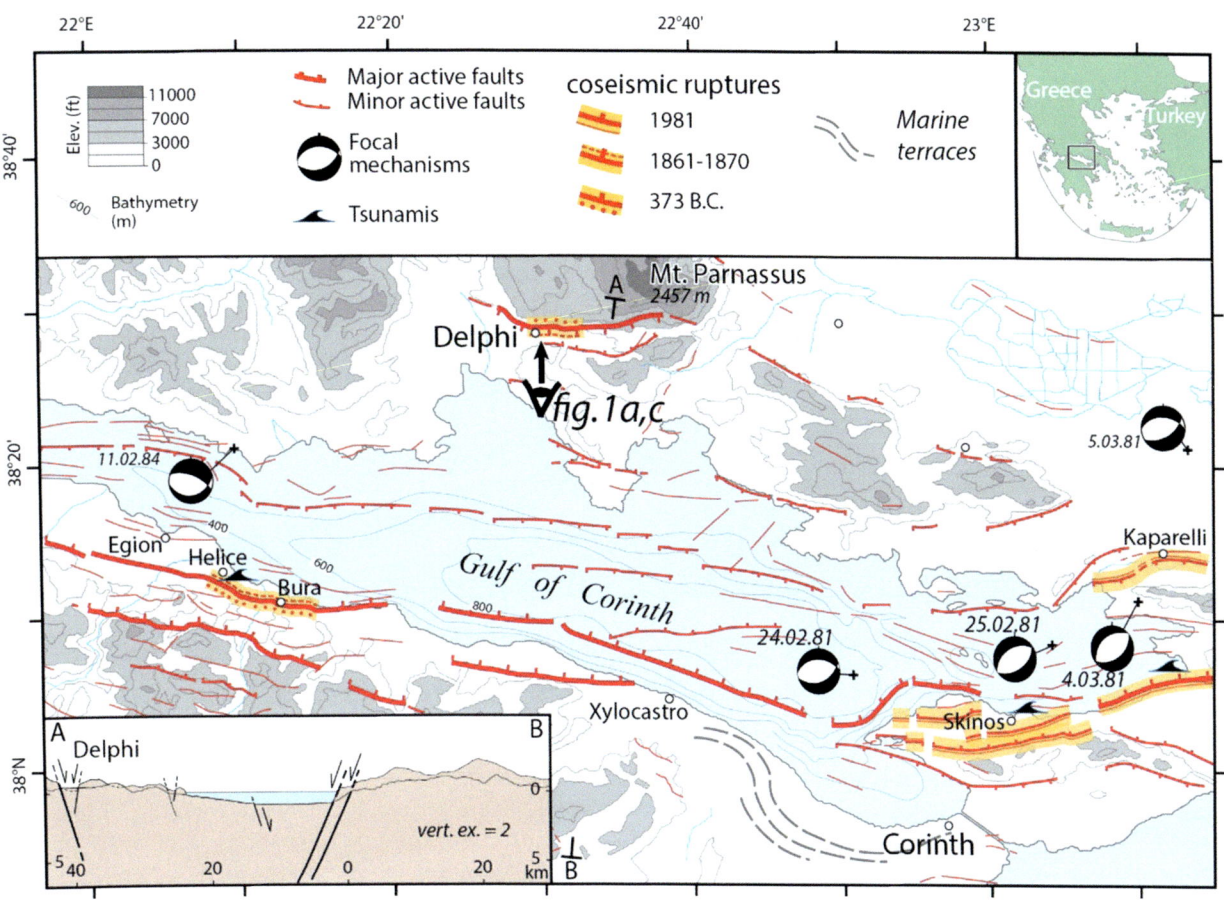

Figure 2: Map of active faults of the Delphi region (modified from Piccardi 2008 and Moretti *et al.* 2003). Dashed lines west of Corinth show the edge of marine terraces that registered the Peloponnese uplift.

of the syn-rift deposits on the southern shore and the shifting deltaic systems northward. Marine terraces of Upper Pleistocene age (*c.* 350 ka) have been exposed especially around the towns of ancient and modern Corinth. Due to hot fluid circulations, travertines are present along some major faults, such as the Xylocastro fault, as well as, at places, quaternary stromatolites all along the faults that border the Perachora peninsula.[15] On the northern shore of the gulf, Oligo-Miocene sediments are mainly absent and Quaternary slope breccias lie unconformably on top of the Cretaceous limestones and Paleogene flysch across most of the region.[16]

The seismicity of the Corinth Gulf represents a major hazard for the inhabitants, with maximum earthquake magnitudes (M) exceeding 6.[17] The occurrence of coseismic surface faulting is a well-known phenomenon in the Gulf of Corinth, and extensional coseismic surface faulting has also been documented by palaeoseismological trenching along the southern margin. Fault rupture at surface has been observed during the 1981 (M = 6.7 and M = 6.4) Corinth seismic sequence, coherent with measured geophysical stress orientation, on conjugate faults at the eastern end of the Corinth Rift.[18] In that sequence, the earthquakes of February 24 and 25, 1981 originated fault ruptures on the south side of the rift, while the shock of March 4, 1981 (Ms = 6.4) ruptured the antithetic fault on the north side of the graben. A rupture (*c.* 13 km long) was described along the Helice active fault in the 1861 Aigion earthquake,[19] and rather similarly the appearance of a large 'chasm' is historically reported for the same fault in the Helice earthquake of 373 BC.

Delphi is located of the northernmost south dipping fault that borders the Corinth Rift (Figures 2, 3). The site of Delphi lies mainly in slope breccias and flysch but Mesozoic carbonate rocks also outcrop and the blocks of this basement cut by the major faults shape the landscape (Figure 3). Three main tectonic units are present in the area of Delphi: 1) the Parnassus Nappe which overlaps (2) the Pindos Nappe to the west and

---

[15] E.g. Armijo *et al.* 1996; Hayward 2003; Kreshaw and Guo 2006.
[16] Solakius *et al.* 1998.
[17] Ambraseys 1996; Ambraseys and Jackson 1990; Mariolakos *et al.* 1989.
[18] Jackson *et al.* 1982.
[19] Mouyaris *et al.* 1992; Schmidt 1881.

Figure 3: View towards the north of Delphi and the southern slope of Mt Parnassus. White triangles indicate the trace of the active fault.

which is overthrust by (3) the Pelagonian Nappe of the Internal Hellenides eastward.[20] The Parnassus Nappe itself is composed of i) Triassic to Upper Cretaceous neritic limestones, ii) Campanian to Maastrichtian pelagic limestones, iii) red Paleocene pelitic rocks and Eocene flysch. The thrusting of the Pindos Nappe toward the west is dated from about 50 myr (Eocene).

North of the sanctuary outcrops the Jurassic limestone of Parnassus (Figure 2) and below the sanctuary, in the hanging wall of the fault, the youngest series are present, dated from the Upper Cretaceous (Maastrichtian, as in the St Elli quarry) to the Eocene. The slope breccia and debris flows, not specifically dated but clearly the product of the activity of the Delphi Fault, and the high of the cliff, largely covers this basement. Another large normal fault, dipping northward, borders the Pleistos Valley, south of Delphi.[21]

The Delphi active fault

Ancient Delphi lies above a seismically active structure, the Delphi Fault, which has been active from at least 1 myr based on calcite cement dating[22] and maybe during the full quaternary as the onshore south dipping faults near Galixidi that affect the full rift sequence based on marine seismic data.[23] The Delphi Fault is a fault system made of parallel slip planes, which bounds the calcareous massif of Mt Parnassus on the south, extending for a distance of c. 28 km from east to west (Figures 2, 3). It is an east/west-trending fault, downthrowing the southern block and with minor right-lateral component of movement, as indicated by the geometry of the fault and slip-parallel lineation on the main fault plane (east of Delphi, the fault plane has an average strike of 260°N, dips 60-65° and is lineated by corrugations and striations with a pitch of about 70°W). For the most part, the main fault plane affects the Mesozoic limestone and Paleogene flysch of the bedrock. Wurmian (< c. 18 ka) slope debris and Holocene (< 10 ka) debris cones are faulted as well. Delphi is positioned at a bend of the fault.

Slip is distributed at the surface on a few parallel subsidiary faults, within a fault-zone several tens of meters wide.[24] Intra-fault hanging wall collapse has created a stepped morphology. This is particularly visible to the east of the Kastalia Spring, east of the Apollo Sanctuary and north of that of Athena Pronaia, where subsidiary faults offset the paleosurface of the Quaternary slope-deposits. Slip during historical times on one of these splay faults may be deduced by archaeological evidence. In the shrine of Athena, for example, one of the segments of the fault cuts and displaces the main temple and altar, built near the end of 6th century BC. An offset of about 30 cm, still evident in the main altar's west wall, caused the break, and tilt, of the basement of the temple.[25]

The Delphi Fault being large, earthquakes generated by its slip can be major, with a magnitude (M) up to

---

[20] Doutsos et al. 2006; Jolivet and Brun 2010; Royden and Papanikolaou 2011; Menant et al. 2016; Nirta et al. 2018.
[21] Péchoux 1977; Piccardi 2000; Piccardi et al. 2008; Valkaniotis et al. 2011.
[22] Causse et al. 2003; Péchoux 1977.
[23] Rohais and Moretti 2017.
[24] Péchoux 1977.
[25] Piccardi 2000.

7. The fault is known to have ruptured a few times in the past: in 373 BC,[26] in AD 551 with an estimated magnitude of M = 7.2.,[27] and more recently in the Ms = 6.7 earthquake of August 1, 1870.[28] The epicentre of the last event was located at Kastri, the village that stood at that time on the main archaeological site of Delphi. The main shock occurred a little to the east, and the two major aftershocks (5.4<M<6.3) occurred west of it.[29] The ejection of sand and water (typically reddish because of the bauxite content of the limestone) from some spring observed after the 1870 earthquake[30] recalls the blood breathed forth by the Homeric dragoness after her death.[31] Other significant earthquakes in the area, although not precisely documented, are indicated in 278 BC (Valkaniotis *et al.* 2011), AD 326 (Marinos and Rondoyanni 2005), in 354–346 and 279 BC, at Galaxidi at the turn of the 11th century (in 976–1025), and in 1580, and at Aigion (1402, 1938, 1965, 1970).

The most famous historical earthquake in this sector of the Corinthian Gulf is that of 373 BC, which destroyed two cities on the southern border of the Gulf. Bura is said to have 'disappeared in a chasm of the Earth' (Strabo, 1.3.18), while Helice was wiped out by 'tidal waves which engulfed the open country and cities' and remained submerged (Diodorus 15.1–4). Mouyaris *et al.* 1992 proposed that the rupture of the Helice Fault, with associated surface faulting, is the most probable source. That earthquake caused massive destruction at Delphi too, although far away from the main-shock epicentre. From archaeological data, Bousquet and Péchoux 1977 attributed a macroseismic intensity of IX MCS to this shock at Delphi, and the 'Catalogue of Strong Earthquakes in the Mediterranean' estimate a Me of 6.4.[32] Integrating tectonic, archaeological and historical data, Piccardi 2000 suggested that during the earthquake of 373 BC, not only was the Helice Fault ruptured on the southern border of the graben, but also the corresponding south dipping fault at Delphi. If this is the case, the seismotectonic mechanism of that earthquake would appear to be the same as that of the 1981 Corinth earthquake.

The description of the geomorphic effects of the M = 6.7, 1870 Kastri earthquake can provide a realistic idea about the coseismic phenomena and about the possible scenario which started the myth of a supernatural chasm in the earth at Delphi.[33] The earthquake caused significant damage to the monuments at Delphi as well as in the nearby towns of Chrisso, Itea, and Arachova. A series of ground ruptures occurred along the trace of the Delphi seismic fault, at the base of the cliff, along a distance of *c.* 18–20 km from east to west, with maximum vertical displacement of 1 m.[34] Just north of the archaeological site: '… large cracks developed at the foot of the cliffs above the village [Kastri] leading in a north-westerly direction', and near Chryso below, 'ground cracks more than one metre wide and 5 metres deep were formed running along the mountain side for a short distance'.[35] Landslides and rock-falls were triggered in many places, being particularly intense in the area of the archaeological site.

Because of the different lithologies on the two sides of the fault, the fault zone determines a threshold for emergence of water circulating in the Parnassus limestone, so that many springs are located in correspondence of this structure, Kerna and Kastalia springs being ones of these. A few instances of gas emissions also have been described by Joseph Fontenrose (1981) and some inhabitants of Delphi, but we did not find traces of such seepages at present. The lack of any geochemical 'anomalies' in springs, soil gases, travertine deposits and local spring-waters, however, indicates that, today, i.e. during the current quiescence period, the Delphi active fault system is not a preferential circulation route for deep fluids, and that the shallow, karstic, hydrological system is isolated from any system at greater depth. The origin of the travertine deposits, covering archaeological relics in places, is definitely related to the superficial circulation of meteoric water within in the Parnassus limestone.[36]

Falling rock

Rockfalls are generated on the steep slope located on the northern side of the archaeological site, the Parnassian Phaedriades (Figures 2, 5a). These almost vertical limestone rock cliffs, rise more than 500 m above the archaeological site (Figure 5a), and consist of limestone that has been intensively fractured into blocks of various dimensions. Rockfalls associated with widespread erosion and detritic slips occur along the borders of the calcareous wall. These phenomena produce toppling and rolling of calcareous blocks as well as recurrent debris flow, and are mainly triggered by seismic phenomena or intense rainfalls and storms,[37] such as recently occurred in September 2009, activated after prolonged rainfall,[38] and as reported at many places during the earthquake of 1870.[39]

---

[26] Piccardi 2000.
[27] Ambraseys and Jackson 1998, Marinos and Rondoyanni 2005; Papazachos and Papazachou 2003.
[28] Ambraseys 1996; Ambraseys and Jackson 1998; Papazachos and Papazachou 2003; Pavlides and Caputo 2004.
[29] Ambraseys and Pantelopoulos 1989.
[30] Ambraseys and Pantelopoulos 1989.
[31] Piccardi 2000.
[32] CFTI5Med, http://storing.ingv.it/cfti/cfti5/; Guidoboni et al. 2019.
[33] E.g. Ambraseys and Pantelopoulos 1989 and 1992; Pavlides and Caputo 2004; Marinos and Rondoyanni 2005; Schmidt 1879.
[34] Pavlides and Caputo 2004.
[35] Schmidt 1879, reported by Ambraseys and Pantelopoulos 1989.
[36] Piccardi et al. 2008.
[37] Lazzari and Lazzari 2012.
[38] Christaras and Vouvalidis 2010.
[39] See Ambraseys and Pantelopoulos 1989.

Figure 4: View towards the south of the main Temple of Athena Pronaia (end of 6th c. BC), with one of the blocks that fell on it in 1905 (photo 2005, since the block has been removed). The figure to the left of the block provides the scale.

Rockfalls are mainly due to the release of blocks individuated by fractures of the rock mass and bedding planes, and the falling blocks vary greatly in size and weight. Fractures which cut the rock mass, limit the falling blocks and contribute to rainwater infiltration. The seepage of groundwater at the interface between the overlying thrusted limestone and the underlying impermeable flysch, softens the flysch, inducing tensile forces from below, enhancing the fracture of the overhanging limestone and rockfall phenomena. The release of 'keyblocks', i.e. blocks which are determinant in the overall equilibrium of other blocks in an intensely fractured area, can sometimes precipitate rockfalls of significant size or, in extreme cases, large-scale slope failures.

The archaeological site of Delphi is so situated in an active scree deposition area. Rockfalls represent the main threat in the area, and have been reported on numerous occasions at or in the vicinity of Delphi, imposing the temporary closing of relevant parts of the archaeological site (e.g. the Stadium, Kastalia Spring, and the Gymnasium). At the western part of the site, major rockfalls occurred repeatedly at the stadium also in recent times, and barrier have been set up for protection. Everywhere in the archaeological site, foundations of several ancient structures have been constructed on large fallen blocks of rock, given the abundance of such blocks in the scree material Many large boulders have fallen after the excavations of the site, and in 1905 a rockfall destroyed a temple in the Sanctuary of Athena, which had remained relatively well preserved until then (fifteen columns were still standing before the rockfall). One of the blocks remained on the temple basement until removed a few years ago (Figure 4).

Other natural risks: waterflow and karstic environment

The Parnassus limestones that form the cliff north of the Sanctuary are subject to dissolution, water

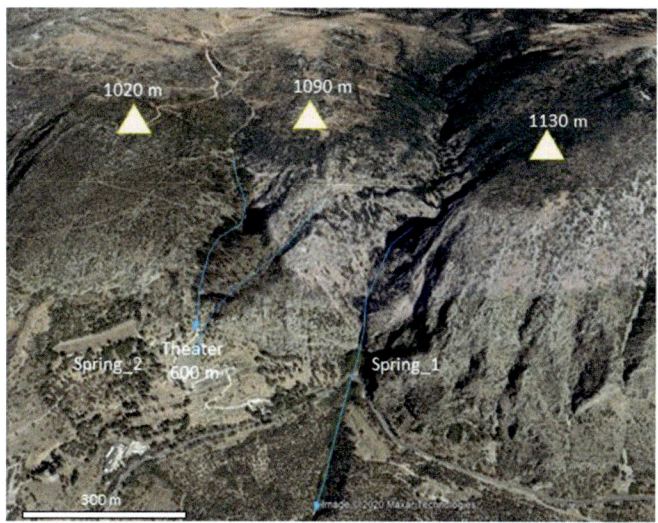 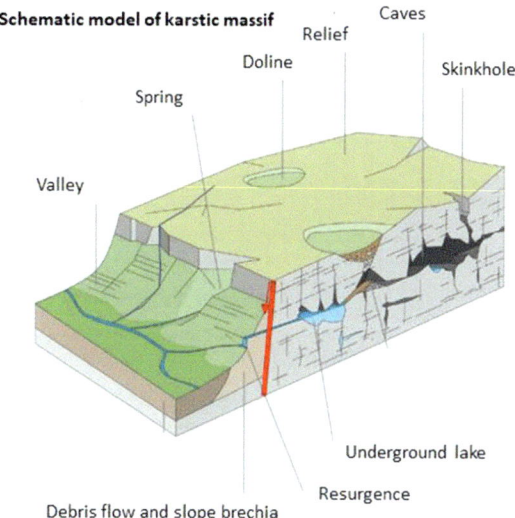

Figures 5a and 5b: Left – 3D view of the reliefs north from Delphi showing some of the seasonal torrent beds as well as the resurgences (springs 1 and 2). Right – schematic behaviours of a karstic relief.

infiltration, development of karstic conditions and resurgence, potentially drastic. Also if limestone may have some primary porosity, circulation of water within these rocks mostly occurs through fractures, and karstification process involves the solution of carbonate minerals (calcite and/or dolomite) due to the acidity of water, enlarging the water conduits. Karstification is driven by climatic conditions because dissolved CO_2 concentrations are controlled by temperature and atmospheric CO_2 partial pressure. If these conditions change, carbonate can be deposited instead of being dissolved: as mentioned above, at Delphi some carbonate, dissolved by meteoric water and circulating within the limestone massif, precipitate creating travertine deposits when emerging. Karst systems are numerous in Greece, and many of them have been extensively studied, especially in central Greece and Crete,[40] but these phenomena have not been studied so far in the area of Delphi and Parnassus. Figures 5a and 5b show the relief around Delphi and propose the presence of a karstic system; since it has not been completely documented, the presented behaviour is generic. The resurgence of water that accumulated in the Parnassus karstic massif leads to torrents of mud and landslides. Historical events are known, and even today, mudslides are still frequent, and rushing waters have often dug natural trenches on the periphery of the archaeological site. In December 1935 one particularly disastrous flood completely buried the site (Figure 6). On a larger scale, the landscape around Delphi allows us to observe slope breccia deposits caused by gravity gliding and, in the Gulf of Corinth itself, seismic data likewise indicate this kind debris flow related to fault activity.[41]

All these natural phenomena, earthquakes, rockfalls, landslides, debris flows and overflowing waters on the surface or in karstic subsurface channels, have affected the area of Delphi and its inhabitants since the first settlement of the region. In the following sections we will study the archaeological traces of these more or less catastrophic events and consider how the ancient Greeks protected themselves.

Traces of natural disasters on the monuments in Delphi

The history of the Sanctuary of Apollo in Delphi could be written as a succession of destructions and reconstructions in the light of the natural disasters it has experienced. In spite of the lack of a stratigraphic record for the Sanctuary of Apollo due to the nature of the 19th-century excavations at the site, part of the chronology of the disasters can be discerned thanks to more recent surveys associated with both the testimony of literary sources and the analysis of monuments and traces of ancient repairs.[42] The oldest examples date back to the Mycenaean and Geometric periods, as detailed below.

The 1938–1939 excavations on Foundations VIII,[43] also known as the Potidaean Treasury, and IX[44] (Figure 7) revealed, under the Geometric layer, a 2-m-thick layer of debris showing that this area had been devastated by a torrent that would have destroyed a modest domestic building from the Late Mycenaean period (1100 BC).[45] In the 1990s, excavations of the Geometric habitation

---
[40] Kazakis et al. 2018.
[41] Moretti et al. 2004.
[42] Perrier 2021.
[43] Bommelaer and Laroche 2015: n° 227.
[44] Bommelaer and Laroche 2015: n° 228.
[45] Amandry 1940-1: 258-259.

149

Figure 6: December, 1935. Deposits left by the catastrophic flood event over the Sanctuary of Apollo (French School at Athens).

area located in the eastern part of the sanctuary (Figure 7) showed that the circulation of underground water was a major risk with which the excavators associated the continual rising of the ground due to flooding and colluvium. The destruction of at least one house (B046) can be explained by the phenomenon of solifluction,[46] in a landslide that separated the soil from the retaining wall. In fact, there are archaeological traces of several landslides at the same place:[47] the wall of this house (B046) shows at least three states of repair. In the 8th century, the village of Delphi was laid out in a multitude of micro-terraces. Houses tended to be built up against retaining walls, which also explains their peculiar plans. According to J.-M. Luce, the laying out of the sanctuary and the first *peribola* in the 6th century BC (c. 585–575) modified the appearance of the site, with the micro-terraces grouped together to form deeper ones,

allowing the emergence of new architectural units.[48] This excavation has clearly shown that the steep natural gradient of the site, combined with significant water circulation, constitutes in itself a major risk for the construction and preservation of the monuments.

The first major natural disaster in Greece for which we have relatively detailed testimonies is the devastating earthquake that struck Sparta c. 464 BC, but it did not affect Phokis.[49] The episode that undoubtedly made the biggest impression on ancient authors, in terms of the extent of the destruction and the political and religious importance of the destroyed sites, was the earthquake, or series of them, dated to 373 BC, as previously mentioned, that affected several sites, including Helice, Bura, Olympia,[50] and Delphi. At Delphi, it also had a significant impact, destroying the Temple of Apollo,

---

[46] Luce 2008: 31.
[47] Luce 2008: 54.
[48] Luce 2008: 94.
[49] Thély 2016: 25.
[50] Partida 2016.

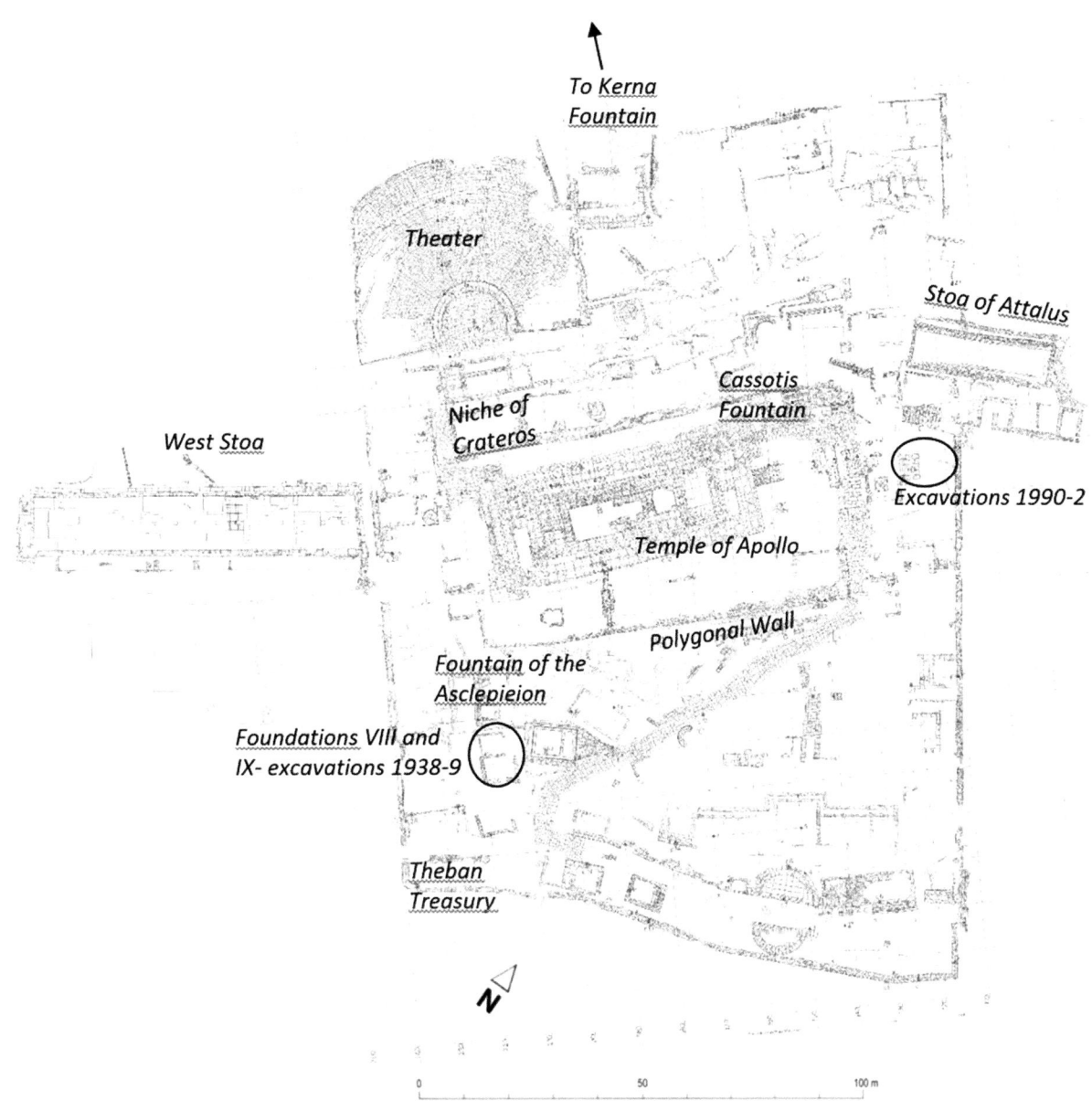

Figure 7: Map of the Sanctuary of Apollo (after Algreen-Ussing and Hansen 1975 and Perrier and Malmary 2013).

and initiating in effect an early 'fundraising campaign' throughout the Greek world for the reconstruction of the temple.[51]

Other rock- and mud-slides occurred from time to time at Delphi, destroying monuments and uncovering ravines. The remains bear witness to several catastrophic episodes that occurred throughout antiquity. The recent survey of the ancient city identified collapses of walls all over the site due to the action of water.[52] The damage caused by the torrents and runoff seem to have been taken very seriously by the ancient caretakers of the site, and several solutions were implemented, especially after the destruction of the temple in the 4th century. The temple terrace was particularly exposed to the effects of water, as the natural pathways of water ran on both the eastern and western sides of the temple.[53]

The construction programme that followed the great catastrophe of 373 BC at Delphi deserves a longer reference, not only because it is very well documented through construction accounts,[54] but because it

---

[51] Herodotus 2.180; Aristotle, *Athenian Constitution*, 19.4; Xenophon, *Hellenica*, 6.4.2; and see Amandry and Hansen 2010.
[52] Luce 2012: 782–783.
[53] Amandry and Hansen 2010: 150, fig. 1.5; 152, fig. 1.7; 153, fig. 1.8.
[54] Bousquet 1988 and 1989.

Figure 8: Temple of Apollo. Distortion of the *euthynteria*.

seems to mark a turning point, from which we see architectural responses to the specific risks at the site. Analysis of the deformation of the present remains has revealed that the 4th-century temple was destroyed, as was probably its predecessor, by a violent landslide due to the underground circulation of water, probably combined with a seismic event. Danish architect Erik Hansen measured a difference in level of more than 53 cm between the northwest and southeast corners of the temple's outer *euthynteria* and a southward shift of 85 cm on the east side to the south.[55] The *euthynteria* on the north side (Figure 8) shows a very pronounced curvature, due to the soil pressure that was partially compensated for, from Antiquity, by the abundant use of cramps of which only the lead gangue gave way. According to this analysis, the deformation of the present ruin is essentially due to a violent episode that took place in Late Antiquity, and not to progressive pressure.[56]

Two elements suggest that the Late Antique event, causing the eastern part to subside and the western part to twist, was the same type of landslide that destroyed the 6th-century temple in 373 BC. On the one hand, the foundations of that temple show at the southwest corner that they suffered a similar collapse and that the courses had to be straightened to accommodate the foundations of the new temple. On the other hand, the measures taken to ensure architectural security and cohesion in the construction of the 4th-century temple let us understand what happened during the destruction of the previous temple. Indeed, the very important reinforcement of the foundations at the external and internal south-western corners suggests that the south-western corner had become dissociated from the north-western one, causing cracks and opening joints. The plan of the current remains shows the effectiveness of the measures implemented during the reconstruction. During the destruction of the 4th-century temple in Late Antiquity, the southwest corner did not move: the rupture occurred a little further east.

The southwest angle had been reinforced, in particular, because this part of the temple is located in a sensitive corridor that experienced major landslides, as shown by the stratigraphy and the remains of other affected structures (Figure 9),[57] such as the Archaic Treasury n° 227 (foundations VIII), whose visible south-eastern corner has moved 130 cm southwards, or the wall of the southern *peribola*, in which blocks from the 6th-century temple were reused in repairs.

---

[55] Amandry and Hansen 2010: 145.
[56] Amandry and Hansen 2010: 147, fig. 1.3–4; 149; 150, fig. 1.5.
[57] Amandry and Hansen 2010: 152, fig. 1.7.

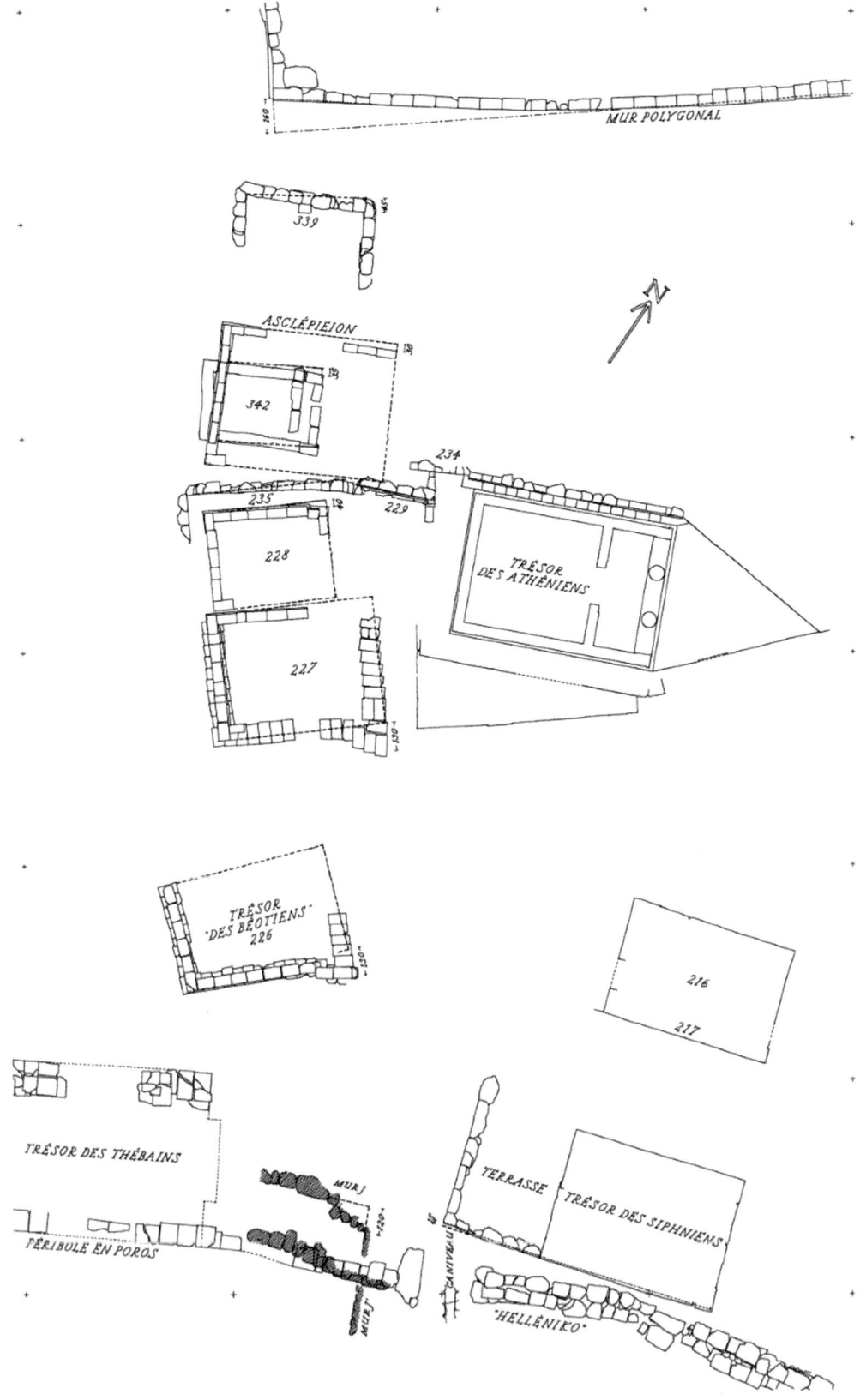

Figure 9: Monuments affected by landslides (after Amandry and Hansen 2010: 152, fig 1.7).

Generally, many retaining walls show deformation due to the effect of soil pressure. Examples include the wall of the Crateros Monument, a prominent 4th-century BC statue niche just above the temple (Figure 7), the wall of the *xystos*, located in the gymnasium near the sanctuary of Athena, on the axis of a fault, and where the same type of deformation related to a catastrophic event can also be seen in the stylobate of the outer colonnade, as well as the retaining wall of the lower terrace of the gymnasium.

Risk prevention and solutions for water management

Preventive measures were implemented as early as the Mycenaean period. These measures are generally of three types: (1) the construction of dams; (2) the creation of a network of hydraulic monuments across the site; and (3) the reinforcement of foundations.

Evidence of the building of dams above the sanctuary spans its entire history. Monumental blocks from the Mycenaean period have been found north of the sanctuary at the site of later dams built in antiquity up to the beginning of the 19th-century excavations.[58] They have been interpreted as the remains of a dam. The remains of a monumental wall from the Mycenaean period found under the Temple of Apollo may have served a similar purpose.[59] This suggests that people from the earliest development of the site may have tried to protect constructions at the site from flash floods from above.

The site of Delphi, especially the Sanctuary of Apollo, has many springs and fountains, whose role exceeded the water needs of the cult and the pilgrims.[60] Hydraulic monuments were constructed on different terraces and operated in part as a network. Water management in Delphi includes fountains, wells, basins, a swimming pool, large cisterns, thermal baths, and an impressive network of underground pipelines. The 4th-century Kassotis Fountain,[61] located in the northeast corner of the temple terrace, was originally probably intended to receive the waters of Kerna,[62] an abundant spring north of the sanctuary, and to channel the waters threatening the temple terrace to the southern perimeter. The dangers of infiltration and solifluction may explain the organisation of an hydraulic network linking several fountains. One of the roles of the Fountain of the Asklepieion (Figure 7) was to drain water from an underground resurgence west of the *opisthodomos* of the temple.[63] Additionally, a fountain located right underneath the temple probably allowed for the drainage of natural springs. From the reconstruction of the temple during the 4th century BC, perhaps already before, the fountains helped divert the overflow of water, if not from the entire sanctuary, then at least from the temple terrace, but they were insufficient to fulfil this role.

Several solutions were adopted to protect certain monuments from the effects of water, in particular by strengthening the masonry of the foundations and the terrace walls, as shown by these few examples. The foundation mass was largely reinforced at the southwest corner of the Temple of Apollo as seen above.[64] The Theban Treasury (Figure 7), built on one of the exposed axes after 371 BC, and thus after the catastrophe of the 4th century, was also later affected by a landslide. Preventive measures had been implemented as early as the construction of the monument – its foundation was reinforced by a wooden frame.[65] The southeast corner of the foundations of the West Stoa, built *c*. 330 BC, were even sealed and separated from the adjacent gutter with lead.[66] This device has not yet been observed in other foundations in Delphi, but the slabs of the northern gutter are themselves well sealed by thick vertical lead joints (Figure 10), a feature little known in Greece but also found in Delphi at the Archaic Fountain of Castalia.[67]

The architectural effort begun in the 4th century to preserve the Sanctuary of Apollo and to protect the temple terrace, in particular, continued with the construction of the theatre, beginning in the middle of the 3rd, or beginning of the 2nd century BC.[68] The monument, located at the northwest corner of the sanctuary, above the temple and south of the Kerna fountains and ancient dams, served as a funnel to collect and channel surface and underground runoff through a system of drains that were built under the steps and the orchestra. The water was collected both in a fountain built on the back wall of the theatre and along the steps. It was then channelled through an hydraulic network organised within the sanctuary, and, at least in the west, evacuated out of the sanctuary by two main conduits through the theatre and into the large western gutter of the West Stoa, and by the channel going from the orchestra to the west of the Crateros Monument.

The majority of the terrace walls in Delphi, up to the Classical period, were built to resist seismic shocks, as well as to the pushing force of earth swollen by water, as shown by the various polygonal masonry methods

---

[58] Homolle 1897: 259.
[59] Darcque 1991: 689.
[60] Perrier 2019.
[61] Bommelaer and Laroche 2015: 222–223, 248.
[62] Bommelaer and Laroche 2015: 259.
[63] Amandry 1942–1943: 342.
[64] Amandry and Hansen 2010: plate XXII (study of deformations on the west side); study of foundation restoration: 161 ff. and conclusions: 467–468.
[65] Michaud and Blécon 1973: 24–25, 93.
[66] Perrier and Malmary 2013: 52–53.
[67] Amandry 1977: 201–203.
[68] Bommelaer 1992: 298–299.

Figure 10: West Stoa: detail of the north gutter sealed with lead; east part of the north gutter.

used for these walls, such as the polygonal masonry with curved joints, known as Lesbian masonry. But after the catastrophe of the 4th century, terrace walls built during the next century have large buttresses, especially the inner buttresses of the two large walls on either side of the temple terrace, outside of the *peribola* (the West Stoa and the 9-E wall located east of the stoa of Attalos).[69]

The multiple architectural solutions implemented over the centuries in Delphi, and in particular from the 4th century BC onwards, reveal an awareness of the natural risks and a technical ingenuity, no doubt progressive and intimately linked to each catastrophic episode.

Evolution of water resources management

From the Archaic to the Protobyzantine period, the approaches to water management in the sanctuary and in the city seem to have evolved, moving from simple risk management, certainly in a religious context where water played a special role, to a real valorisation of the water resource. This phenomenon began as early as the 4th century BC with the development of the swimming pool and basins in the gymnasium,[70] but this is a special case. Except for a few examples, such as the remains of an Hellenistic bath in the northeast of the sanctuary,[71] it was not until Late Antiquity that large-scale water exploitation began to take place.

In fact, from the 3nd century AD onwards, the city of Delphi was transformed with a number of construction projects: i.e. the large villas with baths, such as the Late Antique Southeast Villa;[72] the building of public and private baths and associated reservoirs, with examples known east, south, and west of the sanctuary, especially the East Thermae; and even the adaptation of certain Classical or Hellenistic monuments into baths or cisterns, such as the Stoa of Attalos I (Figure 11), east of the temple terrace, which has been completely enclosed by brick walls and transformed into a huge reservoir (Figure 12), due to the construction of a villa and of the eastern thermal baths. To the west, a large cistern was built with a south wall reinforced with buttresses and a ceiling supported by arches (Figure 13).[73] It is now completely filled with earth and slope debris and has never been excavated. Even inside the heart of the sanctuary, domestic baths have been installed in the Crateros Monument.[74] This transformation can be related to the development of the Protobyzantine city, increasing the water needs, as shown especially by the construction of the large cistern in the new west district,[75] and the end of the distinction between sacred and secular spaces.

The location of these hydraulic installations seems to confirm our preliminary functional analysis of the monuments within the general economy of the site, since some of them are located around the terrace of the temple, as, for example, the Crateros Monument,

---

[69] See Trouki 1993.
[70] Homolle 1899: 581–583; Jannoray 1953.
[71] Ginouvès 1952.
[72] Déroche, Petridis and Badie 2014: 18–20.
[73] Amandry 1981: 733.
[74] Amandry 1981: 736; on baths in Delphi, see Ginouvès 1955.
[75] Petridis 2010: 21.

Figure 11: Main hydraulic structures during Late Antiquity in the area of the Sanctuary of Apollo (from Google Earth 2021).

Figure 12: Eastern cistern in the Stoa of Attalos I.

Figure 13: West cistern.

located exactly between the theatre and the temple, just above the *opisthodomos* square, or the cistern in the Stoa of Attalos I, northeast to the temple (Figure 11).

The observable conduits were reworked in Roman and Protobyzantine times. The network of drains and gutters allowing the circulation of water was adapted and increased. For example, a late 4th-century gutter found during the 2018 excavations to the west of the West Stoa was reworked in Roman times and connected to newly built baths to the west.[76] The so-called Roman Agora, dating to Late Antiquity, is equipped with important underground gutters, which not only drain the waters of the square but also collect the water from the upper terraces, since they are connected to the gutters of the thermal baths and cistern east of the main sanctuary.

An initial assessment of these remains reveals that by then water was no longer considered so much as a danger but as a resource, and one that was particularly abundant in Delphi, which undoubtedly contributed to the success of the site until Late Antiquity. These installations not only promoted new practices but adequately captured and therefore controlled rainwater on the site. It seems that the city, between the 4th and 6th centuries AD, was particularly dynamic and experienced its greatest expansion – as shown, for example, by the number of craft workshops and the construction of luxurious villas.[77]

Conclusion

The impressive landscape of Delphi is strongly marked by rocky escarpments and faults. Although earthquakes and rockfalls are frequent, it seems that the most devastating natural disasters are due to landslides, debris flow, gravity gliding and slope instabilities, from Antiquity to the present day. The remains clearly show that the ancient residents sought solutions and occasionally implemented building strategies to counter the disastrous effects of the passage of water, particularly in the Sanctuary of Apollo. As far as we know, the 4th century BC marked a turning point in the awareness of the risks associated with water, and the large construction site of the temple and its surroundings were redesigned to protect the heart of the sanctuary from this kind of danger. If the Greeks of the Classical period tried to protect themselves from the excess of water, from the Roman period onwards, they took it as an advantage. Water was exploited throughout the Roman Imperial period into Late

---

[76] Perrier *et al.* forthcoming.

[77] Petridis 2010: 142.

Antiquity as a resource in the service of a way of life that transformed the city of Delphi and its sanctuary.

Acknowledgements

The contributors would like to express their gratitude to the French School at Athens and to the Ephorate of Antiquities of Phokis for their support. Thanks are also due to Giovanni Polizzi, Vincent Ollivier and Sophie Bouffier for inviting us to take part in this conference of the WaterTraces Project supported by Aix-Marseille University.

Bibliography

Algreen-Ussing, G. and Hansen, E. 1975. *Atlas de Delphes, Fouilles de Delphes* II 17. Athens: École française d'Athènes.

Amandry, P. 1940-1941. Chronique des fouilles 1940-1941. Delphes. *Bulletin de Correspondance Hellénique* 64-65: 253–266.

Amandry, P. 1942-1943. Chronique des fouilles en 1942. Delphes. Fontaine à l'angle Sud-Ouest du mur polygonal. *Bulletin de Correspondance Hellénique* 66-67: 342–345.

Amandry, P. 1977. Notes de topographie et d'architecture delphique. VI. La fontaine Castalie. Études delphiques. *Bulletin de Correspondance Hellénique Supplément IV*: 179–228.

Amandry, P. 1981. Chronique delphique. *Bulletin de Correspondance Hellénique* 105.2: 673–769.

Amandry, P. and Hansen, E. 2010. *Le temple d'Apollon du IVe siècle. Fouilles de Delphes* II 14. Athens: École française d'Athènes.

Ambraseys, N. 1996. Material for the investigation of the seismicity of Central Greece, in S. Stiros and R.E. Jones (eds) *Archaeoseismology*: 23–36. Fitch Laboratory Occasional Paper 7. Exeter: Exeter University Press.

Ambraseys, N. 2009. Catalogue of earthquakes. *Earthquakes in the Mediterranean and Middle East*: 60–814. Cambridge: Cambridge University Press.

Ambraseys, N. and Pantelopoulos, P. 1992. Long-term seismicity of central Greece. *Mineral Wealth* 76: 23–32.

Ambraseys, N.N. and Jackson, J.A. 1998. Faulting associated with historical and recent earthquakes in the Eastern Mediterranean region. *Geophysical Journal International* 133: 390–406.

Ambraseys, N.N. and Pantelopoulos, P. 1989. The Fokis (Greece) earthquake of 1 August 1870. *European Earthquake Engineering* 1: 10–18.

Armijo, R, Meyer, B., Hubert, A. 1999. Westward propagation of the North Anatolian fault into the northern Aegean; timing and kinematics. *Geology* 27(3): 267–270.

Armijo, R., Meyer, B., King, G.C.P. et al. 1996. Quaternary evolution of the Corinth Rift and its implications for Late Cenozoic evolution of the Aegean. *Geophysical Journal International* 126: 11–53.

Billiris, H., Paradissis, D., Veis, G. et al. 1991. Geodetic determination of tectonic deformation in central Greece from 1900 to 1988. *Nature* 350: 124–129.

Bommelaer, J.-Fr. 1992. Observations sur le théâtre de Delphes, in J.-Fr. Bommelaer (ed.) *Delphes. Centenaire de la « Grande Fouille » réalisée par l'École française d'Athènes (1892-1903). Actes du Colloque Paul Perdrizet. Strasbourg, 6-9 novembre 1991*: 277–300. Leiden: Brill.

Bommelaer, J.-Fr. and Laroche, D. 2015. *Guide de Delphes*. Athens: École française d'Athènes.

Bousquet, B., Dufaure, J.J. and Péchoux, P.-Y. 1977. Le rôle de la géomorphologie dans l'évaluation des déformations néotectoniques en Grèce. *Bulletin de la Société Géologique de France* 19: 685–693.

Bousquet, B. and Péchoux, P.-Y. 1977. La séismicité du Bassin égéen pendant l'Antiquité. Méthodologie et premiers résultats. *Bulletin de la Société Géologique de France* 19: 679–684.

Bousquet, J. 1988. Études sur les comptes de Delphes. *Bibliothèque des Écoles françaises d'Athènes et de Rome*: 267. Athens: École française d'Athènes.

Bousquet, J. 1989. *Les comptes du IVe et du IIIe siècle. Corpus des Inscriptions de Delphes* II. Athens: École française d'Athènes.

Brun, J.-P. and Sokoutis, D. 2007. Kinematics of the Southern Rhodope Core Complex (North Greece). *International Journal of Earth Sciences* 96(6): 1079–1099.

Causse, C., Moretti, I, Ghisetti, F. et al. 2004. Kinematics of the Corinth Gulf inferred from calcite dating and syntectonic sedimentary characteristics. *CRAS* 336: 281–290.

Christaras, B. and Vouvalidis, K. 2010. Rockfalls occurred in the archaeological site of Delphi, Greece. *Proceedings of the 8th International Symposium on the Conservation of Monuments in the Mediterranean Basin*. Patras: **National Technical University of Athens.**

Collier, R.E.L., Pantosti, D., D'Addenzio, G. et al. 1998. Paleoseismicity of the 1981 Corinth earthquake fault: seismic contribution to extensional strain in central Greece and implications for seismic hazard. *Journal of Geophysical Research* 103, no. B12: 30001–30019.

Darcque, P. 1991. Temple d'Apollon : fouille. *Bulletin de Correspondance Hellénique* 115: 689.

Dercourt, J., Zonenshain, L.P., Ricou, L-E et al. 1986. Geological evolution of the Tethys Belt from the Atlantic to the Pamirs since the Lias. *Tectonophysics* 123: 241–315.

Déroche, V., Petridis, P. and Badie, A. 2014. *Le secteur au sud-est du péribole. Fouilles de Delphes* II 15. Athens: École française d'Athènes.

Doutsos, T., Pe-Piper, G., Boronkay, K. and Koukouvelas, I. 1993. Kinematics of the central Hellenides. *Tectonics* 12: 936–953.

Doutsos, T. and Piper, D.J.W. 1990. Listric faulting, sedimentation, and morphological evolution of the quaternary eastern Corinth rift, Greece: first stages of continental rifting. *Geological Society of America Bulletin* 102: 812–829.

Fontenrose, J. 1981. *The Delfic Oracle, its responses and operations*. Berkeley: University of California Press.

Gawthorpe, R.L., Leeder, M.R., Kranis, H. et al. 2018. Tectono- sedimentary evolution of the Plio-Pleistocene Corinth rift, Greece. *Basin Research* 30(3): 448–479.

Ginouvès, R. 1952. Une salle de bains hellénistique à Delphes. *Bulletin de Correspondance Hellénique* 76: 541–561.

Ginouvès, R. 1955. Sur un aspect de l'évolution des bains en Grèce vers le IVe siècle de notre ère. *Bulletin de Correspondance Hellénique* 79: 135–152.

Guidoboni, E., Ferrari, G., Tarabusi, G. et al. 2019. CFTI5Med, the new release of the catalogue of strong earthquakes in Italy and in the Mediterranean area. *Scientific Data* 6, Article number: 80 (2019); doi.org/10.1038/s41597-019-0091-9.

Hayward, C.L. 2003. Geology of Corinth: The Study of a Basic Resource. *American School of Classical Studies at Athens, Corinth, The Centenary: 1896-1996*, Vol. 20: 15–42.

Hellmann, M.-Chr. 1992. Voyageurs et fouilleurs à Delphes, in *La redécouverte de Delphes*: 14–54. Athens: École française d'Athènes.

Homolle, Th. 1897. Topographie de Delphes. *Bulletin de Correspondance Hellénique* 21: 256–420.

Homolle, Th. 1899. Le gymnase de Delphes. *Bulletin de Correspondance Hellénique* 23: 560–583.

Jackson, J.A., Gaignepain, J., Houseman, G. et al. 1982. Seismicity, normal faulting and the geomorphological development of the Gulf of Corinth (Greece): the Corinth earthquakes of February and March 1981. *Earth and Planetary Science Letters* 57: 377–397.

Jannoray, J. 1953. *Le Gymnase. Fouilles de Delphes* II 12. Athens: École française d'Athènes.

Jolivet, L., Brun, J.-P. Brun, Gautier, P. et al. 1994. 3D kinematics of extension in the Aegean region from the early Miocene to the present, insights from the ductile crust. *Bulletin de la Société Géologique de France* 165(3): 195–209.

Jolivet, L., Menant, A., Sternai, P. et al. 2015. The geological signature of a slab tear below the Aegean. *Tectonophysics* 659: 166–182.

Kazakis, N., Chalikakis, K., Mazzilli, N. et al. 2018. Management and research strategies of karst aquifers in Greece: Literature overview and exemplification based on hydrodynamic modelling and vulnerability assessment of a strategic karst aquifer. *Science of the Total Environment* 643: 592–609.

Kortekaas, S., Papadopoulos, G.A., Ganas, A. et al. 2011. Geological identification of historical tsunamis in the Gulf of Corinth, Greece. *Workshop 'Tsunami Deposits and its role in hazard mitigation', June 12-15, 2011*. Washington US.

Krashaw, S. and Guo, L. 2006. Pleistocene calcified cyanobacterial mounds, Perachora Peninsula, central Greece: A controversy of growth and history. *Geological Society London Special Publications* 255(1): 53–69.

Luce, J.-M. 2008. L'aire du Pilier des Rhodiens (*fouille 1990-1992*), à la frontière du profane et du sacré. *Fouilles de Delphes* II 13. Athens: École française d'Athènes.

Luce, J.-M. 2012. La ville de Delphes. *Bulletin de Correspondance Hellénique* 136-7.2: 771–798.

Lyon-Caen, H., Armijo, R., Drakopolous, J. et al. 1988. The Kalamata (South-Peloponnesus) earthquake: detailed study of a normal fault, evidences for East-West extension in the Hellenic Arc. *Journal of Geophysical Research* 93, no. B12: 14967–15000.

Marinos, P. and Rondoyanni, T. 2005. The Archaeological Site of Delphi, Greece: a Site Vulnerable to Earthquakes, Rockfalls and Landslides, in K. Sassa, H. Fukuoka, F. Wang and G. Wang (eds) *Landslides*. Berlin-Heidelberg: Springer.

Mariolakos, E., Bornovas, J. and Mouyiaris, N. (eds) 1989. *Seismotectonic map of Greece, with seismological data, scale 1:500 000*: Athens: Institute of Geology and Mineral Exploration (4 sheets).

McKenzie, D.P. 1978. Active tectonics of the Alpine-Himalayan belt: The Aegean Sea and surrounding regions. *Royal Astronomical Society Geophysical Journal* 55: 217–254.

Menant, A., Jolivet, L. and Vrielynck, B. 2016. Kinematic reconstructions and magmatic evolution illuminating crustal and mantle dynamics of the eastern Mediterranean region since the late Cretaceous. *Tectonophysics* 675: 103–140.

Michaud, J.-P. and Blécon, J. 1973. *Le Trésor de Thèbes. Fouilles de Delphes* II 16. Athens: École française d'Athènes.

Moretti, I., Sakellariou, D., Lykousis, V. et al. 2003. The Gulf of Corinth: an active half graben? *Journal of Geodynamics* 36: 323–340.

Moretti, I., Lykousis, V., Sakellariou, D. et al. 2004. Subsidence rate in the Gulf of Corinth: what we learn from the long piston coring, *CRAS-Structural Geology/Deformation mechanisms* 336: 291–299.

Mouyaris, N., Papastamatiou, D. and Vita-Finzi, C. 1992. The Helice fault? *Terra Nova* 4: 124–129.

Nirta, G., Moratti, G., Piccardi, L. et al. 2015. The Boeotian Flysch revisited: new data from central Greece constraining ophiolite obduction on the Adria plate. *Ofioliti* 40(2): 107–123.

Nirta, G., Moratti, G., Piccardi, L. et al. 2018. From obduction to continental collision: new data from Central Greece. *Geological Magazine* 155(2): 377–421.

Ori, G.G. 1989. Geologic history of the extensional basin of the Gulf of Corinth (Miocene-Pleistocene). *Greece: Geology* 17: 918–921.

Papanikolaou, D.J. and Royden, L.H. 2007. Disruption of the Hellenic arc: Late Miocene extensional

detachment faults and steep Pliocene-Quaternary normal faults – Or what happened at Corinth? *Tectonics* 26(5).
Papazachos, C. and Papazachou, C. 2003. *The earthquakes of Greece*. Thessaloniki: Ziti Edition.
Parke, H.W. and Wormel, D.E.W. 1956. *The Delphic Oracle*. Oxford: Blackwell.
Partida, E. 2016. Post-earthquake architecture at Olympia: The impact of natural phenomena upon the builder's attitude in the 4 c. BC, in E. Partida and D. Katsanopoulou (eds) *ΦΙΛΕΛΛΗΝ. Essays presented to Stephen G. Miller*: 299–316. Athens: The Helike Society.
Pavlides, S. and Caputo, R. 2004. Magnitude versus faults' surface parameters: quantitative relationships from the Aegean Region. *Tectonophysics* 380: 159–188.
Péchoux, P.-Y. 1977. Nouvelles remarques sur les versants Quaternaires du secteur de Delphes. *Revue de Géographie Physique et Géologie Dynamique* 19: 83–92.
Perrier, A. 2019. De l'éloquence et des dangers de l'eau à Delphes. *Les* Études *classiques* 87: 151–170.
Perrier, A. 2021. Ancient repairs and preventive architectural measures in the site of Delphi, in J. Vanden Broeck-Parant and T. Ismaelli (eds) *Ancient architectural restoration in the Greek World. Proceedings of the international workshop held at Wolfson College, Oxford*. Rome: Quasar: 57–65.
Perrier, A. and Malmary, J.-J. 2013. *Le Portique Ouest à Delphes. Étude d'histoire et d'architecture*. Unpublished dissertation. Athens: École française d'Athènes.
Perrier, A., Attuil, R., Camberlein, C. et al. (forthcoming). Fouilles du Portique Ouest de Delphes. *Bulletin des activités archéologiques des Écoles françaises à l'étranger*.
Petridis, P. 2010. *La céramique protobyzantine de Delphes : une production et son contexte*. Athens: École française d'Athènes. Fouilles de Delphes V.
Piccardi, L. 2000. Active faulting at Delphi: seismotectonic remarks and a hypothesis for the geological environment of a myth. *Geology* 28: 651–654.
Piccardi, L., Monti, C., Vaselli, O. et al. 2008. Scent of a myth: tectonics, geochemistry and geomythology at Delphi (Greece). *Journal of the Geological Society* 165: 5–18.
Ricou, L.-E. 1994. Tethys reconstructed: plates, continental fragments and their Boundaries since 260 Ma from Central America to South-eastern Asia. *Geodinamica Acta* 7(4): 169–218.
Ricou, L.-E., Dercourt, J., Geyssant, J. et al. 1986. Geological constraints on the alpine evolution of the Mediterranean Tethys. *Tectonophysics* 123(1): 83–122.
Rohais, S. and Moretti, I. 2017. Structural and Stratigraphic Architecture of the Corinth Rift (Greece): An Integrated Onshore to Offshore Basin-Scale Synthesis. *Lithosphere Dynamics and Sedimentary Basins of the Arabian Plate and surrounding area*: 89–120. New York: Springer.
Royden, L.H. and Papanikolaou, D. 2011. Slab segmentation and late Cenozoic disruption of the Hellenic arc. *Geochemical Geophysics Geosystems* 12: Q03010.
Sakellariou, D., Lykousis, V., Alexandri, S. et al. 2007. Faulting, seismic-stratigraphic architecture and late quaternary evolution of the Gulf of Alkyonides-East Gulf of Corinth, Central Greece. *Basin Research* 19: 273–295.
Schmidt, J. 1879. *Studien über Erdbeden*. Leipzig: Alwin Georgi ed.
Şengör, A.M.C. and Yilmaz, Y. 1981. Tethyan evolution of Turkey: A plate tectonic approach. *Tectonophysics* 75(3): 181–241.
Solakius, N., Carras, N., Mavridis, A. et al. 1998. Late cretaceous to early paleocene planktonic foraminiferal stratigraphy of the Agios Nikolaos sequence, the Parnassus-Ghiona zone, Central Greece. Δελτίον της Ελληνικής Γεωλογικής Εταιρίας 32(2): 13–20.
Taymaz, T., Jackson, J. and McKenzie, D. 1991. Active tectonics of the north and central Aegean Sea. *Geophysical Journal International* 106: 433–490.
Thély, L. 2016. *Les Grecs face aux catastrophes naturelles. Bibliothèque des Écoles françaises d'Athènes et de Rome*, 375. Athens: École française d'Athènes.
Trouki, E. 1993. Αναλήμματα και περίβολοι. Unpublished PhD Thesis, Strasbourg University.
Valkaniotis, S. 2009. *Correlation Between Neotectonic Structures and Seismicity in the broader area of Gulf of Corinth (Central Greece)*. Unpublished PhD Thesis, Aristotle University of Thessaloniki.
Valkaniotis, S., Papathanassiou, G. and Pavlides, S. 2011. Active faulting and earthquake-induced slope failures in archeological sites: case study of Delphi, Greece, in *Proceedings of 2nd INQUA-IGCP-567 International Workshop on Active Tectonics, Earthquake Geology, Archaeology and Engineering, Corinth, Greece, 2011*: 259–262.
Zachos, K. (ed.) 1964. *Geological Map of Greece, Delphi, scale 1:50 000*. Athens: Institute of Geology and Subsurface Research (1 sheet).